MW00837189

Osswald

Understanding Polymer Processing

Tim A. Osswald

Understanding
Polymer Processing

Processes and Governing Equations

HANSER

Hanser Publishers, Munich

Hanser Publications, Cincinnati

The Author:
Professor Dr. Tim A. Osswald
Department of Mechanical Engineering University of Wisconsin-Madison, Madison, WI, USA

Distributed in the USA and in Canada by
Hanser Publications
6915 Valley Avenue, Cincinnati, Ohio 45244-3029, USA
Fax: (513) 527-8801
Phone: (513) 527-8977
www.hanserpublications.com

Distributed in all other countries by
Carl Hanser Verlag
Postfach 86 04 20, 81631 München, Germany
Fax: +49 (89) 98 48 09
www.hanser.de

Library of Congress Cataloging-in-Publication Data

Osswald, Tim A.
Understanding polymer processing : processes and governing equations / Tim A. Osswald.
p. cm.
Includes bibliographical references and index.
ISBN-13: 978-1-56990-472-5 (hardcover)
ISBN-10: 1-56990-472-3 (hardcover)
1. Plastics. 2. Polymers. I. Title.
TP1120.O854 2010
668.4--dc22
2010035726

Bibliografische Information Der Deutschen Bibliothek
Die Deutsche Bibliothek verzeichnet diese Publikation in der Deutschen Nationalbibliografie; detaillierte bibliografische Daten sind im Internet über <http://dnb.d-nb.de> abrufbar.

ISBN 978-3-446-42404-3

Production Management: Steffen Jörg
Coverconcept: Marc Müller-Bremer, www.rebranding.de, München
Coverdesign: Stephan Rönigk
Typeset: Hilmar Schlegel, Berlin
Printed and bound by Kösel, Krugzell
Printed in Germany

In loving memory of Max Robert Osswald

In gratitude to Ronald L. Daggett,
the teacher and the plastics engineering pioneer,
for having the vision to develop and teach a plastics course in
mechanical engineering at the University of Wisconsin-Madison
in the fall of 1946.

Preface

This book provides the background for an understanding of the wide field of polymer processing. It is divided into three parts to give the engineer or student sufficient knowledge of polymer materials, polymer processing and modeling. The book is intended for the person who is entering the plastics manufacturing industry, as well as a textbook for students taking an introductory course in polymer processing.

Understanding Polymer Processing — Materials, Processes and Modeling is based on the 12 year-old Hanser Publisher's book *Polymer Processing Fundamentals*, as well as lecture notes from a 7-week polymer processing course taught at the University of Wisconsin-Madison.

The first three chapters of this book cover essential information required for the understanding of polymeric materials, from their molecule to their mechanical and rheological behavior. The next four chapters cover the major polymer processes, such as extrusion, mixing, injection molding, thermoforming, compression molding, rotomolding and more. Here, the underlying physics of each process is presented without complicating the reading with complex equations and concepts, however, helping the reader understand the basic plastics manufacturing processes. The last two chapters present sufficient background to enable the reader to carry out process scaling and to solve back-of-the-envelope polymer processing models.

I cannot possibly acknowledge everyone who helped in the preparation of this manuscript. First, I would like to thank all the students in my polymer processing course who, in the past two decades, have endured my experimenting with new ideas. I am also grateful to my polymer processing colleagues who taught the introductory polymer processing course before me: Ronald L. Daggett, Lew Erwin, Jay Samuels and Jeroen Rietveld. I thank Nicole Brostowitz for adding color to some of the original graphs, and to Katerina Sánchez for introducing and organizing the equations and for proofreading the final manuscript. I would like to thank Professor Juan Pablo Hernández-Ortiz, of the Universidad Nacional de Colombia, Medellín, for his input in Part III of this book. Special thanks to Wolfgang Cohnen for allowing me to use his photograph of Coyote Buttes used to exemplify the Deborah number in Chapter 3. My gratitude to Dr. Christine Strohm, my editor at Hanser Publishers, for her encouragement, support and patience. Thanks to Steffen Jörg at Hanser Publishers for his help and for putting together the final manuscript. Above all, I thank my wife Diane and my children Palitos and Rudi for their continuous interest in my work, their input and patience.

Summer of 2010
Tim A. Osswald

Contents

Part II Polymer Processes 71

Part I
Polymeric Materials

1 Introduction

Polymers
Macromolecules
Plastics

As the word suggests, polymers[1] are materials composed of molecules of high molecular weight. These large molecules are generally called *macromolecules*. The unique material properties of polymers and the versatility of their processing methods are attributed to their molecular structure. The ease with which polymers and *plastics*[2] are processed makes them, for many applications, the most sought after materials today. Because of their low density and their ability to be shaped and molded at relatively low temperatures, compared to traditional materials, such as metals, plastics. Polymers are the material of choice when integrating several parts into a single component — a design step usually called *part consolidation*. In fact, parts and components, traditionally made of wood, metal, ceramics, or glass, are frequently redesigned with plastics.

This chapter provides a general introduction to polymers and plastics, their molecular structure, their additives, as well as to other relevant topics, such as the plastics industry and plastics processes.

1.1 Historical Background

Natural polymeric materials, such as rubber, have been in use for thousands of years. Natural rubber also known as *caoutchouc* (crying trees) has been used by South American Indians in the manufacture of waterproof containers, shoes, torches, and squeeze bulb pumps. The first Spanish explorers of Haiti and Mexico reported that natives played games on clay courts with rubber balls [1]. Rubber trees were first mentioned in *De Orbe Novo*, originally published in Latin, by Pietro Martire d'Anghiera in 1516. The French explorer and mathematician Charles Maria de la Condamine, who was sent to Peru by the French *Academie des Sciences*, brought caoutchouc from South America to Europe in the 1740s. In his report [2] he mentions several rubber items made by native South Americans, including a piston-less pump composed of a rubber pear with a hole in the bottom. He points out that the most remarkable property of natural rubber is its great elasticity.

The first chemical investigations on *gummi elasticum* were published by the Frenchman Macquer in 1761. However, it was not until the 20th century that the molecular architecture of polymers was well understood. Soon after its introduction to Europe, various uses were found for natural rubber. Gossart manufactured the first polymer tubes in 1768 by wrapping rubber sheets around glass pipes. During the same time period small rubber blocks where introduced to erase lead pencil marks from paper. In fact, the word *rubber* originates from this specific application — *rubbing*.

1) From the Greek, *poli* which means many, and *meros* which means parts.
2) The term plastics describes the compound of a polymer with one or more additives.

These new materials slowly evolved from their novelty status as a result of new applications and processing equipment. Although the screw press, the predecessor of today's compression molding press, was patented in 1818 by McPherson Smith [3], the first documented *polymer processing* machinery dates to 1820 when Thomas Hancock invented a rubber masticator. This masticator, consisting of a toothed rotor in a toothed cylindrical cavity [4], was used to reclaim rubber scraps that resulted from the manual manufacturing process of elastic straps, perhaps the first recycling effort. In 1833 the development of the vulcanization process by Charles Goodyear [5] greatly enhanced the properties of natural rubber, and in 1836 Edwin M. Chaffee invented the two-roll steam heated mill, the predecessor of the calendar. It was used to continuously mix additives into rubber for the manufacture of rubber-coated textiles and leathers. As early as 1845, presses and dies were used to mold buttons, jewelry, dominoes, and other novelties out of shellac and gutta-percha. *Gutta-percha* (rubber clump), a gum found in trees similar to rubber, became the first wire insulation and was used for ocean cable insulation for many years.

The ram-type extruder was invented by Henry Bewley and Richard Brooman in 1845. The first *polymer processing* screw extruder, the most influential equipment in polymer processing, was patented by an Englishman named Mathew Gray in 1879 for the purpose of wire coating. However, the screw pump is attributed to Archimedes and the actual invention of the screw extruder by A. G. DeWolfe of the U. S. dates back to the early 1860s.

Cellulose nitrate plasticized by camphor, possibly the first thermoplastic, was patented by Isaiah and John Hyatt in 1870. Based on experience from metal injection molding, the Hyatt brothers built and patented the first injection molding machine in 1872 to mold cellulose materials [6].

With the mass production of rubber, gutta-percha, cellulose, and shellac articles during the height of the industrial revolution, the polymer processing industry after 1870 saw the invention and development of internal kneading and mixing machines for the processing and preparation of raw materials [7]. A notable invention was the Banbury mixer, developed by Fernley Banbury in 1916. This mixer, with some modifications, is still used for rubber compounding.

Bakelite, developed by Leo Baekeland in 1907, was the first synthetically developed polymer. Bakelite, also known as phenolic, is a thermoset resin that reacts by condensation polymerization occurring when phenol and formaldehyde are mixed and heated.

In 1924, Hermann Staudinger proposed a model that described polymers as linear molecular chains. Once this model was accepted by other scientists, the concept for the synthesis of new materials was realized. In 1927 cellulose acetate and polyvinyl chloride (PVC) [8] were developed. Because of its higher wear resistance, polyvinyl chloride replaced shellac for phonograph records in the early 1930s. Wallace Carothers pioneered condensation polymers such as polyesters and polyamides. It was not until this point that the scientific world was finally convinced of the validity of Staudinger's work. Polyamides, first called Nylon, were set into production in 1938. Polyvinyl acetate, acrylic polymers, polystyrene (PS), polyurethanes, and melamine were also developed in the 1930s [9].

The first single-screw extruder designed for the processing of thermoplastic polymers was built circa 1935 at the Paul Troester Maschinenfabrik [10]. Around that same time period, Roberto Colombo developed a twin-screw extruder for thermoplastics.

World War II and the post-war years saw accelerated development of new polymeric materials. Polyethylene (PE), polytetrafluoroethylene, epoxies, and acrylonitrile-butadiene-styrene (ABS) were developed in the 1940s, and linear polyethylene, polypropylene (PP), polyacetal, polyethylene terephthalate (PET), polycarbonate (PC), and many more materials came in the 1950s. The 1970s saw the development of new polymers such as polyphenylene sulfide and in the 1980s, liquid crystalline polymers were developed.

Developing and synthesizing new polymeric materials has become increasingly expensive and difficult. Developing new engineering materials by blending or mixing two or more polymers or by modifying existing ones with plasticizers is now widely accepted.

The world's annual production of polymer resins has experienced steady growth since the turn of the century, with growth resins to that of steel and aluminum over the past 60 years. Before 1990, the figure depicts the production of plastics in the Western World and after that, when the iron curtain came down, the worldwide production. In developed countries, the growth in annual polymer production has decreased recently. However, developing countries in South America and Asia are now starting to experience tremendous growth.

Of the over 50 million tons of polymers produced in the U.S. in 2008, 90 % were thermoplastics. Figure 1.2 breaks the U.S. polymer production into major polymer

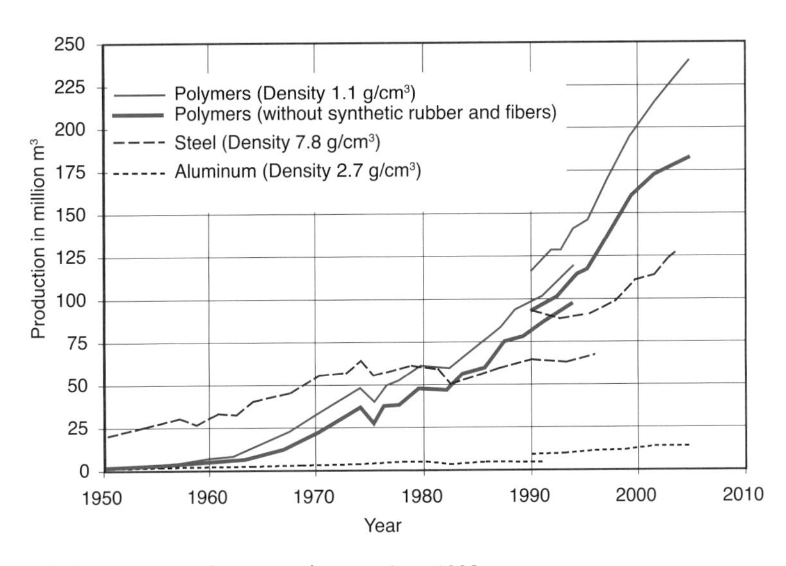

World plastics production

Figure 1.1 World annual plastics production since 1900

categories, including polyethylenes, polypropylene, polystyrene, polyvinyl chloride, and thermosets. Polyethylenes are by far the most widely used polymeric material, accounting for 41 % of the U. S. plastic production.

Thermoplastics – 90 %
Thermosets – 10 %

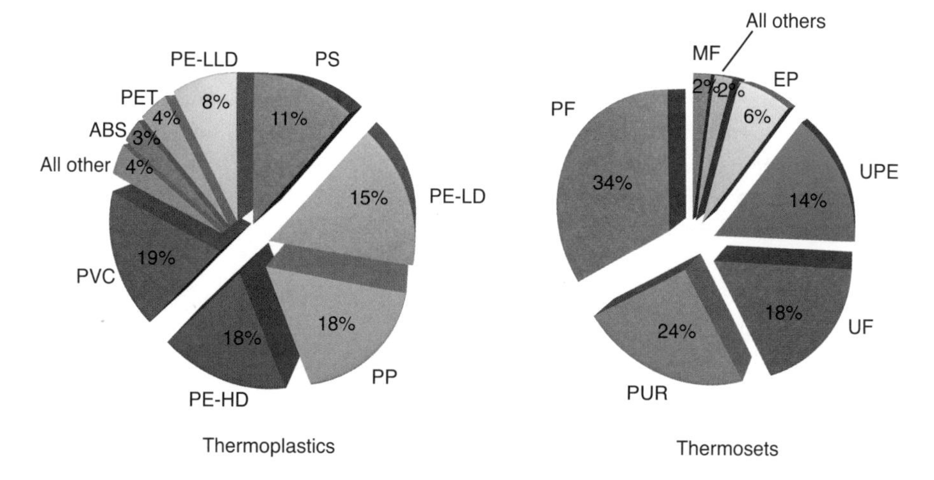

Figure 1.2 Break down of U. S. polymer production into major polymer categories
Source: SPI Committee on Resin Statistics as compiled by Ernst & Young

1.2 General Properties

Any plastic resin can be categorized as either a thermoplastic or thermoset. Thermoplastics are those polymers that solidify as they cool, restricting the motion of the long molecules. When heated, these materials regain the ability to "flow," as the molecules can slide past each other easily. Thermoplastic polymers further are divided into two classes: amorphous and semi-crystalline polymers.

Amorphous thermoplastics have molecules that remain disorderly as they cool, leading to a material with a random molecular structure. An amorphous polymer solidifies, or vitrifies, as it cools below its glass transition temperature, T_g. Semi-crystalline thermoplastics, on the other hand, solidify with a certain order in their molecular structure. Hence, as they cool, they harden when the molecules arrange in a regular order below what is usually called the melting temperature, T_m. The molecules in semi-crystalline polymers not ordered remain in amorphous regions. These regions within the semi-crystalline domains solidify at the glass transition temperature. Most semi-crystalline polymers have a glass transition temperature below the ice point, and behave at room temperature similarly to rubbery or leathery materials. Table 1.1 presents the most common amorphous and semi-crystalline thermoplastics with some of their applications.

On the other hand, thermosetting polymers solidify by a chemical curing process. Here, the long macromolecules crosslink during cure, resulting in a network. The original molecules can no longer slide past each other. These networks prevent "flow" even after re-heating. The high density of crosslinking between the molecules makes thermosetting materials stiff and brittle. Thermosets also exhibit glass transition temperatures, which sometimes exceed thermal degradation temperatures. Some of the most common thermosets and their applications are also found in Table 1.1.

Table 1.1 Common polymers and some of their applications

Polymer	Typical applications
	Thermoplastics
Amorphous	
Polystyrene	Mass-produced transparent articles, packaging, insulation (foamed)
Polymethyl methacrylate	Skylights, airplane windows, lenses, stop lights
Polycarbonate	Helmets, hockey masks, blinker lights, head lights
Unplasticized polyvinyl chloride	Tubes, window frames, siding, bottles, packaging
Plasticized polyvinyl chloride	Shoes, hoses, calendered films and sheets (floors and upholstery)
Semi-crystalline	
High density polyethylene	Milk and soap bottles, mass produced household goods
Low density polyethylene	Mass produced household goods, grocery bags
Polypropylene	Housings for electric appliances, auto battery cases
Polytetrafluoroethylene	Coating of cooking pans, lubricant-free bearings
Polyamide	Gears, bolts, skate wheels, pipes, fishing line, textiles, ropes
	Thermosets
Epoxy	Adhesive, matrix in fiber-reinforced composite parts
Melamine	Decorative heat-resistant surfaces for kitchens and furniture, dishes
Phenolics	Heat-resistant handles for pans, irons and toasters, electric outlets
Unsaturated polyester	Sinks and tubs, automotive body panels (with glass fiber)
	Elastomers
Polybutadiene	Automotive tires, golf ball skin
Ethylene propylene rubber	Automotive radiator hoses and window seals, roof covering
Natural rubber (polyisoprene)	Automotive tires, engine mounts
Polyurethane elastomer	Roller skate wheels, automotive seats (foamed), shoe soles (foamed)
Silicone rubber	Seals, flexible hoses for medical applications
Styrene butadiene rubber	Automotive tire treads

This table can be used to compare plastics

WARNING: The ASTM D638 and ASTM D790 are performed at one rate of deformation. Consequently, the properties measured with these techniques are time independent, and cannot be used for design.

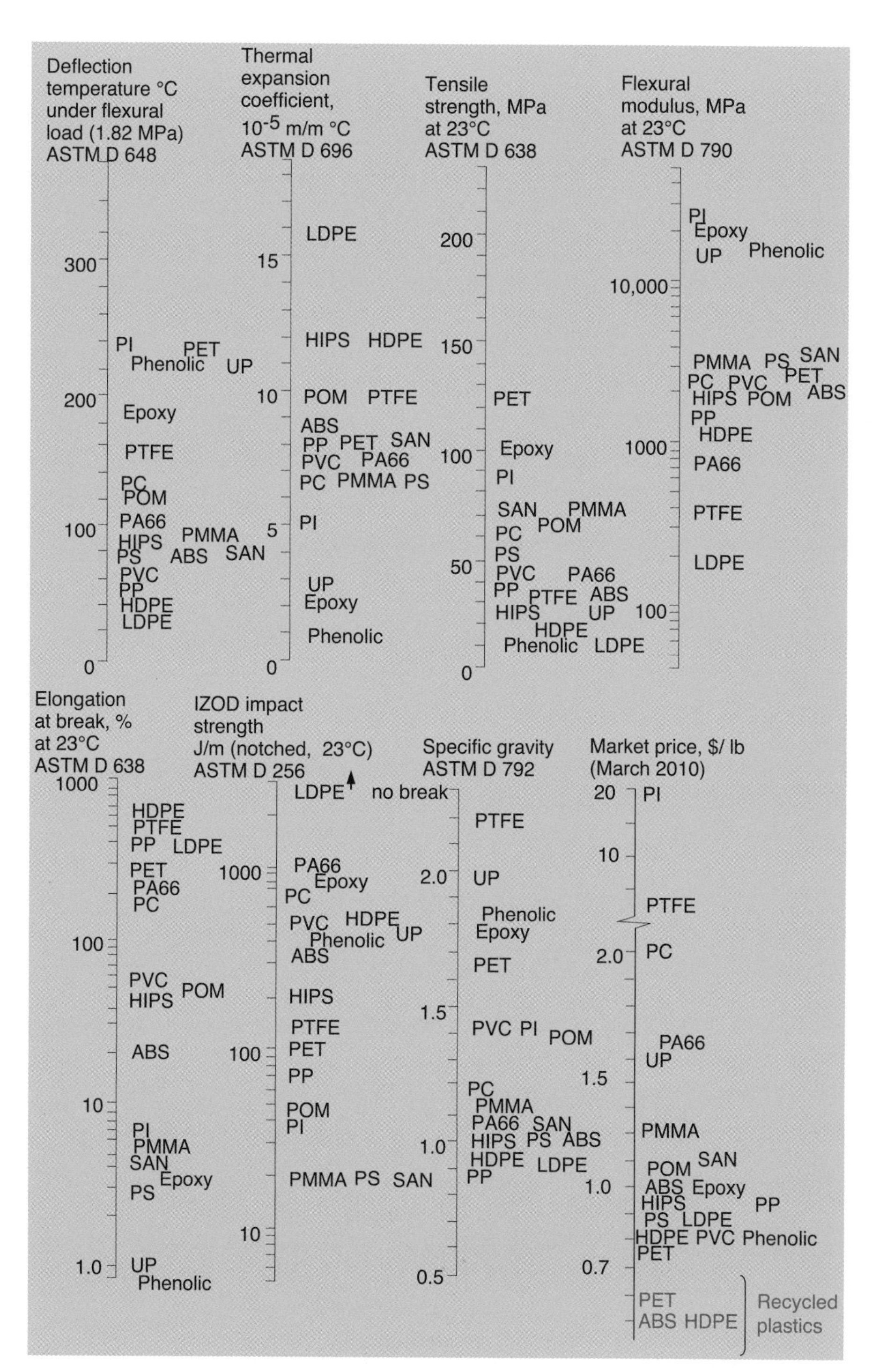

Figure 1.3 Average properties for common polymers

Compared to thermosets, elastomers are only slightly crosslinked, which permits almost full molecular extension. However, the links across the molecules hinder them from sliding past each other, making even large deformations reversible. One common characteristic of elastomeric materials is that the glass transition temperature is much lower than room temperature. Table 1.1 lists the most common elastomers with some of their applications.

As mentioned earlier, there are thousands of grades of polymers available to the design engineer. These cover a wide range of properties, from soft to hard, ductile to brittle, and weak to tough. Figure 1.3 shows this range by plotting important average properties for selected polymers. The abbreviations used in Fig. 1.3 are defined in Table 1.2. The values for each material in Fig. 1.3 are representative averages.

The relatively low stiffness of polymeric materials is attributed to their molecular structure, which allows relative movement with ease while under stress. However, the strength and stiffness of individual polymer chains are much higher than the measured

http://www.campusplastics.com/

Table 1.2 Commonly used acronyms for plastics[a)]

Acronym	Polymer
ABS*	Acrylonitrile-butadiene-styrene
CA	Cellulose acetate
EP*	Epoxy Resin
EPDM	Ethylene propylene diene rubber
EVAC* (EVAL)	Ethylene vinylacetate
LCP*	Liquid crystal polymer
LSR	Liquid silicone rubber
MABS*	Methylmethacrylate acrylonitrile butadiene styrene
MF*	Melamine formaldehyde
NR	Natural rubber
PA	Polyamide
PA6*	Polyamide from e-caprolactam
PA66*	Polyamide from Hexamethylene diamine adipic acid
PAEK*	Polyarylether ketone
PBT*	Polybutylene terephthalate
PC*	Polycarbonate (from bisphenol-A)
PE*	Polyethylene
PE-HD	Polyethylene-high density
PE-LD	Polyethylene-low density
PE-LLD	Polyethylene-linear low density
PE-MD	Polyethylene-medium density
PE-UHMW	Polyethylene-ultra high molecular weight

a) (*) Designated by ISO standards, in conjunction with the materials data bank CAMPUS

Table 1.2 *(continued)* Commonly used acronyms for plastics[a)]

Acronym	Polymer
PE-X	Polyethylene, crosslinked
PEEK	Polyetheretherketone
PEI*	Polyetherimide
PES*	Polyethersulfone
PET*	Polyethylene terephthalate
PET-G*	Polyethylene terephthalate, glycol modified
PF*	Phenolic formaldehyde resin
PI*	Polyimide
PLA	Polylactide
PMMA*	Polymethylmethacrylate
POM*	Polyoxymethylene (polyacetal resin, polyformaldehyde)
PP*	Polypropylene
PPE*	Polyphenylene ether, old notation PPO
PPS*	Polyphenylene sulfide
PPSU*	Polyphenylene sulfone
PS*	Polystyrene
PS-HI	Polystyrene-high impact
PSU*	Polysulfone
PTFE*	Polytetrafluoroethylene
PUR*	Polyurethane
PVC*	Polyvinyl chloride
SAN*	Styrene acrylonitrile
SBR	Styrene butadiene rubber
SI	Silicone, Silicone resin
TPE	Thermoplastic elastomers
TPU* (TPE-U)	Thermoplastic elastomers based on polyurethane. TPA* (polyamide) etc.
UF	Urea formaldehyde resin
UP*	Unsaturated polyester resin

a) (*) Designated by ISO standards, in conjunction with the materials data bank CAMPUS

properties of the bulk. For example, polyethylene, whose molecules have a theoretical stiffness of 300 000 MPa, has a bulk stiffness of only 1000 MPa [12, 13]. By introducing high molecular orientation, the stiffness and strength of a polymer can be substantially increased. In the case of *ultra-drawn, ultra high molecular weight high density polyethylene* (PE-UHMHD), fibers can exceed a stiffness of 200 000 MPa [13].

1.3 Macromolecular Structure of Polymers

Polymers are macromolecular structures generated synthetically or through natural processes. Cotton, silk, natural rubber, ivory, amber, and wood are a few materials that occur naturally with an organic macromolecular structure, whereas natural inorganic materials include quartz and glass. The other class of organic materials with a macromolecular structure is synthetic polymers, which are generated through addition polymerization or condensation polymerization.

In addition polymerization, monomers are added to each other by breaking the double-bonds that exist between carbon atoms, allowing them to link to neighboring carbon atoms to form long chains. The simplest example is the addition of ethylene monomers, schematically shown in Fig. 1.4, to form polyethylene molecules, as shown in Fig. 1.5. The schematic shown in Fig. 1.5 can also be written symbolically as shown in Fig. 1.6. Here, the subscript "n" represents the number of repeat units which determines the molecular weight of a polymer. The number of repeat units is more commonly referred to as the degree of polymerization.

Another technique for producing macromolecular materials is condensation polymerization. Condensation polymerization occurs when two components with end-groups that react with each other are mixed. When they are stoichiometric, these end-groups

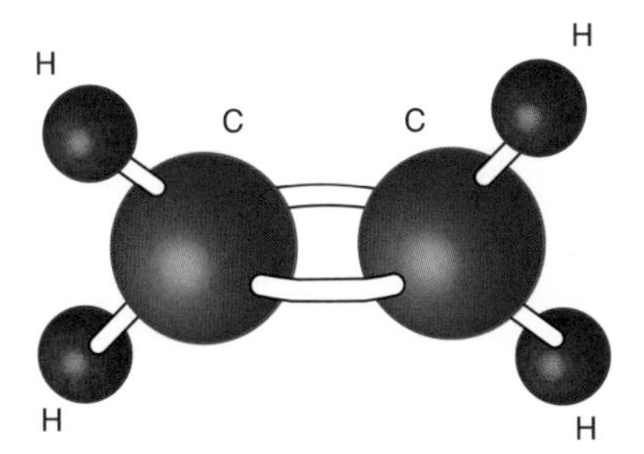

Figure 1.4 Schematic representation of an ethylene monomer

Polyethylene

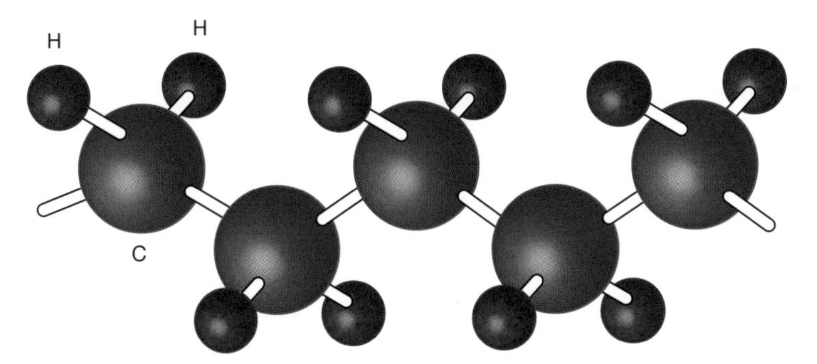

Figure 1.5 A polyethylene molecule

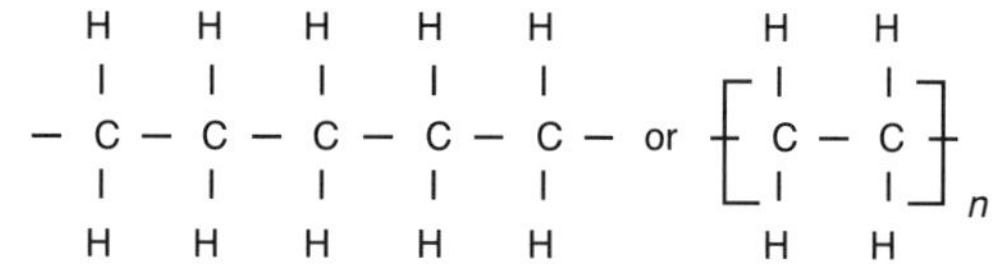

$$-\overset{\displaystyle H}{\underset{\displaystyle H}{C}}-\overset{\displaystyle H}{\underset{\displaystyle H}{C}}-\overset{\displaystyle H}{\underset{\displaystyle H}{C}}-\overset{\displaystyle H}{\underset{\displaystyle H}{C}}-\overset{\displaystyle H}{\underset{\displaystyle H}{C}}- \quad \text{or} \quad \left[\overset{\displaystyle H}{\underset{\displaystyle H}{C}}-\overset{\displaystyle H}{\underset{\displaystyle H}{C}}\right]_n$$

Figure 1.6 Symbolic representation of a polyethylene molecule

Polyamide (nylon)

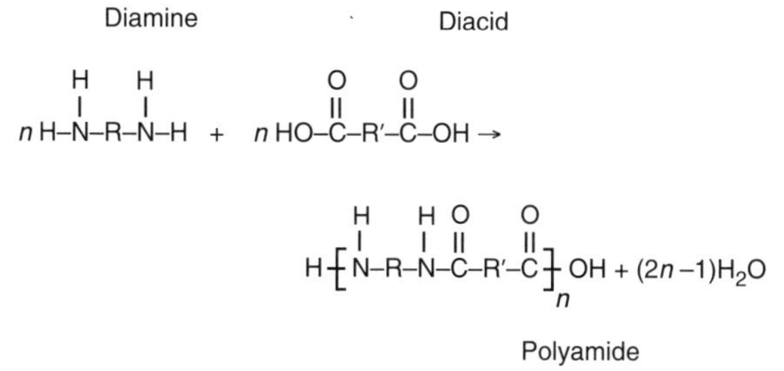

Diamine Diacid

$$n\,H-\overset{\displaystyle H}{N}-R-\overset{\displaystyle H}{N}-H \;+\; n\,HO-\overset{\displaystyle O}{\overset{\|}{C}}-R'-\overset{\displaystyle O}{\overset{\|}{C}}-OH \rightarrow$$

$$H\left[\overset{\displaystyle H}{N}-R-\overset{\displaystyle H}{N}-\overset{\displaystyle O}{\overset{\|}{C}}-R'-\overset{\displaystyle O}{\overset{\|}{C}}\right]_n OH + (2n-1)H_2O$$

Polyamide

Figure 1.7 Symbolic representation of the condensation polymerization of polyamide

react, linking them to chains and leaving a by-product such as water. A common polymer made by condensation polymerization is polyamide, where diamine and diacid groups react to form polyamide and water, as shown in Fig. 1.7.

In the molecular level, there are several forces that hold a polymeric material together. The most basic forces are the covalent bonds, which hold the polymer backbone together, such as the —C—C— or —C—N— bond.

1.4 Molecular Weight

A polymeric material usually consists of polymer chains of various lengths. With the exception of some naturally occurring polymers, most polymers have a molecular weight distribution such as the one shown in Fig. 1.8 and the molecular weight is described by a set of averages. The properties of a polymeric material are strongly linked to the molecular weight of the polymer, as shown schematically in Fig. 1.9. A polymer, such as polystyrene, is stiff and brittle at room temperature with a degree of polymerization of 1000. However, at a degree of polymerization of 10, polystyrene is sticky and soft at room temperature. The stiffness properties reach an asymptotic maximum, whereas the transition temperatures, and consequently the viscosity, increase with molecular weight. One must find the molecular weight of a polymer that renders ideal mechanical properties while maintaining flow properties that make it easy to shape the material during manufacturing.

Polymer chain branching, which occurs due to irregularities during polymerization, also influences the final structure, crystallinity, and properties of the polymeric material. Figure 1.10 shows the molecular architecture of high density, low density, and linear low density polyethylenes (PE-LLD). The high density polyethylene (PE-HD)

Number average molecular weight

Weight average molecular weight

Viscosity average molecular weight

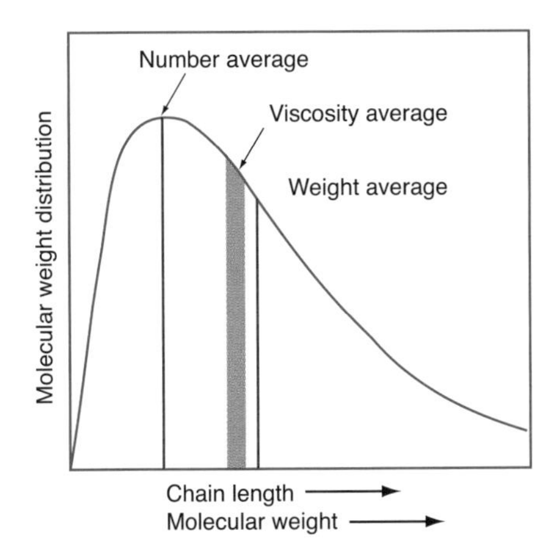

Figure 1.8 Molecular weight distribution of a typical thermoplastic

Mechanical properties and molecular weight

Processing and molecular weight

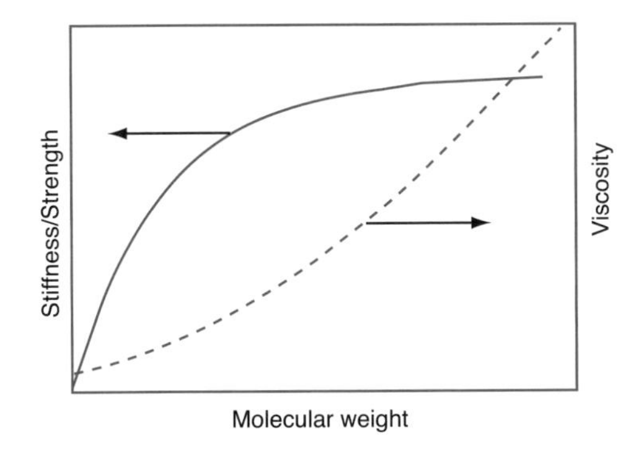

Figure 1.9 Influence of molecular weight on mechanical properties

Branched polymer molecules

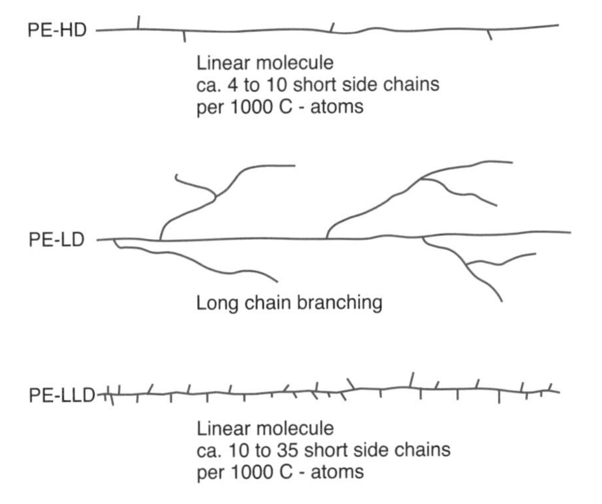

Figure 1.10 Schematic of the molecular structure of different polyethylenes

has between 5 and 10 short branches per every 1000 carbon atoms. The low density material (PE-LD) has the same number of branches as PE-HD, however, they are much longer and are themselves usually branched. The PE-LLD has between 10 and 35 short chains every 1000 carbon atoms. Polymer chains with fewer and shorter branches can crystallize more easily, resulting in higher degree of crystallinity, and therefore density.

Example 1.1 Molecular weight of a polymer

Estimate the degree of polymerization of a polypropylene with an average molecular weight of 100 000.

As shown in the diagram below, the repeat unit of a polypropylene molecule contains 3 carbon and 6 hydrogen atoms.

From the diagram we can see that each repeat unit's molecular weight is 6(1) + 3(12) = 42. Thus, a molecule with a molecular weight of 100 000 is formed by 100 000/42 = 2381 repeat units.

■

1.5 Arrangement of Polymer Molecules

From basic molecular standpoint, polymers are either un-crosslinked or cross-linked. However, polymeric materials are categorized as either thermoplastics, thermosets, or elastomers. Thermoplastics can re-melt after solidification, while thermosets and elastomers solidify via a chemical reaction that causes polymer molecules to crosslink. For elastomers, the crosslinking process is referred to as vulcanization. These crosslinked materials cannot be re-melted after solidification.

As thermoplastic polymers solidify, they take on two different types of structure: amorphous and semi-crystalline. Amorphous polymers are those where the molecules solidify in a random arrangement, whereas the molecules in semi-crystalline polymers align with their neighbors, forming regions with a three-dimensional order.

1.5.1 Thermoplastic Polymers

The formation of macromolecules from monomers occurs if there are unsaturated carbon atoms (carbon atoms connected with double or triple bonds), or if there are monomers with reactive end-groups. For example, in an ethylene monomer a double bond is split, which frees two valence electrons per monomer and leads to the formation of a macromolecule such as polyethylene. This process is called polymerization.

Similarly, two complementing monomers (R and R′) that each possess two reactive end groups (bifunctional) can react with each other, also leading to the formation of a polymer chain.

1.5.2 Amorphous Thermoplastics

Amorphous thermoplastics, with their randomly arranged molecular structure, are analogous to spaghetti. Because of their structure, the characteristic size of the largest ordered region is the length of a carbon-carbon bond. This dimension is much smaller than the wavelength of visible light and so generally makes amorphous thermoplastics very clear, and in most cases close to transparent.

Amorphous Polymers

Glass transition temperature

Transparent

Figure 1.11 [14] shows the shear modulus, G', versus temperature for polystyrene, one of the most common amorphous thermoplastics. The figure, which was obtained through a dynamic-mechanical test, shows two general regions: one where the modulus appears fairly constant and one where the modulus drops significantly with increasing temperature. With decreasing temperatures, the material enters the glassy region where the slope of the modulus approaches zero. At high temperatures, the modulus is negligible and the material is soft enough to flow. Although there is no clear transition between "solid" and "liquid", the temperature dividing the two states in an amorphous thermoplastic is called the *glass transition temperature*, T_g. For the polystyrene in Fig. 1.11, the glass transition temperature is about 110 °C. Although data is usually presented in the form shown in Fig. 1.11, the curve shown in the figure was measured at a constant frequency. If the test frequency is increased — reducing the time scale — the curve shifts

DMA test

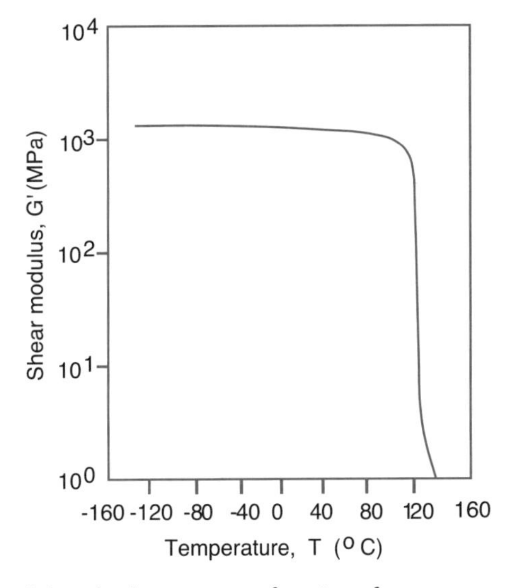

Figure 1.11 Shear modulus of polystyrene as a function of temperature

pvT behavior

Glass transition temperature increases with increasing pressure

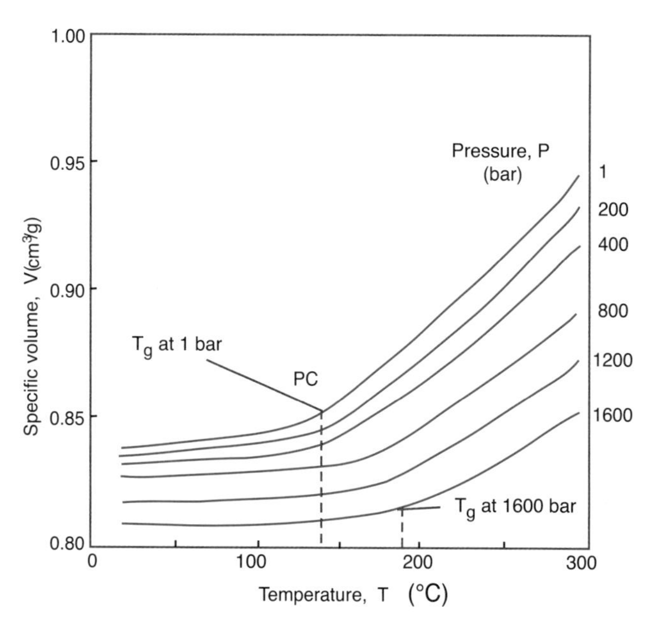

Figure 1.12 pvT behavior of a PC

to the right, because higher temperatures are required to achieve molecular motion at higher frequencies.

Furthermore, the glass transition temperature is also affected by pressure, as becomes evident when observing the specific volume of a polymer as a function of temperature at various test pressures. Figure 1.12 presents a so-called *pressure-volume-temperature diagram*, or pvT diagram, for a polycarbonate. As can be seen in the graph, the glass transition temperature shifts approximately 50 K when increasing the pressure from 1 bar to 1600 bar.

1.5.3 Semi-Crystalline Thermoplastics

The molecules in semi-crystalline thermoplastic polymers align in an ordered crystal structure, as shown for polyethylene in Fig. 1.13. The schematic shows the general structure and hierarchical arrangement in semi-crystalline materials. The crystalline structure is part of a *lamellar crystal*, which in turn forms the *spherulites*. The sperulitic structure is the largest domain with a specific order and has a characteristic size of 50 to 500 μm. This is much larger than the wavelength of visible light, making semi-crystalline materials translucent, not transparent.

However, the crystalline regions are tiny, with molecular chains comprised of both crystalline and amorphous regions. The degree of crystallinity in a typical thermoplastic varies from grade to grade. For example, in polyethylene, the degree of crystallinity depends on the branching and the cooling rate. LDPE with its long branches (Fig. 1.10)

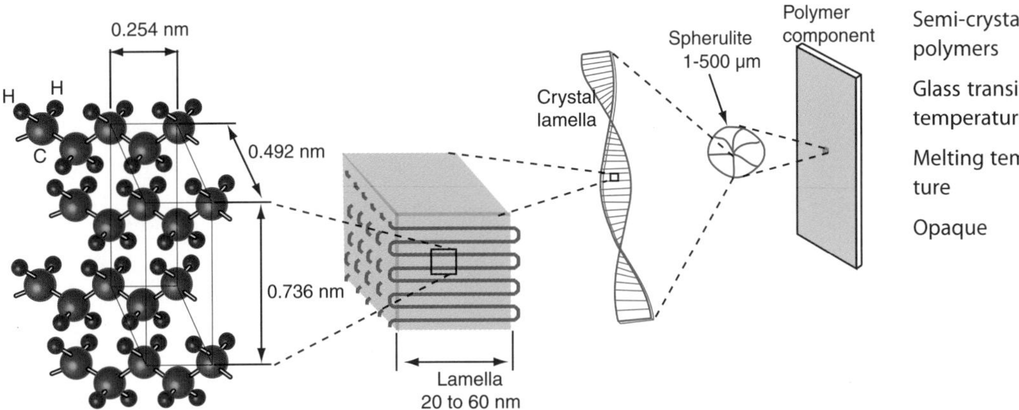

Figure 1.13 Schematic representation of the general molecular structure and arrangement of typical semi-crystalline materials

Semi-crystalline polymers

Glass transition temperature

Melting temperature

Opaque

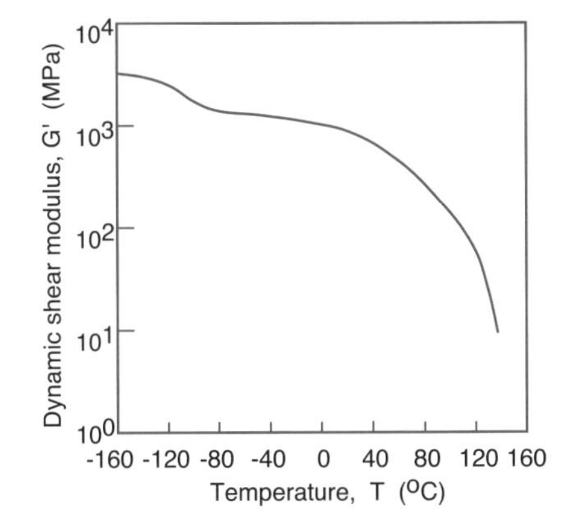

Figure 1.14 Shear modulus of HDPE versus temperature

DMA test

can only crystallize to about 40 to 50 %, whereas a HDPE crystallizes to up to 80 %. The density and strength of semi-crystalline thermoplastics increase with the degree of crystallinity.

Figure 1.14 [14] shows the dynamic shear modulus versus temperature for a HDPE, the most common semi-crystalline thermoplastic. Again, this curve presents data measured at one test frequency. The figure clearly shows two distinct transitions: one at about –110 °C, the *glass transition temperature*, and another near 140 °C, the *melting temperature*. Above the melting temperature, the shear modulus is negligible and the material flows. Crystalline arrangement begins to develop as the temperature dips below the melting point. Between the melting and glass transition temperatures the material behavior is leathery. As the temperature dips below the glass transition temperature, the amorphous regions within the semi-crystalline structure solidify, forming a glassy, stiff, and sometimes brittle polymer.

pvT behavior

Melting temperature increases with increasing pressure

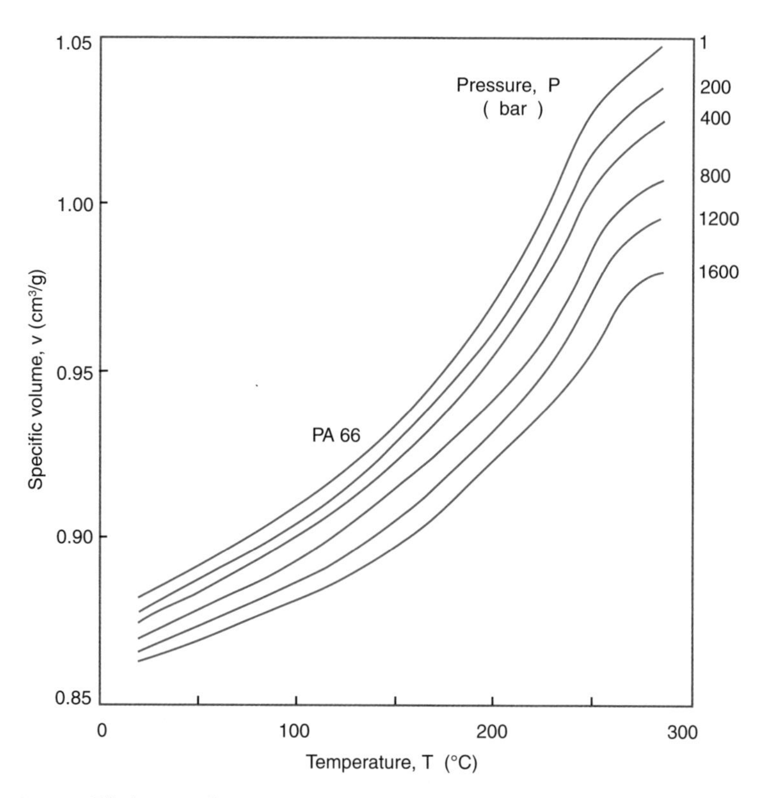

Figure 1.15 pvT behavior of a PA66

The pvT diagram for a PA66, a typical semi-crystalline polymer, is presented in Fig. 1.15.

1.5.4 Thermosets and Crosslinked Elastomers

Thermosets and some elastomers are polymeric materials that can crosslink. The cross-linking causes the material to resist heat after solidification. A more in-depth explanation of the crosslinking chemical reaction that occurs during solidification is given in Section 8.3.

Crosslinking is usually a result of double bonds breaking, allowing molecules to link with their neighbors. One of the oldest thermosetting polymers is phenol-formaldehyde, or phenolic. Figure 1.16 shows the chemical symbol representation of the reaction. The phenol and formaldehyde molecules react to create a three-dimensional cross-linked network that is stiff and strong. The by-product of this chemical reaction is water.

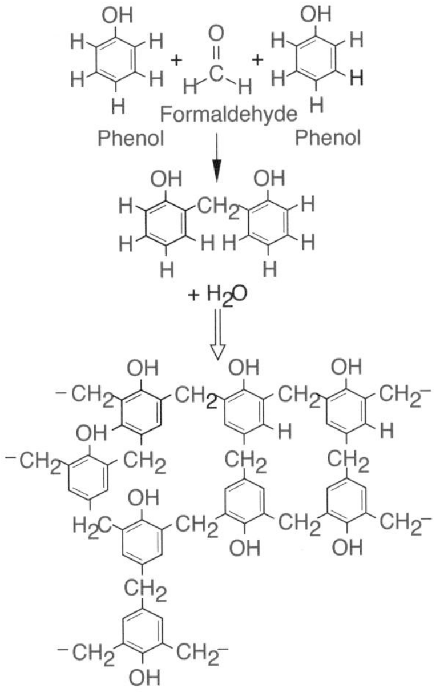

Phenolic-condensation reaction

Figure 1.16 Symbolic representation of the condensation polymerization of phenol-formaldehyde resins

1.6 Copolymers and Polymer Blends

Copolymers are polymeric chains of materials with two or more monomer types. A copolymer that is composed of two monomer types is called a *bipolymer*, and one formed from three different monomer groups is called a *terpolymer*. Depending on how the different monomers are arranged in the polymer chain, we distinguish between *random*, *alternating*, *block*, or *graft* copolymers. These copolymer types are schematically represented in Fig. 1.17.

A widely used copolymer is high impact polystyrene (PS-HI), formed by grafting polystyrene to polybutadiene. If styrene and butadiene are randomly copolymerized, the resulting material is an elastomer called styrene-butadiene-rubber (SBR). Another classic example of copolymerization is the terpolymer acrylonitrile-butadiene-styrene (ABS).

Polymer blends belong to another family of polymeric materials made by mixing or blending two or more polymers to enhance the physical properties of each individual component. Common polymer blends include PP+PC, PVC+ABS, PE+PTFE, and ABS+PC.

Randon copolymers

Alternating copolymers

Block copolymers

Graft copolymers

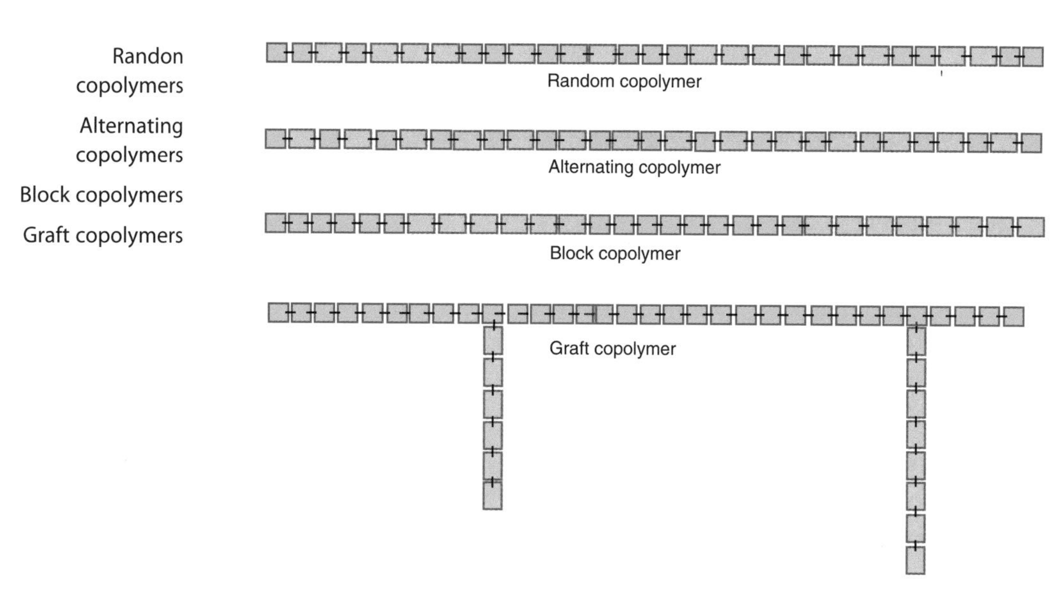

Figure 1.17 Schematic representation of different copolymers

1.7 Polymer Additives

A polymer seldom is sold as a pure material. More often a polymer contains several additives to aid during processing, add color, or enhance the mechanical properties [15].

1.7.1 Plasticizers

Plasticizers

Solvents, commonly called plasticizers, are sometimes mixed into a polymer to dramatically alter its rheological or mechanical properties. Plasticizers are used as processing aids because they have the same effect as raising the temperature of the material. The resulting lowered viscosities reduce the risk of thermal degradation during processing. For example, cellulose nitrite thermally degrades if it is processed without a plasticizer.

Plasticizers are more commonly used to alter a polymer's mechanical properties, such as stiffness, toughness, and strength. For example, adding a plasticizer such as dioctylphthalate (DOP) to PVC can reduce its stiffness by three orders of magnitude and can lower its glass transition temperature by 35 °C. In fact, highly plasticized PVC is rubbery at room temperature.

1.7.2 Flame Retardants

Flame retardants

Since polymers are organic materials, most of them are flammable. The flammability of polymers has always been a serious technical problem. However, some additives that contain halogens, such as bromine, chlorine, or phosphorous, reduce the possibility of either ignition within a polymer component, or once ignited, flame spread is limited. Bromine is a more effective flame retardant than chlorine [16].

1.7.3 Stabilizers

Stabilizers

The combination of heat and oxygen can result in thermal degradation in a polymer. Heat or energy produce free radicals which react with oxygen to form carbonyl compounds, giving rise to yellow or brown discolorations in the final product.

Thermal degradation can be suppressed by adding stabilizers, such as antioxidants or peroxide decomposers. These additives do not eliminate thermal degradation, but slow it down. Once the stabilizer has been consumed by reaction with oxygen, the polymer is no longer protected against thermal degradation.

Polyvinyl chloride is probably the polymer most vulnerable to thermal degradation. In polyvinyl chloride, scission of the C—Cl bond occurs at the weakest point of the molecule. The chlorine radicals react with their nearest CH group, forming HCl and creating new weak C—Cl bonds. A stabilizer must therefore be used to neutralize HCl and stop the autocatalytic reaction, as well as prevent corrosion of processing equipment.

1.7.4 Antistatic Agents

Antistatic agents

Since polymers have such low electrical conductivity, they can easily build up electric charges. The amount of charge build-up is controlled by the rate at which the charge is generated compared to the charge decay. The rate of charge generation at the surface of the component can be reduced by reducing the intimacy of contact, whereas the rate of charge decay is increased through surface conductivity. Hence, a good antistatic agent should be an ionizable additive that allows the charge to migrate to the surface. At the same time, it should be creating bridges to the atmosphere through moisture in the surroundings. Typical antistatic agents are nitrogen compounds, such as long-chain amines, and amides and polyhydric alcohols.

1.7.5 Fillers

Fillers

Fillers can be classified three ways: those that reinforce the polymer and improve its mechanical performance; those used to take up space and so reduce the amount of resin to produce a part — sometimes referred to as *extenders;* and those, less common, that are dispersed through the polymer to improve its electric conductivity.

Polymers that contain fillers that improve their mechanical performance are often referred to as reinforced plastics or *composites.* Composites can be furthermore divided into composites with *high performance* reinforcements and composites with *low performance* reinforcements. The high performance composites are those where the reinforcement is placed inside the polymer so that optimal mechanical behavior is achieved, such as unidirectional glass fibers in an epoxy resin. High performance composites usually have 50 to 80 % reinforcement by volume and usually have a laminated tubular shape containing braided reinforcements. The low performance composites are those where the reinforcement is small enough that it can be well dispersed into the matrix. These materials can be processed the same way as their unreinforced counterparts.

The most common filler used to reinforce polymeric materials is glass fiber. However, wood fiber, which is commonly used as an extender, also increases the stiffness and mechanical performance of some thermoplastics. To improve the bonding between the polymer matrix and the reinforcement, *coupling agents* such as *silanes* and *titanates* are often added.

Extenders, used to reduce the cost of the component, are often particulate fillers. The most common of these are calcium carbonate, silica flour, clay, and wood flour or fiber. As mentioned earlier, some fillers, such as clay, silica flour, and wood fiber, also slightly reinforce the polymer matrix. Polymers with extenders often have significantly lower toughness than when unfilled.

1.7.6 Blowing Agents

The task of blowing or foaming agents is to produce cellular polymers, also called expanded plastics. The cells can be completely enclosed (closed cell) or can be interconnected (open cell). Polymer foams are produced with densities between 1.6 kg/m^3 and 960 kg/m^3. There are many reasons for using polymer foams, such as their high strength to weight ratio, excellent insulating and acoustic properties, and high energy and vibration absorbing properties.

Polymer foams can be made by mechanically whipping gases into the polymer, or by either chemical or physical means. The basic steps of the foaming process are (1) cell nucleation, (2) expansion or growth of the cells, and (3) stabilization of the cells. Cell nucleation occurs when, at a given temperature and pressure, the solubility of a gas is reduced, leading to saturation, and expelling the excess gas to form bubbles. Nucleating agents are used for initial formation of a bubble. The bubble reaches an equilibrium shape when the pressure in the bubble balances the surface tension.

1.8 Plastics Recycling

Plastics recycling

We can divide plastics recycling into two major categories: industrial and postconsumer plastic scrap recycling. Industrial scrap is rather easy to recycle and re-introduce into the manufacturing stream, either within the same company as a regrind or sold to third parties as a homogeneous, reliable, and uncontaminated source of resin. Post-consumer plastic scrap recycling requires the material to go through a full life cycle prior to being reclaimed. This life cycle can be from a few days for packaging material to several years for electronic equipment housing material. The post-consumer plastic

SPI recycling codes

Figure 1.18 SPI resin identification codes

scrap can come from commercial, agricultural, and municipal waste. Municipal plastic scarp primarily consists of packaging waste, but also plastics from de-manufactured retired appliances and electronic equipment.

Post-consumer plastic scrap recycling requires collecting, handling, cleaning, sorting, and grinding. Availability and collection of post-consumer plastic scrap is perhaps one of the most critical aspects. Today, the demand for recycled plastics is higher than the availability of these materials. Although the availability of HDPE from bottles has seen a slight increase, the availability of recycled PET bottles has decreased in the past years. One of the main reasons for the decrease of PET is the fact that single serving PET bottles are primarily consumed outside of the home, making recycling and collection more difficult. On the other hand, HDPE bottles, which come from milk containers, soap and cleaning bottles, are consumed in the home and are therefore thrown into the recycling bin by the consumer. A crucial issue when collecting plastic waste is identifying the type of plastic used to manufacture the product. Packaging is often identified with the standard SPI identification symbol, which contains the triangular-shaped recycling arrows and a number between 1 and 7. Often, this is accompanied by the abbreviated name of the plastic. Table 1.3 and Fig. 1.18 [11] present the seven commonly recycled plastics in the United States along with the characteristics of each plastic, the main sources or packaging applications and the common applications for the recycled materials. Electronic housings are often identified with a molded-in name of the polymer used, such as ABS, as well as an identifier that reveals if a flame retardant was used, such as ABS-FR. When a product is not identified, various simple techniques exist, such as water or burning tests. The water test simply consists in determining if a piece of plastic floats or sinks after having added a drop of soap to the a container filled with water. If a part floats, it is either a polyethylene, a polypropylene, or an expanded or foamed plastic. Most of the remaining polymers will likely sink.

Table 1.3: Packaging plastics, characteristics, applications and use after recycling

Codes	Characteristics	Packaging applications	Recycled Products
(1) PET	Clarity, strength, toughness, barrier to gas and moisture, resistance to heat.	Plastic soft drink, water, sports drink, beer, mouthwash, catsup and salad dressing bottles; peanut butter, pickle, jelly and jam jars; heatable film and food trays.	Fiber, tote bags, clothing, film and sheet, food and beverage containers, carpet, strapping, fleece wear, luggage and bottles.

Codes	Characteristics	Packaging applications	Recycled Products
(2) PE-HD	Stiffness, strength, toughness, resistance to chemicals and moisture, permeability to gas, ease of processing, and ease of forming.	Milk, water, juice, shampoo, dish and laundry detergent bottles; yogurt and margarine tubs; cereal box liners; grocery, trash and retail bags.	Liquid laundry detergent, shampoo, conditioner and motor oil bottles; pipe, buckets, crates, flower pots, garden edging, film and sheet, recycling bins, benches, dog houses, plastic lumber, floor tiles, picnic tables, fencing.
(3) PVC	Versatility, clarity, ease of blending, strength, toughness, resistance to grease, oil and chemicals.	Clear food and nonfood packaging, medical tubing, wire and cable insulation, film and sheet, construction products such as pipes, fittings, siding, floor tiles, carpet backing and window frames.	Packaging, loose-leaf binders, decking, paneling, gutters, mud flaps, film and sheet, floor tiles and mats, resilient flooring, cassette trays, electrical boxes, cables, traffic cones, garden hose, mobile home skirting.
(4) PE-LD	Ease of processing, strength, toughness, flexibility, ease of sealing, barrier to moisture.	Dry cleaning, bread and frozen food bags, squeezable bottles, e. g., honey, mustard.	Shipping envelopes, garbage can liners, floor tile, furniture, film and sheet, compost bins, paneling, trash cans, landscape timber, lumber.
(5) PP	Strength, toughness, resistance to heat, chemicals, grease and oil, versatile, barrier to moisture.	Catsup bottles, yogurt containers and margarine tubs, medicine bottles.	Automobile battery cases, signal lights, battery cables, brooms, brushes, ice scrapers, oil funnels, bicycle racks, rakes, bins, pallets, sheeting, trays.
(6) PS	Versatility, insulation, clarity, easily formed.	Compact disc jackets, food service applications, grocery store meat trays, egg cartons, aspirin bottles, cups, plates, cutlery.	Thermometers, light switch plates, thermal insulation, egg cartons, vents, desk trays, rulers, license plate frames, foam packing, foam plates, cups, utensils.
(7) Other	Dependent on resin or combination of resins.	Three and five gallon reusable water bottles, some citrus juice and catsup bottles.	Bottles, plastic lumber applications.

1.9 The Plastics and Rubber Industries

After the discovery of the vulcanization process, the rubber industry experienced significant growth. However, in the late 1860s the rubber industry saw one of its main applications, dentures, threatened by celluloid, a new material that would render better texture and color, giving people a much nicer smile. In view of loosing a rather profitable business to these new materials, referred to as plastics, the rubber industry started a propaganda campaign against cellulose in all major US newspapers. They falsely claimed that celluloid dentures could easily explode in ones mouth when coming in contact with hot food. This not only cheated people of a better smile, but also started a rivalry between the two industries which has caused them to maintain as completely separate entities to this day (Fig. 1.19). In fact, despite of the materials and processing similarities between plastics and rubber, the plastics industry and the rubber industry have completely separate societies and technical journals. A plastics engineer is likely to be found in meetings organized by the *Society of Plastics Engineers* (SPE) or the *Society of the Plastics Industry* (SPI), while a member of the rubber industry will attend meetings organized by their own society, the *Rubber Division of the American Chemical Society*.

The plastics industry versus the rubber industry

More importantly, each industry utilizes its own sets of standards to evaluate the materials. Furthermore, whole companies concentrate on either one or the other industry.

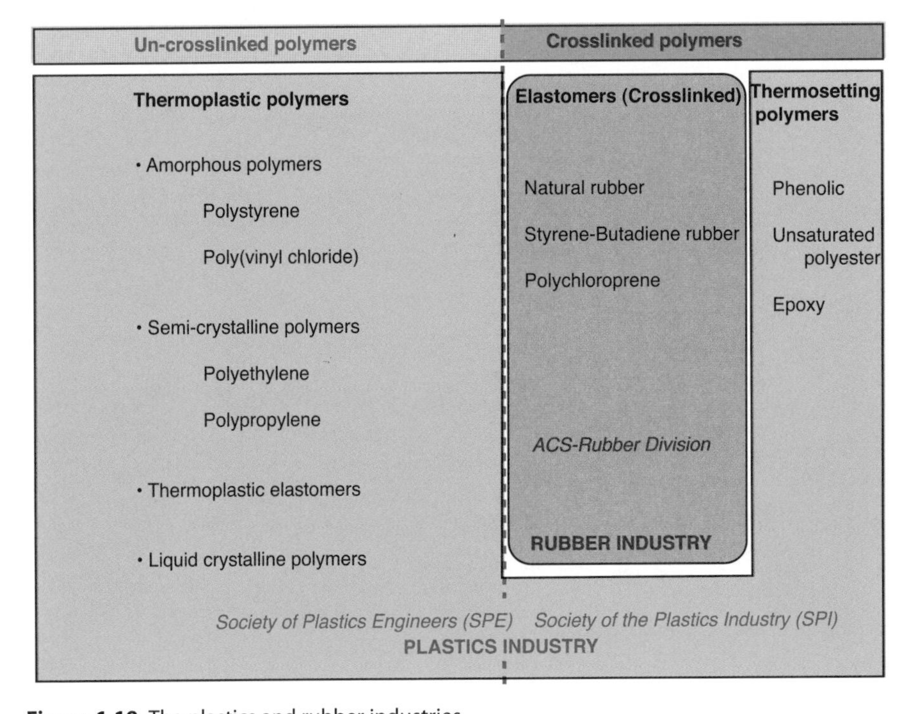

Figure 1.19 The plastics and rubber industries

1.10 Polymer Processes

The two main plastics processing techniques are the *extrusion* and *injection molding processes.* As covered in detail in Chapters 4 and 6, the fundamental element of both

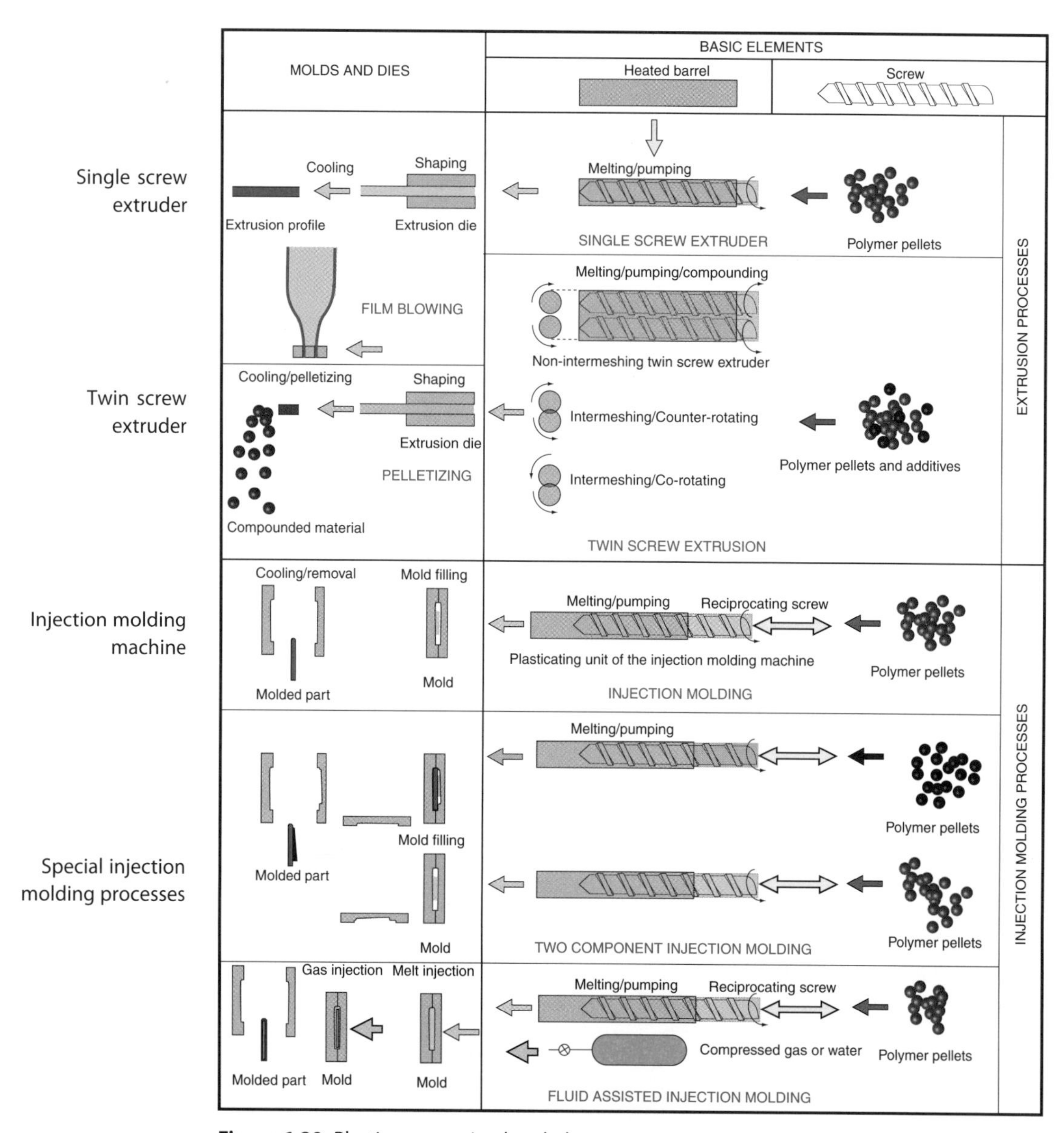

Figure 1.20 Plastics processing break down

these manufacturing methods is the screw and heated barrel system, as depicted in Fig. 1.20.

The right column in the figure first presents extrusion, with its two major categories, the *single screw extruder* and the *twin screw extruder.* The twin screw system is an arrangement of two screws inside a double or twin barrel, primarily used for mixing (*compounding*). Mixing and compounding is covered in Chapter 5 of this book. Both, the single and twin screw systems, are used for melting the resin, as well as for pumping the polymer melt through the extrusion die. The pumping action is accomplished by generating the pressure required to push the melt through the die. The die in the extrusion systems are used to shape the material into a continuous product, such as a film, a plate, a tube, a strand, or any desired profile. In most cases, twin screw extruder dies produce strands that are cut into pellets of the compounded plastic or *resin*, which is used in subsequent extrusion or injection molding processes. The pellets that result from a twin screw compounding process are typically pellets of *polymer blends.*

Injection molding is perhaps the most widely and versatile process in the plastics industry. Figure 1.20 presents the process with some of its variations, such as *two-component injection molding*, where one plastic is over-molded on a pre-injected substrate. Related to the two-component injection process is the *co-injection molding process*, where one resin displaces another creating a product with a skin made up of the first resin, and a central core of the second material. If the second material is a fluid, such as nitrogen or water, one creates a hollow part. This type of process can be referred to as *fluid assisted injection molding.* The two commercial fluid assisted processes are the *gas-assisted injection molding* and the *water assisted injection molding processes.*

References

1. Osswald, T. A., and Menges, G., *Materials Science of Polymers for Engineers*, Hanser Publishers (2003), Munich
2. de la Condamine, C. M., *Relation Abregee D'un Voyage Fait Dans l'interieur de l'Amerique Meridionale*, Academie des Sciences (1745), Paris
3. DuBois, J. H., *Plastics History U. S. A.*, Cahners Publishing Co., Inc. (1972), Boston
4. Tadmor, Z., and Gogos, C. G., *Principles of Polymer Processing*, John Wiley & Sons (2006), New York
5. McPherson, A. T., and Klemin, A., *Engineering Uses of Rubber*, Reinhold Publishing Corporation (1956), New York
6. Sonntag, R., *Kunststoffe* (1985), *75*, 4
7. Herrmann, H., *Kunststoffe* (1985), *75*, 2
8. Regnault, H. V. Liebigs Ann. (1835), 14, 22
9. Ulrich, H., *Introduction to Industrial Polymers*, 2nd Ed., Hanser Publishers (1993), Munich
10. Rauwendaal, C., *Polymer Extrusion*, 4th ed., Hanser Publishers (2001), Munich

11. Osswald, T. A., Baur, E., Brinkmann, S. and Schmachtenberg, E., *International Plastics Handbook*, Hanser Publishers (2006), Munich

12. Campo, E. A., *Industrial Polymers*, Hanser Publishers (2008), Munich

13. Ehrenstein, G. W., *Polymeric Materials: Structure-Properties-Applications*, Hanser Publishers (2001), Munich

14. Domininghaus, H., *Plastics for Engineers*, Hanser Publishers (1993), Munich

15. Zweifel, H., Maier, R. and Schiller, M., *Plastics Additives Handbook*, Hanser Publishers, (2009) Munich

16. Weil, E. D. and Levchik, S. V. *Flame Retardants for Plastics and Textiles — Practical Applications, Hanser Publishers* (2009), Munich

2 Mechanical Behavior of Polymers

The mechanical properties of a polymeric component are dominated by its viscoelasticity. This is reflected by the time-dependency of the mechanical response of a component during loading. Hence, a polymer behaves differently if subjected to short term or long term loads. This chapter briefly explains polymer viscoelasticity and covers the short and long term mechanical behavior of polymers.

2.1 Viscoelastic Behavior of Polymers

A polymer, at a specific temperature and molecular weight, may behave as a liquid or a solid, depending on the speed (time scale) at which its molecules deform. This behavior, which ranges between liquid and solid, is generally referred to as the viscoelastic behavior or material response. This discussion is limited to *linear viscoelasticity*, which is valid for polymer systems undergoing *small or slow deformations*. *Non-linear viscoelasticity* is required when modeling *large rapid deformations*, such as those encountered in flowing polymer melts.

In linear viscoelasticity, the *stress relaxation test* is often used, along with the *time-temperature superposition principle* and the *Boltzmann superposition principle*, to explain the behavior of polymeric materials during deformation.

2.1.1 Stress Relaxation

In a stress relaxation test, a polymer specimen is suddenly deformed by a fixed amount, ϵ_0, and the stress required to hold that amount of deformation is recorded over time. This test is cumbersome to perform, so the design engineer and the material scientist have tended to ignore it. In fact, the standard relaxation test, ASTM D2991, was dropped by ASTM in the early 1990s. Rheologists and scientists, however, have consistently used the stress relaxation test to interpret the viscoelastic behavior of polymers.

Figure 2.1 [1] presents the stress relaxation modulus measured for polyisobutylene[1] at various temperatures. Here, the stress relaxation modulus is defined by

Stress relaxation

$$E_r = \frac{\sigma(t)}{\varepsilon_0} \tag{2.1}$$

Relaxation modulus decreases with time

where ε_0 is the applied strain and $\sigma(t)$ the stress being measured. From the test results, stress relaxation is clearly time and temperature dependent, especially around the glass transition temperature where the curve is steepest. For the polyisobutylene in Fig. 2.1,

1) Better known as the matrix material for chewing gum.

Relaxation time

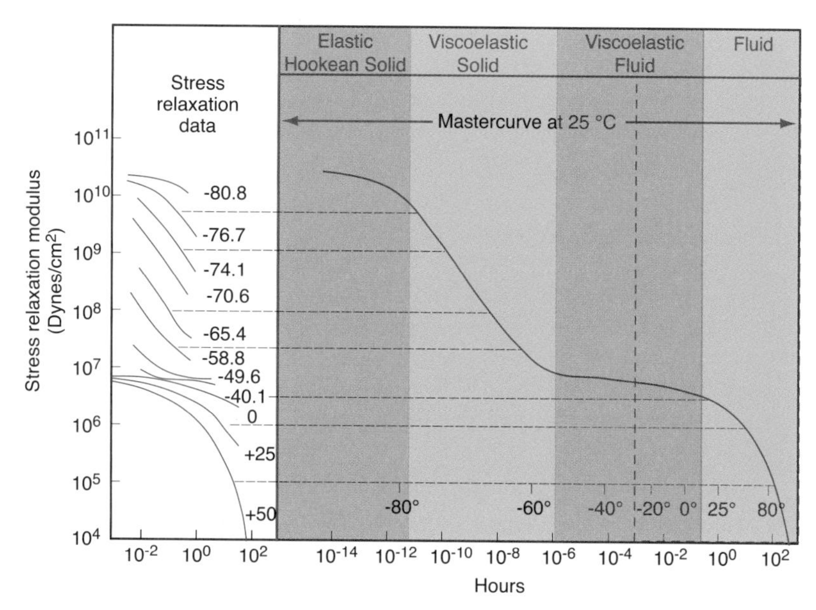

Figure 2.1 Relaxation modulus curves for polyisobutylene and corresponding master curve at 25 °C

the glass transition temperature is about –70 °C. The measurements were completed in an experimental time window between a few seconds and one day. The tests performed at lower temperatures were used to record the initial relaxation while the ones at higher temperatures only captured the end of relaxation of the rapidly decaying stresses. The tests at various temperatures can be used to assemble a master curve at one of the test temperatures, as is shown for 25 °C on the right side of Fig. 2.1.

The time it takes for the imposed stresses to relax is governed by the relaxation time, λ. For the 25 °C case, the relaxation time is about 100 hours. High temperatures lead to small molecular relaxation times and low temperatures lead to materials with long relaxation times. When changing temperature, the shape of relaxation test results remains the same, except for a horizontal shift to the left or right, representing lower or higher response times, respectively.

2.1.2 Time-Temperature Superposition

The time-temperature equivalence seen in stress relaxation test results can be used to reduce data at various temperatures to one general *master curve* for a reference temperature, T. To generate a master curve at any reference temperature, the curves shown on the left of Fig. 2.1 must be shifted horizontally, holding the reference curve fixed. The master curve for the data in Fig. 2.1 is on the right of the figure. Each curve was shifted horizontally until the ends of all the curves superposed. The amount that each curve was shifted can be plotted with respect to the temperature difference taken

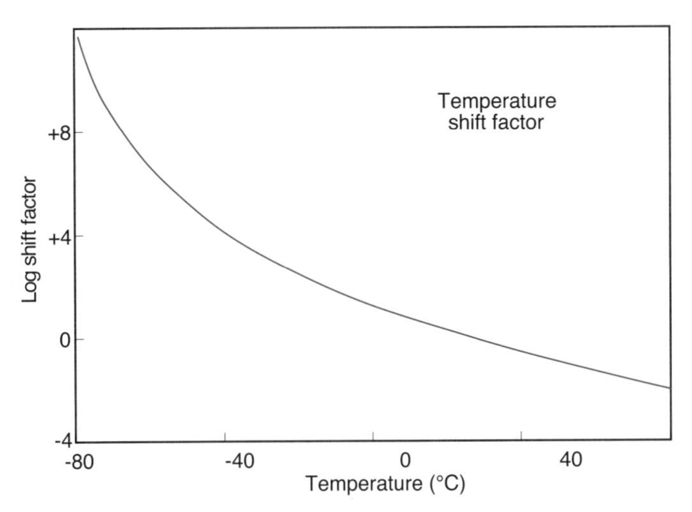

Figure 2.2 Shift factor as a function of temperature used to generate the master curve plotted in Fig. 2.1

from the reference temperature. For the data in Fig. 2.1, the shift factor is shown in Fig. 2.2. It is important to point out here that the relaxation master curve represents a material at a single temperature, but depending on the time scale, it can be regarded as a Hookean solid, or a viscous fluid. In other words, if the material is loaded for a short time, the molecules are not allowed to move and slide past each other, resulting in a perfectly elastic material. In such a case, the deformation is fully recovered. However, if the test specimen is maintained deformed for an extended period of time, such as 100 hours for the 25 °C case, the molecules will have enough time to slide and move past each other, fully relaxing the initial stresses, resulting in permanent deformation. For such a time scale the material can be regarded as a fluid.

WLF Equation [2]

The amount relaxation curves must be shifted in the time axis to line-up with the master curve at a reference temperature is represented by

Williams-Landel-Ferry (WLF) equation

$$\log t - \log t_{\mathrm{ref}} = \log\left(\frac{t}{t_{\mathrm{ref}}}\right) = \log a_T \qquad (2.2)$$

Although the results in Fig. 2.2 where shifted to a reference temperature of 298 K (25 °C), Williams, Landel and Ferry [2] chose $T_{\mathrm{ref}} = 243$ K for

$$\log a_T = \frac{-8.86\,(T - T_{\mathrm{ref}})}{101.6 + T - T_{\mathrm{ref}}} \qquad (2.3)$$

which holds for nearly all amorphous polymers if the chosen reference temperature is 45 K above the glass transition temperature. In general, the horizontal shift, $\log a_T$, between the relaxation responses at various temperatures to a reference temperature can be computed using the well known Williams-Landel-Ferry [2] (WLF) equation. The WLF equation is

$$\log a_T = \frac{-C_1(T - T_{\text{ref}})}{C_2 + T - T_{\text{ref}}} \tag{2.4}$$

where C_1 and C_2 are material dependent constants. It has been shown that with $C_1 = 17.44$ and $C_2 = 51.6$, Eq. 2.4 fits well for a wide variety of polymers as long as the glass transition temperature is chosen as the reference temperature. These values for C_1 and C_2 are often referred to as universal constants. Often, the WLF equation must be adjusted until it fits the experimental data. Master curves of stress relaxation tests are important because the polymer's behavior can be traced over much longer periods than those that can be determined experimentally.

Boltzmann Superposition Principle

In addition to the *time-temperature superposition principle* (WLF), the *Boltzmann superposition principle* is of extreme importance in the theory of linear viscoelasticity. The Boltzmann superposition principle states that the deformation of a polymer component is the sum or superposition of all strains that result from various loads acting on the part at different times. This means that the response of a material to a specific load is independent of pre-existing loads. Hence, we can compute the deformation of a polymer specimen upon which several loads act at different points in time by simply adding all strain responses.

2.2 The Short-Term Tensile Test

The most commonly used mechanical test is the short-term stress-strain tensile test. Stress-strain curves for selected polymers are displayed in Fig. 2.3 [3]. For comparison, the figure also presents stress-strain curves for copper and steel. Although they have much lower tensile strengths, many engineering polymers exhibit much higher strains at break than metals.

The next two sections discuss the short-term tensile test for cross-linked elastomers and thermoplastic polymers separately. The main reason for separating these two polymers is that the deformation of a crosslinked elastomer and an uncrosslinked thermoplastic differ greatly. The deformation in a crosslinked polymer is generally reversible, while the deformation in typical uncross-linked polymers is associated with molecular chain relaxation, making the process time-dependent and irreversible.

Comparing stress-strain behavior

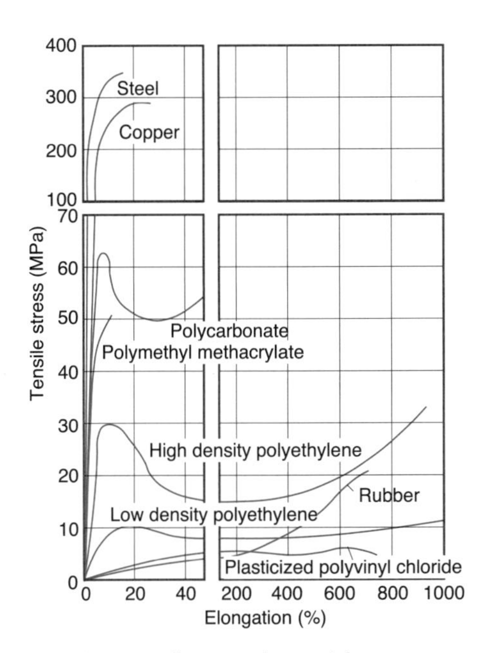

Figure 2.3 Tensile stress-strain curves for several materials

2.2.1 Elastomers

The main feature of cross-linked elastomeric materials is that they can undergo large, reversible deformations. This is because the curled polymer chains stretch during deformation but are hindered from sliding past each other by the crosslinks between the molecules. Once a load is released, most molecules recoil. As an elastomeric polymer component is deformed, the slope in the stress-strain curves drops significantly because the uncurled molecules provide less resistance and entanglement, allowing them to move more freely. Eventually, at deformations of about 400 %, the slope starts to increase because the polymer chains are fully stretched. This stretch is followed by polymer chain breakage or crystallization, ending with the fracture of the component. Stress-deformation curves for natural rubber (NR) [4] and a rubber compound [5] composed of 70 parts of styrene-butadiene-rubber (SBR) and 30 parts of natural rubber are shown in Fig. 2.4.

Rubber elasticity

Because of the large deformations, typically several hundred percent, the stress-strain data is usually expressed in terms of the extension ratio, λ, defined by

Extension and strain

$$\lambda = \frac{l}{l_0} \tag{2.5}$$

where l and l_0 are the instantaneous and initial lengths of the specimen, respectively.

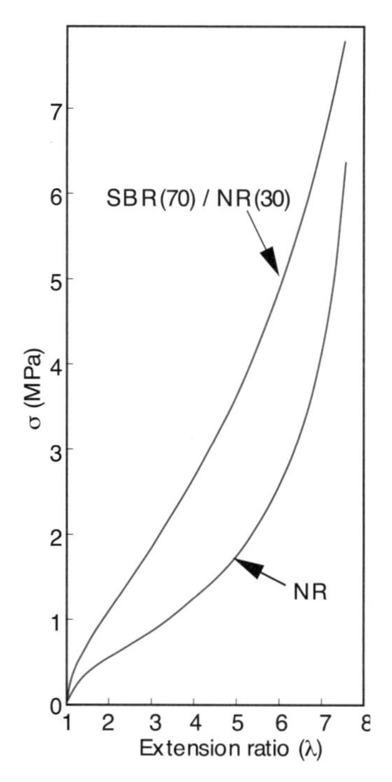

Figure 2.4 Experimental stress-extension curves for NR and a SBR/NR compound

For a component in uniaxial extension, or compression, the stress can be computed as

$$\sigma = G_0 \left(\lambda - \frac{1}{\lambda^2} \right) \tag{2.6}$$

where G_0 is the shear modulus at zero extension, which for rubbers can be approximated by

$$G_0 = \frac{E_0}{3} \tag{2.7}$$

with E_0 as the elastic tensile modulus at zero extension. The model agrees with experiments up to about 50 % extension ($\lambda = 1.5$). For compression, the model agrees much better with experiments, as shown for natural rubber in Fig. 2.5 [4]. Fortunately, rubber products are rarely deformed more than 25 % in compression or tension, allowing the use of the above model.

The corresponding model for equibiaxial extension (inflation) of thin sheets is:

$$\sigma = G_0 \left(\lambda^2 - \frac{1}{\lambda^4} \right) \tag{2.8}$$

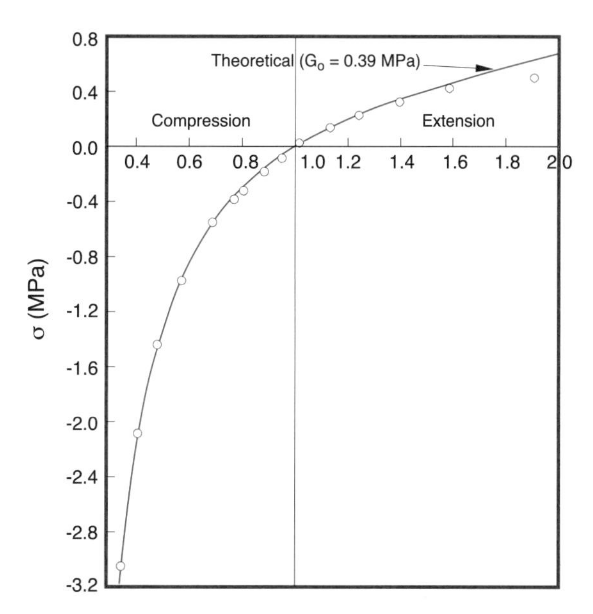

Rubber elasticity

Figure 2.5 Experimental and theoretical stress-extension and compression curves for natural rubber

2.2.2 Thermoplastic Polymers

Of all the mechanical tests done on thermoplastic polymers, the tensile test is the least understood, and the results are often misinterpreted and misused. Because the test was inherited from other materials that have linear elastic stress-strain responses, it is often inappropriate for testing polymers.

A typical test performed on PMMA at various strain rates at room temperature is shown in Fig. 2.6. The increased curvature in the results with slow elongational speeds suggests that stress relaxation plays a significant role during the test. Here, the high rates of deformation reduce the loading time, allowing higher stress to build-up during the test.

The ASTM D638 stress-strain test should only be used to compare one material with another. The resulting modulus, ultimate strength and maximum strain at failure are only valid for the rate of deformation used during the test

The standard test used to measure the stress-strain behavior of thermoplastic polymers is the widely accepted ASTM D638 test and its ISO counterpart the ISO 527 test. Because the standard test uses a single deformation speed, the resulting moduli, maximum strain and ultimate stress, can only be used when comparing one material against another, and not for design purposes.

The stiffness and strength of polymers and rubbers is increased by filling them with solid particles, such as calcium carbonate and carbon black. The most common expression for describing the effect of carbon black content on the modulus of rubber was originally derived by Guth and Simha [6, 7] for the viscosity of particle suspensions, and later used by Guth to predict the modulus of filled polymers. The Guth equation

Effect of rate of deformation on mechanical behavior of thermoplastics

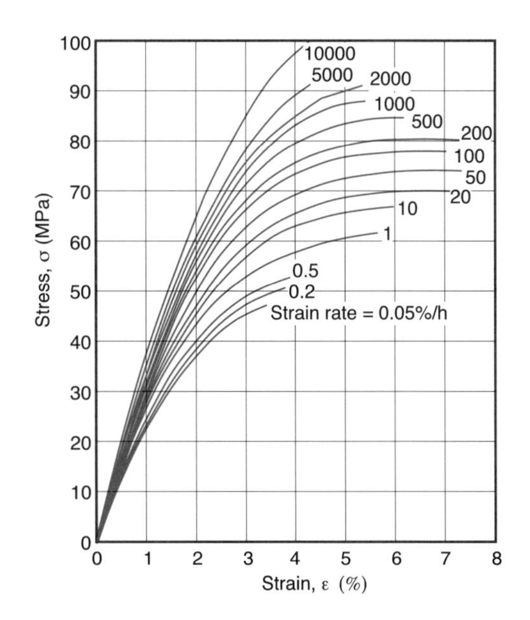

Figure 2.6 Stress-strain behavior of PMMA at various strain rates

Effect of filler fraction on modulus

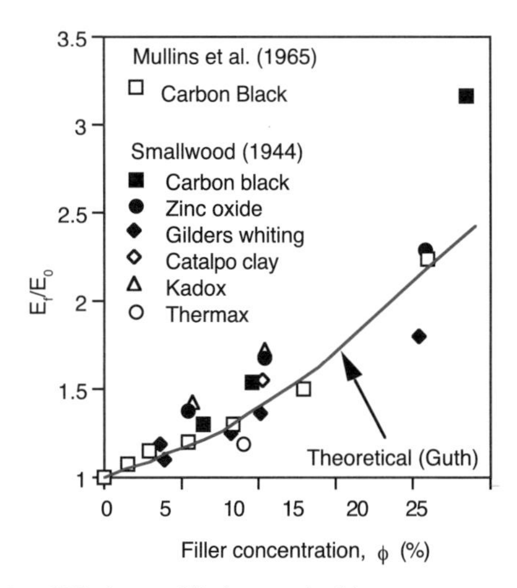

Figure 2.7 Modulus ratio of filled to unfilled natural rubber

can be written as

Effect of filler fraction on modulus

$$\frac{G_f}{G_0} = 1 + 2.5\varphi + 14.1\varphi^2 \tag{2.9}$$

where G_f is the shear modulus of the filled material and φ is the volume fraction of particulate filler. The above expression is compared to experiments [8, 9] in Fig. 2.7.

2.3 Long-Term Tests

The *stress relaxation* and the *creep* test are well-known long-term tests. The stress relaxation test is difficult to perform and is, therefore, often approximated by data acquired through the more commonly used creep test. The stress relaxation of a polymer is often considered the inverse of creep.

The creep test, which can be performed either in shear, compression, or tension, measures the flow of a polymer component under a constant load. It is a common test that measures the strain, ε, as a function of stress, time, and temperature. Standard creep tests such as DIN 53444 and ASTM D2990 can be used. Creep tests are performed at a constant temperature, using a range of applied stress, as shown in Fig. 2.8 [10], where the creep responses of a polypropylene copolymer are presented for a range of stresses in a graph with a log-scale for time. If plotting creep data in a log-log graph, in the majority of the cases, the creep curves reduce to straight lines. Hence, the creep behavior of most polymers can be approximated with a power law:

Creep model

$$\varepsilon(t) = k(T)\,\sigma^m\,t^n \tag{2.10}$$

where $k(T)$, m and n are material dependent properties.

As for the stress relaxation test, the creep behavior of a polymer depends heavily on the material temperature during testing, with the highest rates of deformation near the glass transition temperature.

Processing creep data

Isochronous curves

Isometric curves

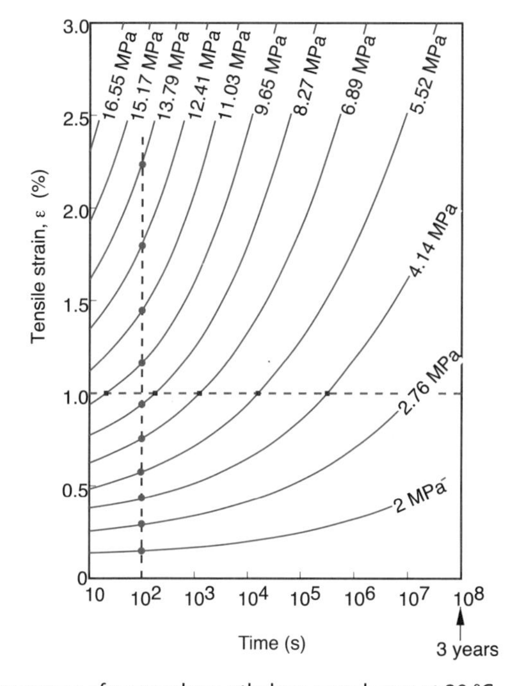

Figure 2.8 Creep response of a propylene-ethylene copolymer at 20 °C

Creep data is often presented in terms of the creep modulus, E_c, defined by

$$E_c = \frac{\sigma_0}{\varepsilon(t)} \tag{2.11}$$

2.3.1 Isochronous and Isometric Creep Plots

Typical creep test data, shown in Fig. 2.8, can be manipulated to be displayed as short-term stress-strain tests or as stress relaxation tests. These manipulated creep curves are called *isochronous* and *isometric* graphs.

An isochronous plot of the creep data is generated by cutting sections through the creep curves at constant times and plotting the stress as a function of strain, as shown in Fig. 2.8. The isochronous curves of the creep data in Fig. 2.8 are shown in Fig. 2.9 [10, 11]. Similar curves can also be generated by performing a series of *short creep tests*, where a specimen is loaded at a specific stress for a short period, typically around 100 seconds. The load is then removed, and the specimen relaxes for a period of at least four times the loading interval. The specimen is then reloaded at a different stress, and the test is repeated until there are sufficient points for an isochronous graph. This procedure is more expedient than the regular creep test and is often used to predict the short-term behavior of polymers. However, it should be pointed out that the short-term tests described in Section 2.2.2 are more accurate, quicker, and less expensive to run.

The isometric or "equal size" plots of the creep data are generated by taking constant strain sections of the creep curves and by plotting stress versus time. Isometric curves of the polypropylene creep data in Fig. 2.8 are shown in Fig. 2.10 [10]. This plot resembles the stress relaxation test results and is often used similarly. When we divide the stress axis by the strain, we can also plot modulus versus time.

2.3.2 Creep Rupture

A creeping polymer component eventually undergoes catastrophic failure, generally called creep rupture or static fatigue. The standard test for creep rupture is the same as the creep test discussed earlier. Results from creep rupture tests are usually graphs of applied stress versus the logarithm of time to rupture. An example of a creep rupture test that ran for several decades, starting in 1958, is shown in Fig. 2.11 [12, 13]. Here, the creep rupture of HDPE pipes under internal pressures was tested at different temperatures. Two general regions with different slopes become obvious in the plots. The curves to the left of the knee correspond to ductile failure, whereas those to the right correspond to brittle failure. Generating a graph such as the one presented in Fig. 2.11 is lengthy, taking several years of testing. Once the steeper slope, typical of brittle fracture, has been reached, extrapolation with some confidence can be used to estimate future creep rupture times.

Isochronous and isometric curves

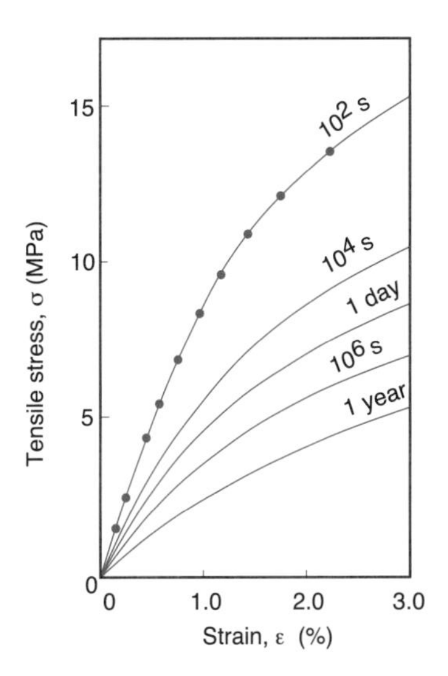

Figure 2.9 Isochronous stress-strain curves for the creep responses in Fig. 2.8

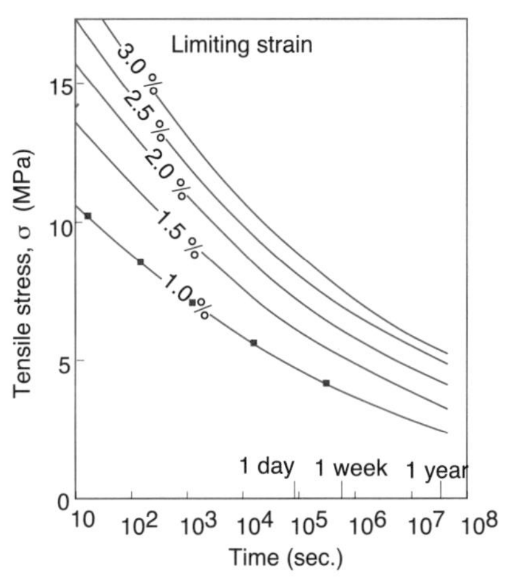

Figure 2.10 Isometric stress-time curves for the creep responses in Fig. 2.8

Creep rupture curves

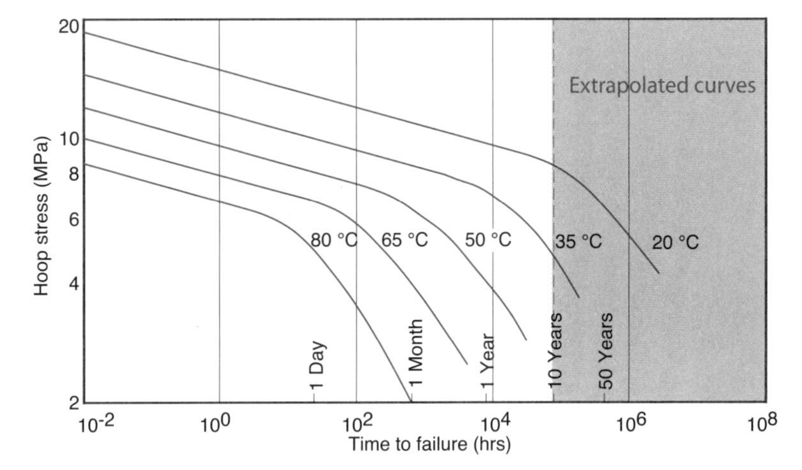

Figure 2.11 Creep rupture behavior as a function of temperature for a HDPE

2.4 Dynamic Mechanical Tests

The sinusoidal oscillatory test, also called the dynamic-mechanical-analysis (DMA) test, is one of the most useful mechanical tests for polymers. Here, a specimen is excited with a low frequency stress input, which is recorded along with the strain response. The shapes of the test specimen and the testing procedure vary significantly from test to test. The various tests and their corresponding specimen shapes are described by ASTM D4065 and the terminology is described by ASTM D4092. The typical responses measured in these dynamic tests are a storage modulus, G', and a loss modulus, G''. The storage modulus is related to the elastic modulus of the polymer at the loading frequency and the loss modulus to the damping or dissipative component observed during loading. The loss modulus can also be written in terms of loss tangent (tan δ) or logarithmic decrement, Δ. The latter is related to the damping of a freely oscillating specimen.

DMA tests

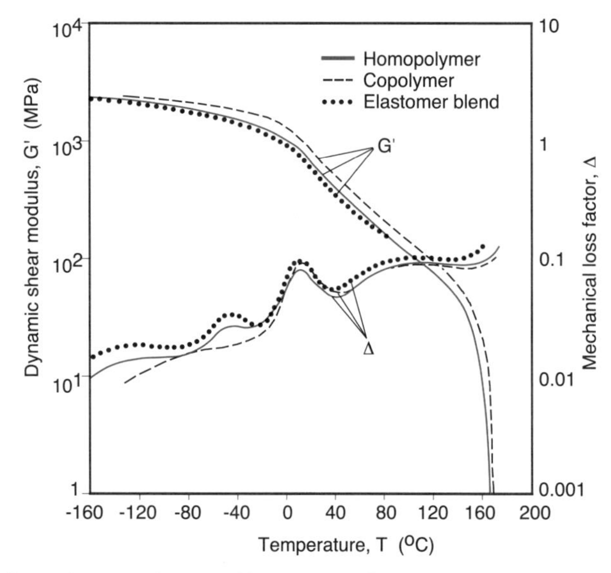

Figure 2.12 Elastic shear modulus and loss tangent for various polypropylene grades

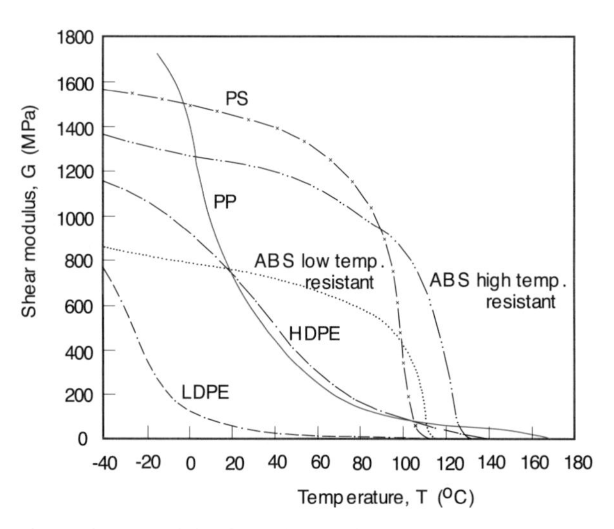

Figure 2.13 Elastic shear modulus for various polymers

Figure 2.12 [3] shows the elastic shear modulus and the logarithmic decrement or loss factor for various polypropylene grades. Here, the glass transition temperatures and the melting temperatures can be seen for the various polypropylene grades. The vertical scale in plots such as in Fig. 2.12 is usually logarithmic. However, a linear scale better describes the mechanical behavior of polymers in design aspects. Figure 2.13 [3] presents the elastic shear modulus versus temperature on a linear scale for several thermoplastic polymers.

If the test specimen in a sinusoidal oscillatory test is linearly elastic, the strain input and stress response would be in phase

Elastic response – stress response is in sync with the strain input

$$\gamma = \gamma_0 \sin(\omega t) \tag{2.12}$$

$$\tau = \tau_0 \sin(\omega t) \tag{2.13}$$

For an ideally viscous test specimen (Newtonian), the stress response lags $\frac{\pi}{2}$ radians behind the strain input

Viscous response – stress response lags behind the strain input by $\frac{\pi}{2}$ radians

$$\gamma = \gamma_0 \sin(\omega t) \tag{2.14}$$

$$\tau = \tau_0 \sin\left(\omega t - \frac{\pi}{2}\right) \tag{2.15}$$

Polymers behave somewhere in between these cases and their response is described by

Viscoelastic response – stress response lags behind the strain input

$$\gamma = \gamma_0 \sin(\omega t) \text{ and} \tag{2.16}$$

$$\tau = \tau_0 \sin(\omega t - \delta) \tag{2.17}$$

which results in a complex shear modulus

Complex modulus

$$G^* = \frac{\tau(t)}{\gamma(t)} = \frac{\tau_0 e^{i\delta}}{\gamma_0} = \frac{\tau_0}{\gamma_0}(\cos\delta + i\sin\delta) = G' + G'' \tag{2.18}$$

where G' is usually called the *storage modulus* and G'' is the *loss modulus*. The ratio of loss modulus to storage modulus is called the *loss tangent:*

$$\tan\delta = \frac{G''}{G'} \tag{2.19}$$

Although the elastic shear modulus, G' and the loss modulus G'', are sufficient to characterize a material, one can also compute the logarithmic decrement, Δ, or loss factor by using

Logarithmic decrement – damping

$$\Delta = \frac{G''\pi}{G'} \tag{2.20}$$

The logarithmic decrement can also be written in terms of *loss tangent*, $\tan\delta$, where δ is the mechanical loss angle. The loss tangent is then

$$\tan\delta = \frac{G''}{G'} = \frac{\Delta}{\pi} \tag{2.21}$$

2.5 Mechanical Behavior of Filled and Reinforced Polymers

Fillers are materials intentionally inserted in polymers to make them stronger, lighter, electrically conductive, or less expensive. Any filler affects the mechanical behavior of a polymeric material. For example, long fibers make the polymer stiffer but usually denser, whereas foaming makes it more compliant and much lighter. On the other hand, a filler, such as calcium carbonate, decreases the polymer's toughness, while making it considerably cheaper to produce.

Reinforced plastics are matrix polymers whose properties have been enhanced by introducing a reinforcement (fibers) of higher stiffness and strength. Such a material is usually called a *fiber reinforced polymer* (FRP) or a *fiber reinforced composite* (FRC). The purpose of introducing a fiber into a matrix is to transfer the load from the weaker material to the stronger one. This load transfer occurs over the length of the fiber as shown in Fig. 2.14.

A matrix completely transfers the load to the fiber

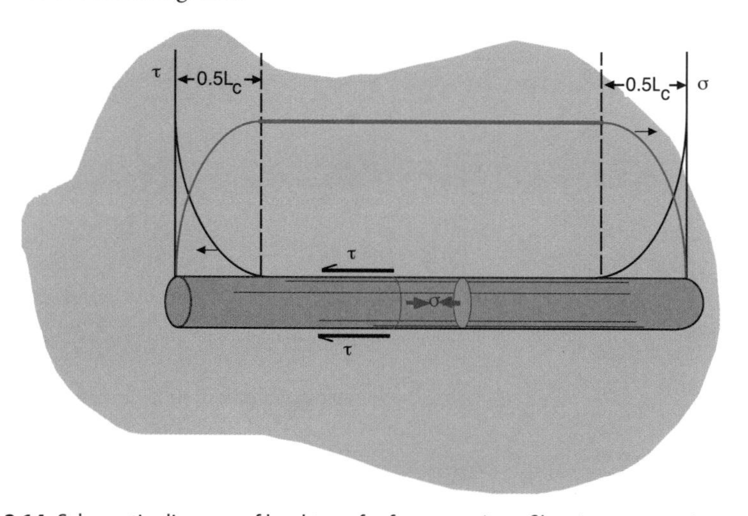

Figure 2.14 Schematic diagram of load transfer from matrix to fiber in a composite

The length to complete the load transfer from the matrix to the fiber, without fiber or matrix fracture, is usually called the critical length, L_C. For the specific case, where there is perfect adhesion between fiber and matrix, experimental evidence suggests that aspect ratios of 100 or higher are required for maximum strength [14]. If composites have fibers that are shorter than their critical length, they are referred to as *short fiber composites*. If the fibers are longer, they are called *long fiber composites* [15].

Halpin and Tsai [16] developed a widely used model to predict the mechanical properties of aligned fiber reinforced composite laminates. With the notation in Fig. 2.15, where f and m represent the fiber and matrix, respectively; L the fiber length; D the fiber diameter; φ the volume fiber fraction, the longitudinal (L) and transverse (T)

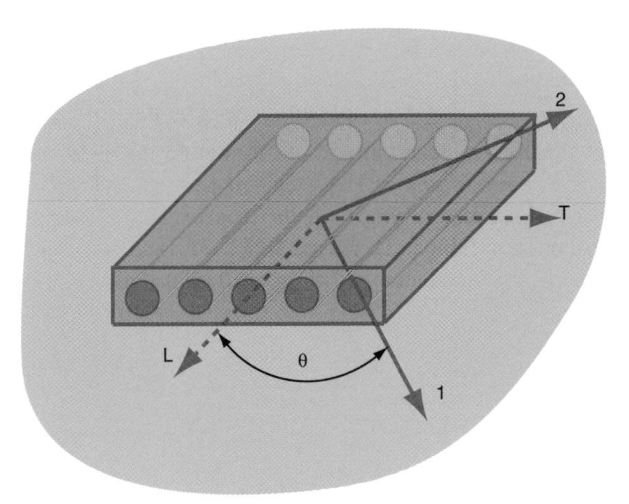

Laminate layer model

Figure 2.15 Schematic diagram of unidirectional, continuous fiber reinforced laminated structure

properties can be predicted using

$$E_{\mathrm{L}} = E_{\mathrm{m}}\left(\frac{1+\xi\,\eta\,\varphi}{1-\eta\,\varphi}\right) \tag{2.22}$$

$$E_{\mathrm{T}} = E_{\mathrm{m}}\left(\frac{1+\eta\varphi}{1-\eta\varphi}\right) \tag{2.23}$$

$$G_{\mathrm{LT}} = G_{\mathrm{m}}\left(\frac{1+\lambda\varphi}{1-\lambda\varphi}\right) = G_{\mathrm{m}}\frac{\nu_{\mathrm{LT}}}{\nu_{\mathrm{m}}} \tag{2.24}$$

where,

Halpin-Tsai model

$$\eta = \frac{\left(\dfrac{E_{\mathrm{f}}}{E_{\mathrm{m}}}-1\right)}{\left(\dfrac{E_{\mathrm{f}}}{E_{\mathrm{m}}}+\xi\right)} \tag{2.25}$$

$$\lambda = \frac{\left(\dfrac{G_{\mathrm{f}}}{G_{\mathrm{m}}}-1\right)}{\left(\dfrac{G_{\mathrm{f}}}{G_{\mathrm{m}}}+1\right)} \tag{2.26}$$

$$\xi = 2\frac{L}{D} \tag{2.27}$$

Most models accurately predict the longitudinal modulus, as shown in Fig. 2.16 [17]. However, differences do exist between models when predicting the transverse modulus, as shown in Fig. 2.17 [17].

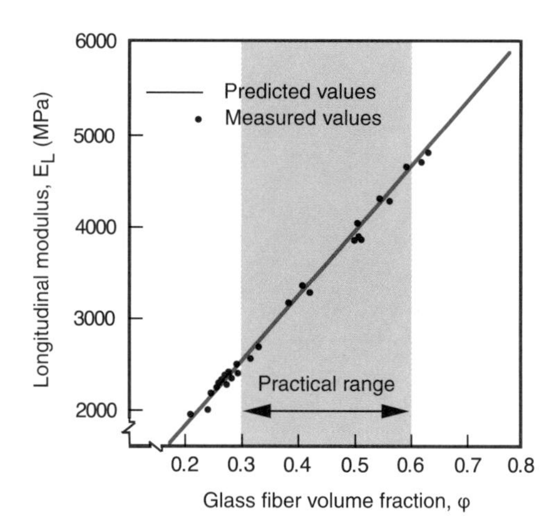

Figure 2.16 Measured and predicted longitudinal modulus for an unsaturated polyester/aligned glass fiber composite laminate as a function of volume fraction of glass content

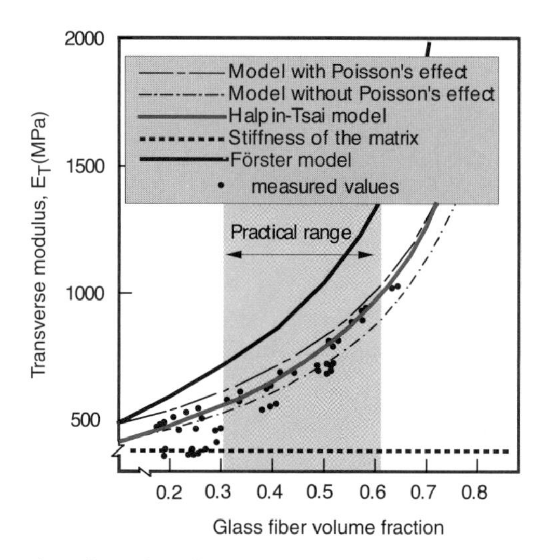

Figure 2.17 Measured and predicted transverse modulus for an unsaturated polyester/aligned glass fiber composite laminate as a function of volume fraction of glass content

Figure 2.18 [18] shows how the stiffness decreases as one rotates away from the longitudinal axis for an aligned fiber reinforced composite with different volume fraction fiber content.

For high volume fraction fiber contents, only a slight misalignment of the fibers from the loading direction results in drastic property reductions.

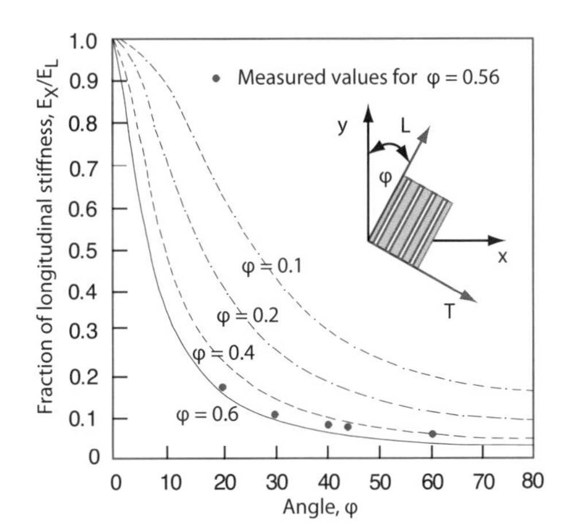

Mechanical properties decrease rapidly with misalignment

Figure 2.18 Measured and predicted elastic modulus in a unidirectional fiber reinforced laminate as a function of angle between loading and fiber direction

The stiffness in a long fiber reinforced composite with a random planar orientation, such as encountered in sheet molding compound (SMC) charges, can be estimated using

$$E_{11} = E_{22} = E_{\text{random}} = \left(\frac{3}{8}\frac{1}{E_L} + \frac{3}{8}\frac{1}{E_T} - \frac{2}{8}\frac{\nu_{LT}}{E_L} + \frac{1}{8}\frac{1}{G_{LT}} \right)^{-1} \qquad (2.28)$$

2.6 Impact Strength

In practice, nearly all polymer components are subjected to impact loads. Since many polymers are tough and ductile, they are often well suited for this type of loading. However, under specific conditions even the most ductile materials, such as polypropylene, can fail in a brittle manner at low strains. These types of failure are prone to occur at low temperatures during high deformation rates.

According to several researchers [19, 20], a significantly high rate of deformation leads to complete embrittlement of polymers, resulting in a lower threshold of elongation at break. Menges and Boden designed a special high-speed elongational testing device to measure the minimum work required to break specimens. The minimum strain, ϵ_{min}, which can be measured with such a device, is a safe value to use in design calculations. One should always assume that if this minimum strain is exceeded anywhere in the component, initial fracture has already occurred. Table 2.1 [21] presents minimum elongation at break values for selected thermoplastics under impact loading.

On the other hand, the stiffness and the stress at break of the material under consideration increases with the rate of deformation. Table 2.2 [21] presents data for the stress at break, σ_{min}, for selected thermoplastics under impact loading. This stress corresponds to the point where the minimum elongation at break has just been reached.

Even ductile polymers become brittle during impact

Table 2.1 Minimum elongation at break under impact loading

Polymer	ε_{min} (%)
HMW-PMMA	2.2
PA6 + 25 % SFR	1.8
PP	1.8
uPVC	2.0
POM	4.0
PC + 20 % SFR	4.0
PC	6.0

Table 2.2 Minimum stress at break on impact loading

Polymer	σ_{min} (MPa)
HMW-PMMA	135
PA6 + 25 % Short fiber reinforced (SFR)	175
uPVC	125
POM	> 130
PC + 20 % SFR	> 110
PC	> 70

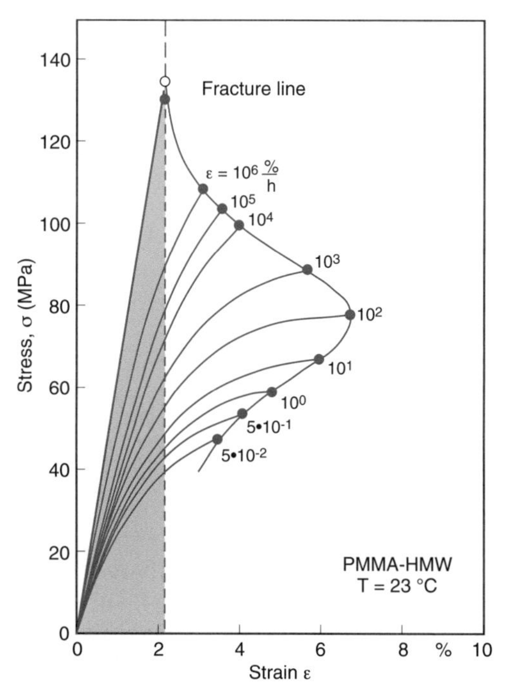

Figure 2.19 Stress-strain behavior of HMW-PMMA at various rates of deformation

Figure 2.19 summarizes the stress-strain and fracture behavior of a HMW-PMMA tested at various rates of deformation. The area under the stress-strain curves represents the *volume-specific energy to fracture* (w). For impact, the elongation at break of 2.2 % and the stress at break of 135 MPa represent a minimum of volume-specific energy, because the stress increases with higher rates of deformation, but the elongation at break remains constant. Hence, if we assume a linear behavior, the *minimum volume-specific energy absorption* to fracture can be calculated using

Minimum volume-specific energy absorption during impact

$$w_{min} = \frac{1}{2}\sigma_{min}\varepsilon_{min} \tag{2.29}$$

If the stress-strain distribution in the polymer component is known, one can estimate the minimum energy absorption capacity using w_{min}. It can be assumed that failure occurs if w_{min} is exceeded anywhere in the loaded component. This minimum volume-specific energy absorption, w_{min}, can be used as a design parameter. It can also be used for fiber reinforced polymeric materials [22].

2.7 Fatigue

Dynamic loading of any material that leads to failure after a certain number of cycles is called *fatigue* or *dynamic fatigue*. Dynamic fatigue is of extreme importance because a cyclic or fluctuating load causes a component to fail at much lower stresses than it does under monotonic loads [23]. Dynamic fatigue is of extreme importance since a cyclic or fluctuating load causes a component to fail at much lower stresses than it does under monotonic loads.

Fatigue test results are plotted as stress amplitude versus number of cycles to failure. These graphs are usually called *S-N curves*, a term inherited from metal fatigue testing. Figure 2.20 [24] presents S-N curves for several thermoplastic and thermoset polymers tested at a frequency of 30 Hz and about a zero mean stress, σ_m.

Fatigue in plastics is strongly dependent on the environment, the temperature, the frequency of loading, and surface finish. For example, because of surface irregularities and scratches, crack initiation at the surface is more likely in a polymer component that has been machined than in one that was injection molded. An injection molded article is formed by several layers of different orientation. In such parts, the outer layers act as a protective skin that inhibits crack initiation. In an injection molded part, cracks are more likely to initiate inside the component by defects such as weld lines and filler particles. The gate region is also a prime initiator of fatigue cracks. Corrosive environments also accelerate crack initiation and failure via fatigue.

Fatigue tests must be done at low frequencies to avoid overheating

Temperature increases during testing is one of the main causes of failure when experimentally testing thermoplastic polymers under cyclic loads. The temperature rise during testing is caused by the combination of internal frictional or hysteretic heating and low thermal conductivity. At low frequency and stress, the temperature in the polymer specimen will rise and can eventually reach thermal equilibrium when the

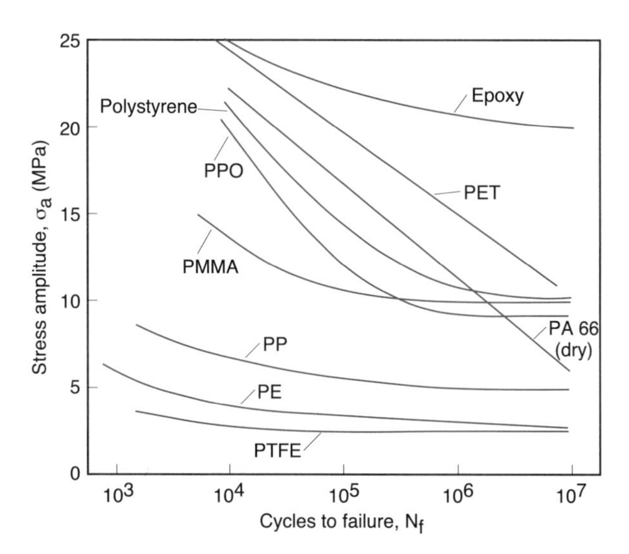

Figure 2.20 Stress-life (S-N) curves for several thermoplastic and thermoset polymers tested at a 30 Hz frequency about a zero mean stress

heat generated by hysteretic heating equals the heat removed from the specimen by conduction. As the frequency is increased, viscous heat is generated faster, causing the temperature to rise even further. After thermal equilibrium has been reached, a specimen eventually fails by conventional brittle fatigue, assuming the stress is above the endurance limit. However, if the frequency or stress level is increased even further, the temperature rises to the point that the test specimen softens and ruptures before reaching thermal equilibrium. This mode of failure is usually referred to as *thermal fatigue.*

2.8 Weathering

When exposed to the elements, polymeric materials can exhibit environmental cracks, which lead to failure at stress levels significantly lower than the ones determined under specific lab conditions. Ultraviolet radiation, moisture, and extreme temperatures harm the mechanical properties of plastic parts.

The strength losses and discoloration from weathering are mainly attributed to the ultra-violet rays in sunlight. This can be demonstrated by plotting properties as a function of sunlight exposure instead of total time exposed. Figure 2.21 [25] is a plot of percent of initial impact strength for an ABS as a function of sunlight exposure in three different locations: Florida, Arizona, and West Virginia. The curve reveals that by "normalizing" the curves with respect to exposure to sunshine, the three different sites with three completely different weather conditions lead to the same relationship between impact strength and sunlight exposure.

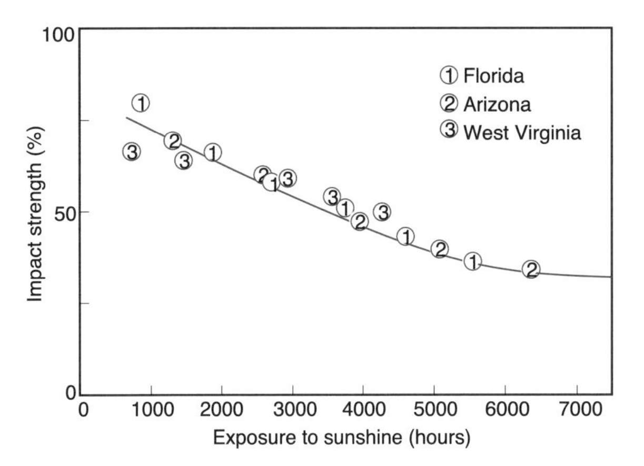

Weathering is primarily caused by sunlight

Figure 2.21 Impact strength of an ABS as a function of hours of actual sunlight exposure

The effect of weathering can often be mitigated with pigments, such as TiO_2 or carbon black, which absorb ultraviolet radiation, preventing penetration through the polymer component surface.

Example 2.1 **Stress Relaxation**

For the poly-α-methylstyrene stress relaxation data in Fig. 2.22 [26], create a master creep curve at T_g (204 °C). Identify the glassy, rubbery, viscous, and viscoelastic regions of the master curve. Identify each region with a spring-dashpot model diagram.

The master creep curve for the above data is generated by sliding the individual relaxation curves horizontally until they match with their neighbors, if the scale is fixed for a hypo-

Generating a stress relaxation master curve

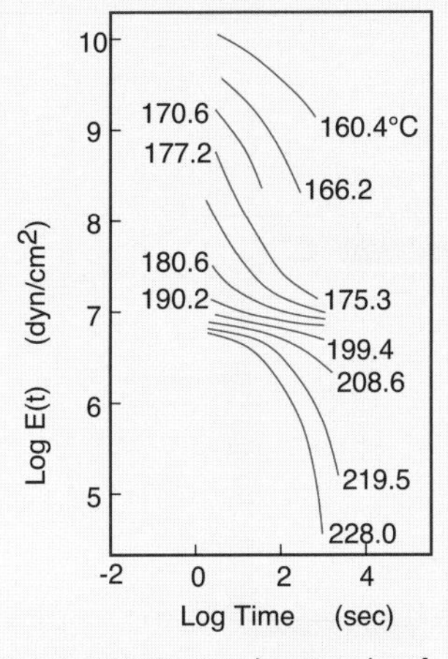

Figure 2.22 Stress relaxation data for poly-α-methylstyrene

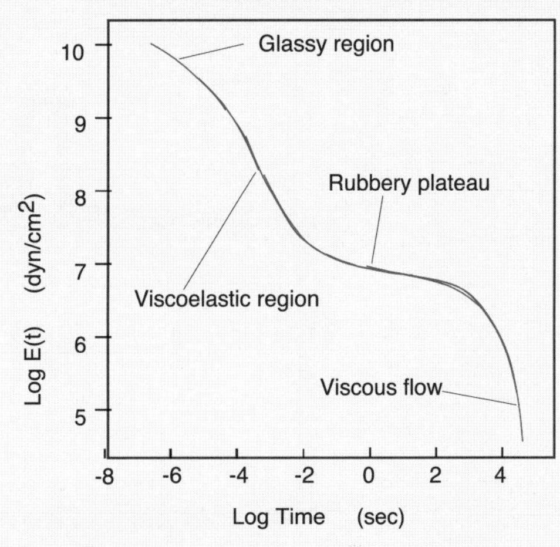

Figure 2.23 Master curve for poly-α-methylstyrene at 204 °C

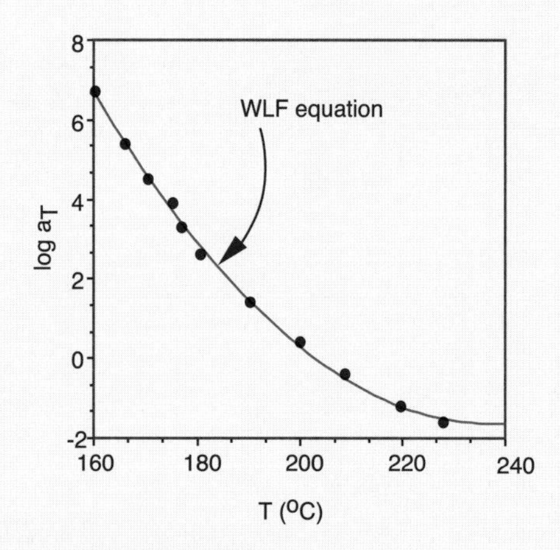

Figure 2.24 Shift factor and WLF for $T_{ref} = 204$ °C

thetical curve at 204 °C. Since a curve does not exist for the desired temperature, we can interpolate between 208.6 °C and 199.4 °C. The resulting master curve is plotted in Fig. 2.23. The amount each curve must be shifted to meet the master curve is the shift factor, $\log a_T$. Figure 2.24 represents the shift factor versus temperature. The solid line indicates the shift factor predicted by the WLF equation. Good agreement can be seen.

Example 2.2 Creep – Isochronous Curves

In a special laboratory experiment, a PMMA pipe is used to cap a tank that is pressurized at 2 MPa, as shown in Fig. 2.25. The 3 mm thick pipe has a 50 mm internal diameter and

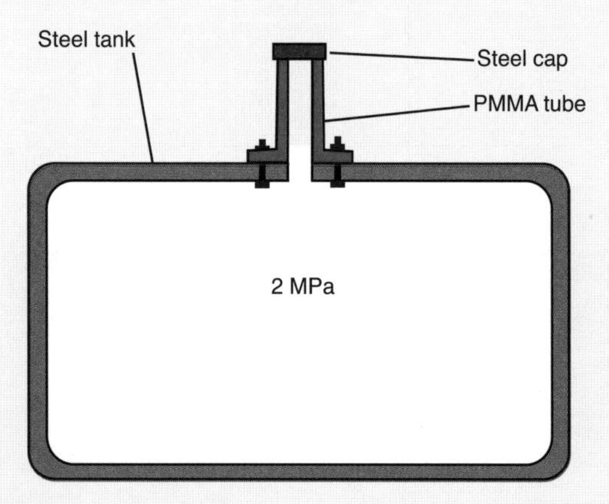

Figure 2.25 Schematic of the laboratory set-up

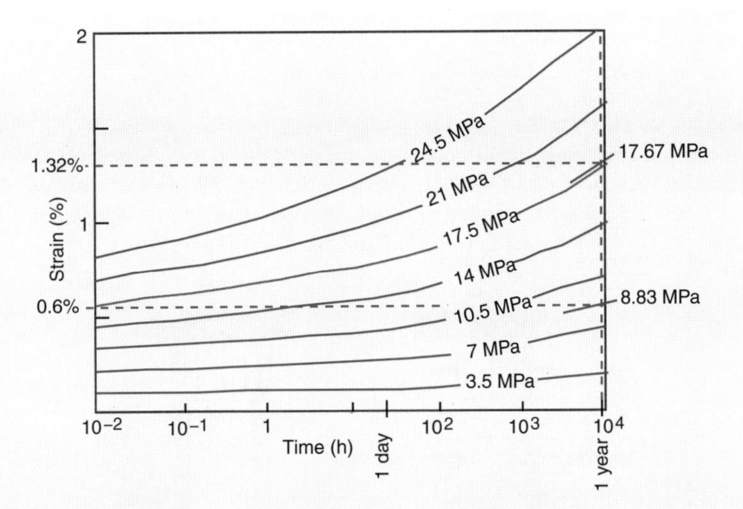

Figure 2.26 PMMA creep data

Example using isochronous curves — creep

is 300 mm long. Estimate the diameter change of the pipe after one year of testing. Use the creep data given below (Fig. 2.26) [7].

To solve this problem, we can use the thin pressure vessel approximation, working with an average diameter, $\overline{D} = 53$ mm. This is a case of biaxial stress, composed of a hoop stress, σ_H, and an axial stress, σ_A, defined by

$$\sigma_H = \frac{p\overline{D}}{2h} \quad \text{and} \quad \sigma_A = \frac{p\overline{D}}{4h}$$

where, p is the pressure and h is the thickness. These stresses are constant and will cause the pipe to creep. Using the PMMA creep data we can generate a 1-year isochronous curve, or we can determine the strains directly from the creep data. The strains corresponding to stresses of 17.67 MPa and 8.83 MPa are $\varepsilon_H = 1.32\,\%$ and $\varepsilon_A = 0.60\,\%$, respectively. Since this is a biaxial case, we must correct the hoop strain using Poisson's effect, before computing the diameter change. For this we use

$$\varepsilon_H^{\text{corrected}} = \varepsilon_H - \nu\varepsilon_A$$

However, since we were not given Poisson's ratio, ν, we assume a value of ⅓. Thus,

$$\varepsilon_H^{\text{corrected}} = 1.38\,\% - (0.33)(0.6\,\%) = 1.12\,\%.$$

To compute the change in diameter we use

$$\varepsilon_H^{\text{corrected}} = \frac{\Delta D}{\overline{D}}$$

Hence,

$$\Delta D = (53\text{ mm})(0.0112) = 0.594\text{ mm}$$

■

Example 2.3 Creep – Isometric Curves

Example using isometric curves – Stress relaxation

In the assembly shown in Fig. 2.27, a tubular polypropylene feature is pressed on a 15 mm long metal stud. The inner diameter of the 1 mm thick PP tubular element is 10 mm. The metal stud is slightly oversized with a diameter of 10.15 mm. With a coefficient of friction $\mu = 0.3$ estimate the force required to disassemble the parts shortly after assembly, and after one year. Use the creep data given in Fig. 2.8.

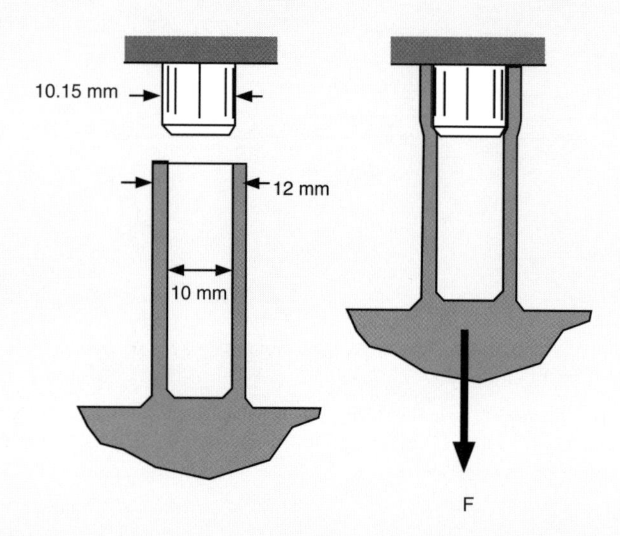

Figure 2.27 Press-fit assembly

This is a classic constant strain, ε_0, stress relaxation problem. The initial hoop stress that holds the assembly together can be quite high. However, as time passes, the hoop stress relaxes and it becomes easier to disassemble the two components. The strain in the system after assembly is computed using

$$\varepsilon_0 = \frac{\Delta D}{\bar{D}} = \frac{0.15\,\text{mm}}{11.0\,\text{mm}} = 0.0145 \rightarrow 1.45\%$$

In order to follow the hoop stress history after assembly, we generate a 1.45 % isometric curve, which is shown in Fig. 2.28.

From the isochronous curve we can deduce that the hoop stress, σ_H, is 13.7 MPa shortly after assembly and about 5 MPa one year after assembly. The pressure acting on the metal stud, due to the hoop stress, can be computed using

$$p = \frac{2h\,\sigma_H}{\bar{D}}$$

which gives a 2.49 MPa pressure right after assembly and 0.91 MPa after one year. From the pressure and the friction we can calculate the disassembly force with

$$F = \mu p (\pi D L)$$

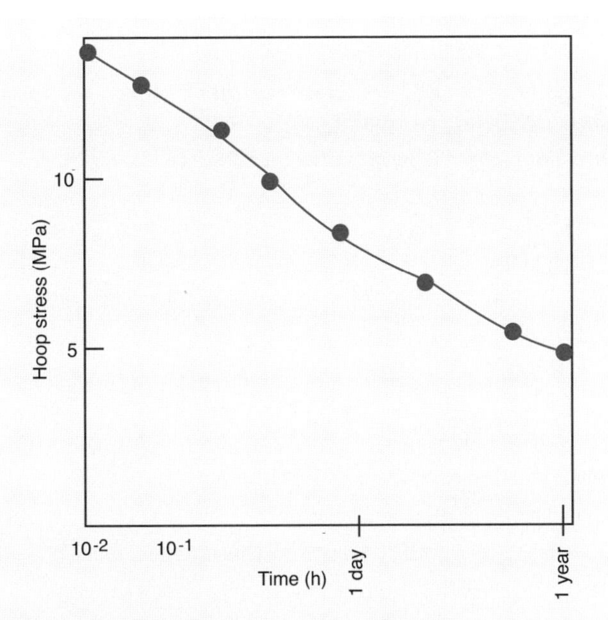

Figure 2.28 Isometric curve for a 1.45 % strain, derived from Fig. 2.8

where $L = 15$ mm is the length of the stud. Using the above equation, the computed force necessary to pull the two components apart is 358 N (80 lb) after assembly and 130 N (29 lb) after one year.

■

References

1. Castiff, E. and A.V.J. Tobolsky, *Colloid Sci.* (1955), *10*, 375
2. Williams, M.L., R.F. Landel, and J.D. Ferry, *J. Amer. Chem. Soc.* (1955), *77*, 3701
3. Osswald, T.A., E. Baur, S. Brinkmann and E. Schmachtemberg, *International Plastics Handbook*, Hanser Publishers (2006), Munich
4. Treloar, L.R.G., *The Physics of Rubber Elasticity*, 3rd Ed. (1975), Clarendon Press, Oxford
5. Courtesy ICIPC, Medellín, Colombia.

   6. Guth, E., and R. Simha, *Kolloid-Zeitschrift* (1936), *74*, 266
6. Osswald, T.A., and G. Menges, *Materials Science of Polymers for Engineers*, Hanser Publishers (2003) Munich
7. Smallwood, H.M., *J. Appl. Phys.* (1944), *15*, 758
8. Mullins, L., and N.R. Tobin, *J. Appl. Polym. Sci.* (1965), *9*, 2993
9. ASTM, Plastics (II), 08.02, ASTM (1994), Philadelphia
10. Erhard, G., *Designing with Plastics*, Hanser Publishers (2006), Munich
11. Richard, K., E. Gaube and G. Diedrich, *Kunststoffe* (1959), *49*, 516

12. Gaube, E. and H. H. Kausch, *Kunststoffe* (1973), *63*, 391
13. Nielsen, L. E., and R. F. Landel, *Mechanical Properties of Polymers and Composites*, 2nd Ed., Marcel Dekker (1994), Inc., New York
14. Loos, R. D. A., *Processing of Composites*, Hanser Publishers (2000), Munich
15. Tsai, S. W., J. C. Halpin, and N. J. Pagano, *Composite Materials Workshop*, Technomic Publishing Co. (1968), Stamford
16. Brintrup, H., Ph.D. Thesis (1974), IKV, RWTH-Aachen, Germany
17. Menges, G., *Kunststoffverarbeitung III*, 5, Lecture notes (1987), IKV, RWTH-Aachen
18. Boyer, R. F., *Polymer Eng. Sci.* (1968), *8*, 161
19. Ehrenstein, G. W., *Polymeric Materials: Structure-Properties-Applications*, Hanser Publishers (2001), Munich
20. Menges, G., and H.-E. Boden, *Failure of Plastics*, Chapter 9, in W. Brostow, and R. D. Corneliussen (Eds.), Hanser Publishers (1986), Munich
21. Rest, H., Ph.D. Thesis (1984), IKV-Aachen
22. Erhard, G., *Designing with Plastics*, Hanser Publishers (2006), Munich
23. Riddell, M. N., *Plast. Eng.* (1974), *40*, 4, 71
24. Naranjo, A., M. Noriega, T. Osswald, A. Roldán and J. D. Sierra, *Plastics Testing and Characterization* (2008), Munich
25. Fujimoto, T., Ozaki, M. and Nagasawa, M., *J. Polymer Sci.* (1968), 2, 6, 129

3 Melt Rheology

Rheology is the science of fluid behavior during flow-induced deformation. Among the variety of materials that rheologists study, polymers have been found to be the most interesting and complex. Polymer melts are shear thinning and viscoelastic and their flow properties are temperature dependent. This chapter discusses the phenomena that are typical of polymer melts and covers the basic properties used to represent the flow behavior of polymers. The chapter also introduces rheometry. For further reading on rheology of polymer melts consult references 1 to 6.

3.1 Introduction to Rheology

Viscosity is the most widely used material parameter when determining the behavior of polymers during processing. Because the majority of polymer processes are shear rate dominated, the viscosity of the melt is commonly measured using shear deformation measurement devices. For example, the simple shear flow generated in the sliding plate rheometer [1], shown in Fig. 3.1, exhibits a deviatoric stress defined by

$$\tau_{xy} = \eta(T, \dot{\gamma})\dot{\gamma}_{xy} \tag{3.1}$$

where the shear stress, τ_{xy}, is the ratio of the force it takes to move the plate and the area of the plate, F/A, $\eta(T, \dot{\gamma})$ is the viscosity and $\dot{\gamma}_{xy}$ the shear rate defined by u/h. For the flow in Fig. 3.1, the magnitude of the rate-of-deformation tensor, $\dot{\gamma}$, is $\dot{\gamma}_{xy}$. There are also polymer processes, such as blow molding, thermoforming, and fiber

Sliding plate rheometer

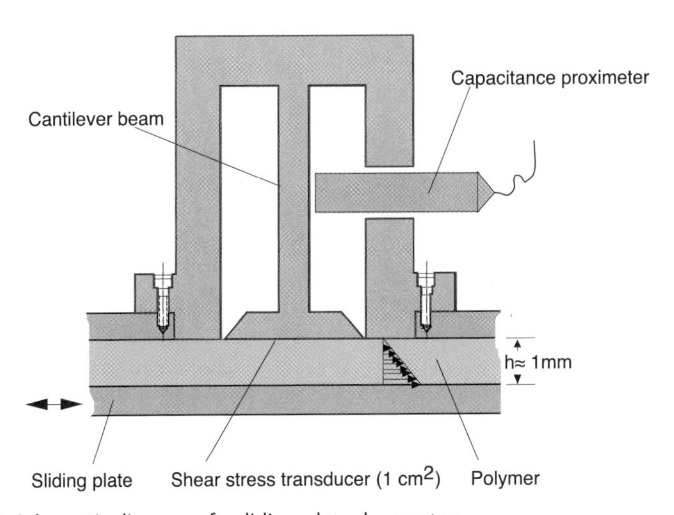

Figure 3.1 Schematic diagram of a sliding plate rheometer

Rate of deformation and temperature effect

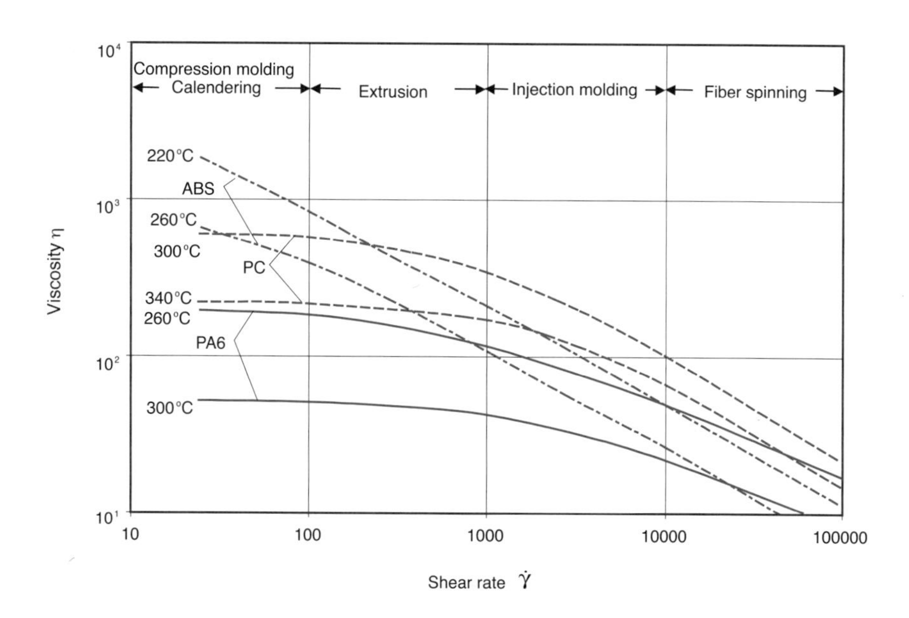

Figure 3.2 Viscosity curves for a selected number of thermoplastics

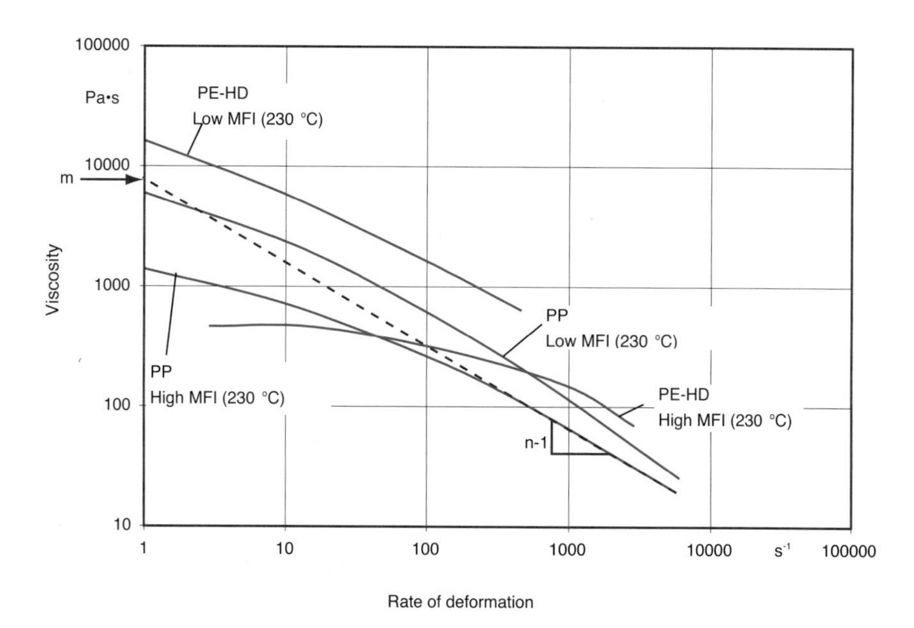

Figure 3.3 Viscosity curves for PE-HD and PP with low and high MFI

spinning, which are dominated by either elongation or by a combination of shear and elongational deformation. In addition, some polymer melts exhibit significant elastic effects during deformation. The equation presented above is generally referred to as a constitutive equation. Equation 3.1 reflects a Generalized Newtonian Fluid constitutive equation, however, in the field of polymer rheology some constitutive equations can be rather complex.

3.1.1 Shear Thinning Behavior of Polymers

Most polymer melts are *shear thinning fluids*. The shear thinning effect is the reduction in viscosity at high rates of deformation, as shown in Fig. 3.2 for various polymers. This phenomenon occurs because at low rates of deformation, the polymer molecules are entangled and at high rates of deformation, the molecules are stretched out and disentangled. The disentangled molecules can slide past each other more easily, thus lowering the bulk melt viscosity.

The power-law model proposed by Ostwald [2] and de Waale [3] is the simplest model that accurately represents the shear thinning region in the viscosity versus strain rate curve, but overshoots the Newtonian plateau at small strain rates. The power-law model is:

Power law model of the shear thinning region

$$\eta = m(T)\dot{\gamma}^{n-1} \tag{3.2}$$

where m is called the *consistency index* and n the *power law index*. Figure 3.3 presents two types of PE-HD and PP (low and high melt flow indices) with a sample slope $(n-1)$ and intercept (m). The consistency index may include the temperature dependence of the viscosity and can be represented as

$$m(T) = m_0 e^{-a(T-T_0)} \tag{3.3}$$

Modeling the temperature effect

Power-law constants, as used in Eqs. (3.2) and (3.3) are presented in Table 3.1 for common thermoplastics.

Table 3.1 Power-law indices, consistency indices, and temperature dependence constants for common thermoplastics

Polymer	m (Pa·s^n)	n	a (1/°C)	T_0 (°C)
Polystyrene	2.80×10^4	0.28	−0.025	170
High density polyethylene	2.00×10^4	0.41	−0.002	180
Low density polyethylene	6.00×10^3	0.39	−0.013	160
Polypropylene	7.50×10^3	0.38	−0.004	200
Polyvinylchloride	1.70×10^4	0.26	−0.019	180

Example 3.1 **Power Law Model Constants**

Determine the power law model constants of the sample in Fig. 3.3

The power law index n is found from the slope of the viscosity–rate of deformation curve, plotted on a log-log scale. Here, the value of $n - 1$ is approximately $-2/3$, resulting in a power law index of 0.33. The consistency index m is found at the position where $\dot{\gamma} = 1s^{-1}$. The consistency index here is 7100 Pa·s^n.

3.1.2 Normal Stresses in Shear Flow

The tendency of polymer molecules to "curl-up" while they are being stretched in shear flow results in normal stresses in the fluid. For example, the shear flow presented in Eq. 3.1 has measurable normal stress differences, N_1 and N_2, which are called the *first* and *second normal stress differences*, respectively. The first and second normal stress differences are material dependent and are defined by

Normal stresses arise during shear deformation

$$N_1 = \tau_{xx} - \tau_{yy} = \psi_1(T, \dot{\gamma})\dot{\gamma}_{xy}^2 \tag{3.4}$$

$$N_2 = \tau_{yy} - \tau_{zz} = \psi_2(T, \dot{\gamma})\dot{\gamma}_{xy}^2 \tag{3.5}$$

The material functions, ψ_1 and ψ_2, are called the primary and secondary normal stress coefficients, and are also functions of the magnitude of the strain rate tensor and temperature. The first and second normal stress differences do not change sign when the direction of the strain rate changes. This is reflected in Eqs. 3.4 and 3.5. Figures 3.4

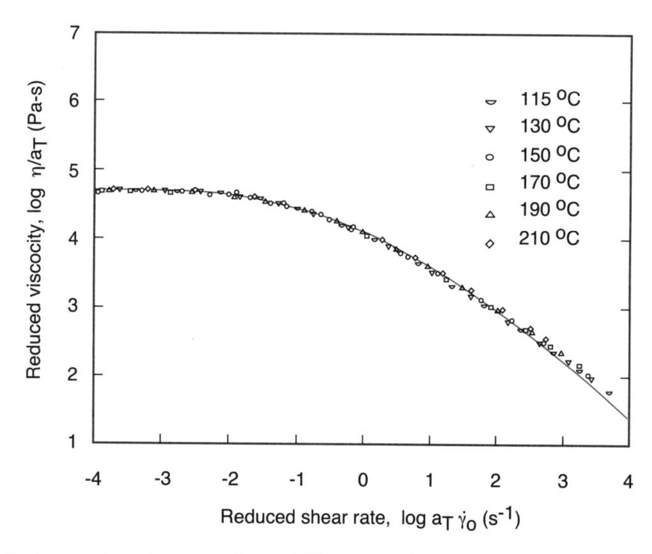

Figure 3.4 Reduced viscosity curve for an LDPE at a reference temperature of 150 °C

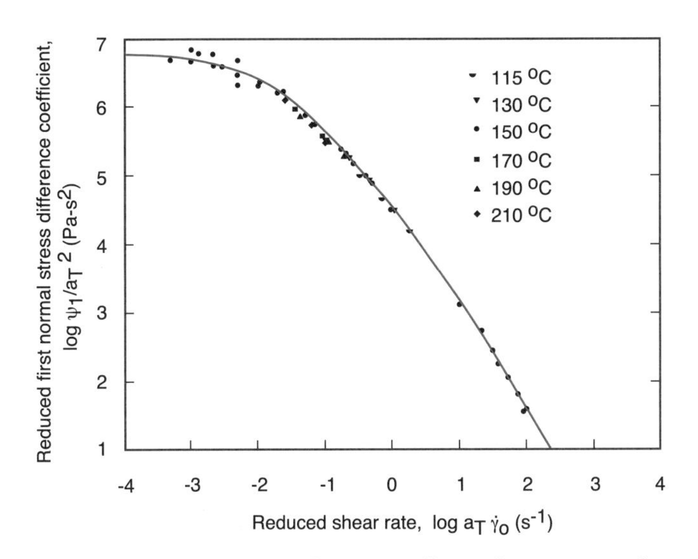

Figure 3.5 Reduced first normal stress difference coefficient for an LDPE melt at a reference temperature of 150 °C

and 3.5 [4] present the viscosity and first normal stress difference coefficient, respectively, for an LDPE melt at a reference temperature of 150 °C. The second normal stress difference is difficult to measure and is often approximated by

$$\psi_2 \approx -0.1\psi_1 \tag{3.6}$$

The normal stress differences play significant roles during processing. For example, the first normal stress difference is partly responsible for *extrudate swell* (Fig. 3.6) at the exit of the die. The second normal stress differences help diminish the eccentricity of a wire in the die during the wire coating process [5].

Extrudate swell

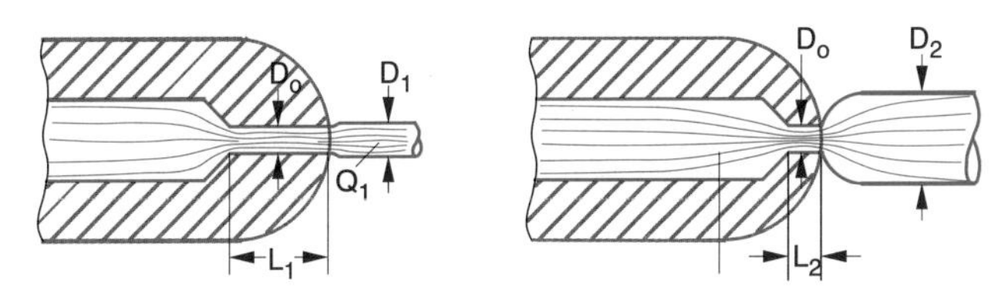

Figure 3.6 Schematic diagram of extrudate swell during extrusion

3.1.3 Deborah Number

A useful parameter for estimating the elastic effects during flow is the Deborah number:

$$De = \frac{\lambda}{t_{\text{process}}} \tag{3.7}$$

where λ is the relaxation time of the polymer and $t_{process}$ is a characteristic process time. The characteristic process time can be defined by the ratio of characteristic die dimension and average speed through the die. A Deborah number of zero represents a viscous fluid and a Deborah number of ∞ an elastic solid. As the Deborah number exceeds one, the polymer does not have enough time to relax during the process, resulting in *extrudate swell,*[1)] *shark skin,* or even *melt fracture.*

Although many factors affect the amount of extrudate swell, fluid "memory" and normal stress effects are most significant. However, abrupt changes in boundary conditions, such as the separation point of the extrudate from the die, also play a role in the swelling or cross section reduction of the extrudate. In practice, the fluid memory contribution to die swell can be mitigated by lengthening the land length of the die. This is schematically depicted in Fig. 3.6. A long die land separates the polymer from the manifold long enough to allow it to "forget" its past shapes.

Typically, melt fracture occurs when the stress at the wall exceeds 0.1 MPa

Waves in the extrudate may also appear as a result of high speeds during extrusion, where the polymer cannot relax. This phenomenon is generally called *shark skin* and is shown for an HDPE in Fig. 3.7(a) [6]. Polymers can be extruded at such high speeds that an intermittent separation of melt and inner die walls occurs, as shown in Fig. 3.7(b). This phenomenon is often called *stick-slip* or *spurt flow* and is attributed to high shear stresses between the polymer and the die wall. This phenomenon occurs when the shear stress is near the critical value of 0.1 MPa [7–9]. If the speed is further

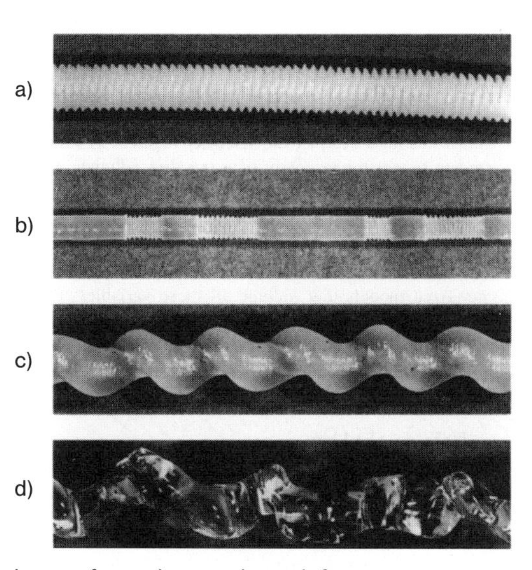

Figure 3.7 Various shapes of extrudates under melt fracture

1) Newtonian fluids, which do not experience elastic or normal stress effects, also show some extrudate swell or reduction. A Newtonian fluid extruded at high shear rates reduces its cross section to 87 % of the diameter of the die. This swell is the result of inertia effects caused by the rearrangement from the parabolic velocity distribution inside the die to the flat velocity distribution of the extrudate. If the polymer is extruded at very low shear rates it swells to 113 % of the diameter of the die.

increased, a helical geometry is extruded, as shown for a polypropylene extrudate in Fig. 3.7(c). Eventually the speeds are so high that a chaotic pattern develops, such as the one shown in Fig. 3.7(d). This well known phenomenon is called *melt fracture*. Shark skin is frequently absent and spurt flow seems to occur only with linear polymers.

The critical shear stress has been reported to be independent of the melt temperature, but inversely proportional to the weight average molecular weight [9, 10]. However, Vinogradov et al. [7] presented results where the critical stress was independent of molecular weight except at low molecular weights. Dealy and co-workers [9] and Denn [11] give an extensive overview of various melt fracture phenomena. Both references are recommended.

Song of Deborah

M. Reiner is credited with naming the Deborah Number after the song of Deborah, Judges 5:5 — "The mountains flowed before the Lord" (Fig. 3.8). It was first mentioned in his article "The Deborah Number" in the January 1964 issue of Physics Today.

Figure 3.8 Coyote Buttes North 1 Second Wave, Arizona. Courtesy of Wolfgang Cohnen (© 1998)

3.1.4 Rheology of Curing Thermosets

A curing thermosetting polymer exhibits a conversion or cure dependent viscosity that increases as the molecular weight of the reacting polymer increases. For the vinyl ester,

Effect of cure on viscosity

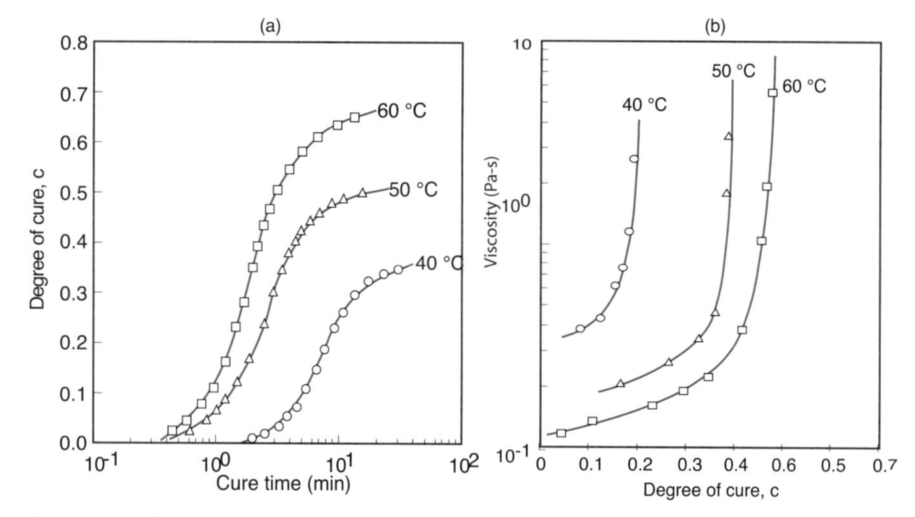

Figure 3.9 Effect of cure on viscosity (a) Cure as a function of time (b) Viscosity as a function of cure

whose curing history is shown in Fig. 3.9 [12], the viscosity behaves as shown in Fig. 3.10 [11]. Hence, a complete model for viscosity of a reacting polymer must contain the effects of strain rate, $\dot{\gamma}$, temperature, T, and degree of cure, c, such as

$$\eta = \eta(\dot{\gamma}, T, c) \tag{3.8}$$

There are no generalized models that include all these variables for thermosetting polymers. However, extensive work has been done on the viscosity of polyurethanes [12, 13] used in the reaction injection molding process. An empirical relation which models the viscosity of these mixing activated polymers, given as a function of temperature and degree of cure, is written as

$$\eta = \eta_0 e^{E/RT} \left(\frac{c_g}{c_g - c} \right)^{C_1 + C_2 c}$$

Where E is the activation energy of the polymer, R is the ideal gas constant, T is the temperature, c_g is the gel point, c the degree of cure, and C_1 and C_2 are constants that fit the experimental data. The gel point is the degree of cure when the molecular weight goes to infinity or when the molecules become interconnected.

3.1.5 Suspension Rheology

Particles suspended in a material, such as in filled or reinforced polymers, have a direct effect on the properties of the final article and on the viscosity during processing. The model that best fits experimental data is the one given by Guth [14]:

$$\frac{\eta_f}{\eta_0} = 1 + 2.5\varphi + 14.1\varphi^2 \tag{3.9}$$

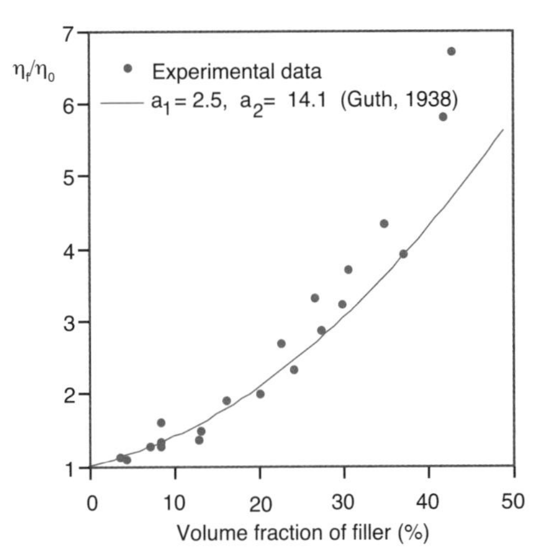

Effect of filler fraction on viscosity

Figure 3.10 Viscosity increase as a function of volume fraction of filler for polystyrene and LDPE

Figure 3.10 compares experimental data to Guth's equation. The experiments were performed on polyethylene and polystyrene containing different fill factors of spherical glass particles ranging in size between 36 μm and 99.8 μm in diameter. The model agrees well with the experimental data up to volume fractions of 30 %.

3.1.6 Viscoelastic Flow Models

Viscoelasticity has already been introduced in Chapter 2 based on linear viscoelasticity. However, in polymer processing large deformations are imposed on the material, requiring the use of non-linear viscoelastic models. There are two types of general non-linear viscoelastic flow models, the differential model and the integral model.

An overview of numerical simulation of viscoelastic flow systems and an extensive literature review on the subject was given by Keunings [15] and details on numerical implementation of viscoelastic models are given by Crochet et al. [16] and Debbaut et al. [17]. As an example of the application of differential models to predict flow of polymeric liquids, Dietsche and Dooley [18] recently evaluated the White–Metzner, the Phan-Thien Tanner-1 and the Giesekus models by comparing finite element and experimental results of the flow inside multi-layered coextrusion dies. Figure 3.11 [19] presents the progression of a matrix of dyed circular polystyrene strands flowing in an identical polystyrene matrix down a channel with a square cross section of 0.95 × 0.95 cm. The cuts in the figure are shown at intervals of 7.6 cm. The circulation pattern caused by the secondary normal stress differences inside non-circular dies was captured well by the Phan-Thien Tanner and Giesekus models but, as expected, not by the White–Metzner model. Figure 3.12 also presents flow patterns predicted by the Phan-Thien Tanner model. The shape of the circulation patterns was predicted accurately. The flow simu-

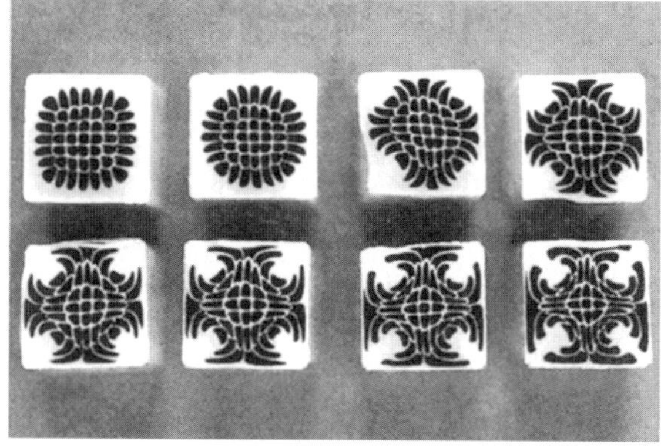

Figure 3.11 Simulated flow patterns and polystyrene strand profile progression in a square die

lation of the square die predicted a velocity on the order of 10^{-5} m/s along the diagonal of the cross section, which agreed with the experimental results. Also, Baaijens [20] recently evaluated the Phan-Thien Tanner models 1 and 2, and the Giesekus models. He compared finite element results to measured isochromatic birefringence patterns using complex experiments with polymer melts and solutions. His simulation results predicted the general shape of the measured birefringence patterns. He found that at high Deborah numbers, the Phan-Thien Tanner models converged more easily than the Giesekus model.

3.2 Rheometry

In industry, there are various ways to qualify and quantify the properties of the polymer melt. The techniques range from simple analyses for checking the consistency of the material at certain conditions, to more difficult fundamental measurements to evaluate viscosity and normal stress differences. This section includes three such techniques to give the reader a general idea of current measuring methods.

3.2.1 The Melt Flow Indexer

The melt flow indexer is often used in industry to characterize a polymer melt and as a simple and quick means of quality control. It takes a single point measurement using standard testing conditions specific to each polymer class on a ram type extruder or extrusion plastometer, as shown in Fig. 3.12. The standard procedure for testing the flow rate of thermoplastics using an extrusion plastometer is described in the ASTM D1238 test [21]. During the test, a sample is heated in the barrel and extruded from a short cylindrical die using a piston actuated by a weight. The weight of the polymer in grams extruded during the 10-minute test is the melt flow index (MFI) of the polymer. A melt flow indexer is an indispensable materials quality control device in any plastics manufacturing facility. This tool helps making sure that the same material is used in a process, thus, resulting in reproducible parts using the same processing conditions. A variation in MFI can have profound differences in a process, but not in the final mechanical properties of the part. Figure 3.3 shows two PE-HD and PP resins with low and high MFI's.

The melt flow indexer can be used for material quality control

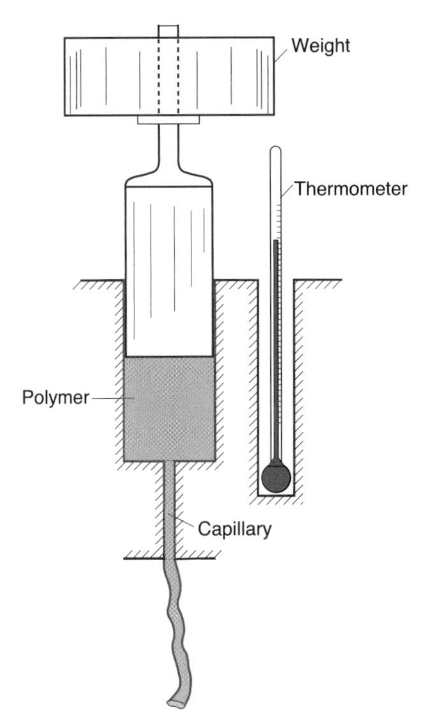

Figure 3.12 Schematic diagram of an extrusion plastometer used to measure melt flow index

3.2.2 The Capillary Viscometer

The most common and simplest device for measuring viscosity is the capillary viscometer. Its main component is a straight tube or capillary, and it was first used to measure the viscosity of water by Hagen [22] and Poiseuille [23]. A capillary rheometer has a pressure driven flow for which the shear rate is maximum at the wall and zero at the center of the flow, making it a non-homogeneous flow.

Since pressure driven viscometers employ heterogeneous flows, they can only measure steady shear functions such as viscosity, $\eta(\dot{\gamma})$. However, they are widely used because they are relatively inexpensive and simple to operate. Despite their simplicity, long capillary viscometers provide the most accurate viscosity data available. Another major advantage is that the capillary rheometer has no free surfaces in the test region, unlike other types of rheometers, such as the cone and plate rheometer discussed next. When the strain rate dependent viscosity of polymer melts is measured, capillary rheometers are capable of obtaining such data at shear rates greater than 10 s^{-1}. This is important for processes with higher rates of deformation such as mixing, extrusion, and injection molding. Because its design is basic and it only needs a pressure head at its entrance, the capillary rheometer can easily attach to the end of a screw- or ram-type extruder for online measurements. This makes the capillary viscometer an efficient tool for industry. The basic features of the capillary rheometer are shown in Fig. 3.13.

A capillary tube of radius R and length L is connected to the bottom of a reservoir. Pressure drop and flow rate through this tube are used to determine the viscosity. At

The capillary viscometer can be used to measure viscosity as a function of rate of deformation and temperature

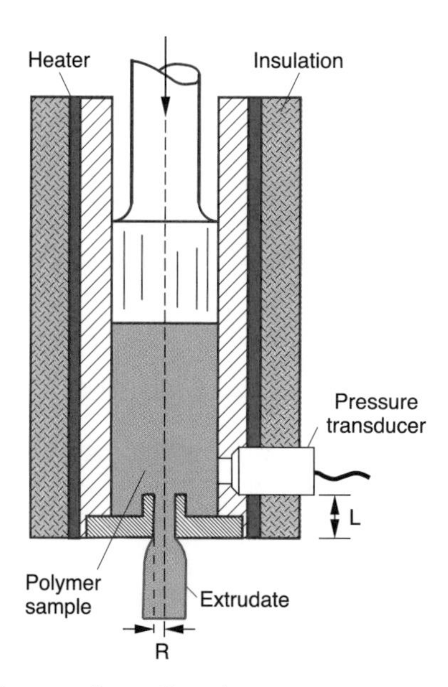

Figure 3.13 Schematic diagram of a capillary rheometer

the wall, the shear stress is:

$$\tau_w = \frac{R(p_0 - p_L)}{2L} = \frac{R\Delta p}{2L} \tag{3.10}$$

Equation 3.11 requires that the capillary be long enough to assure fully developed flow, where end effects are insignificant. However, because of entrance effects, the actual pressure profile along the length of the capillary exhibits curvature. The effect is shown schematically in Fig. 3.14 [24] and was corrected by Bagley [25] using the end correction e:

$$\tau_w = \frac{R(p_0 - p_L)}{2(L/D + e)} \tag{3.11}$$

The correction e at a specific shear rate can be found by plotting pressure drop for various capillary L/D ratios, as shown in Fig. 3.15 [24].

The equation for shear stress is then

$$\tau_{rz} = \frac{r}{R}\tau_w \tag{3.12}$$

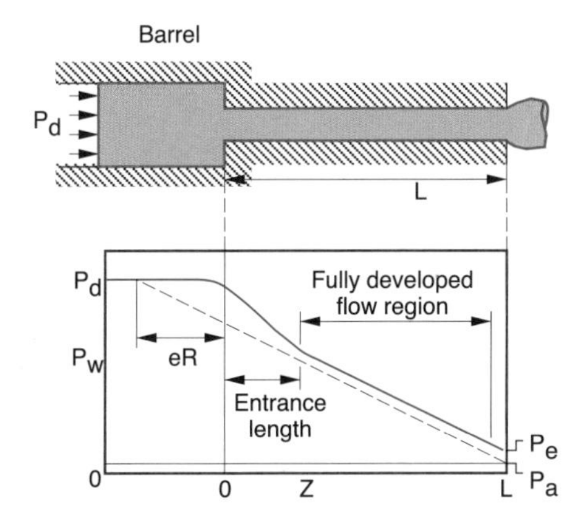

Figure 3.14 Entrance effects in a typical capillary viscometer

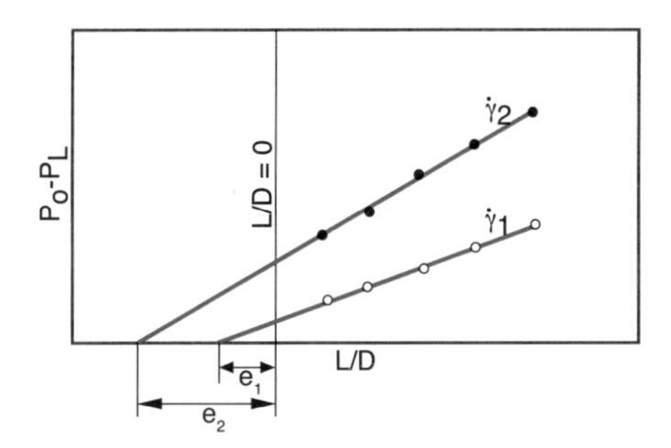

Figure 3.15 Bagley plots for two shear rates

To obtain the shear rate at the wall the Weissenberg–Rabinowitsch [26] equation can be used

$$\dot{\gamma}_{\mathrm{w}} = \frac{1}{4}\dot{\gamma}_{\mathrm{aw}}\left(3 + \frac{d(\ln Q)}{d(\ln \tau)}\right) \tag{3.13}$$

where $\dot{\gamma}_{\mathrm{aw}}$ is the apparent or Newtonian shear rate at the wall and is written as

$$\dot{\gamma}_{\mathrm{aw}} = \frac{4Q}{\pi R^3} \tag{3.14}$$

The shear rate and shear stress at the wall are now known. Therefore, using the measured values of the flow rate, Q, and the pressure drop, $p_0 - p_L$, the viscosity is calculated using

$$\eta = \frac{\tau_{\mathrm{w}}}{\dot{\gamma}_{\mathrm{w}}} \tag{3.15}$$

3.2.3 The Cone-and-Plate Rheometer

The cone-and-plate rheometer can be used to measure viscosity as well as first normal stress difference

The cone-and-plate rheometer is another rheological measuring device widely accepted in the polymer industry. Here, a disk of polymer is squeezed between a plate and a cone, as shown in Fig. 3.16. When the disk is rotated, the torque and the rotational speed are related to the viscosity and the force required to keep the cone at the plate is related to the first normal stress difference. The secondary normal stress difference is related to the pressure distribution along the radius of the plate.

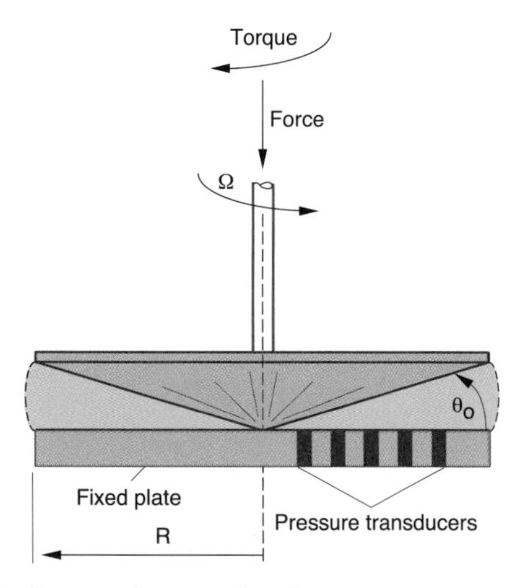

Figure 3.16 Schematic diagram of a cone-plate rheometer

References

1. Giacomin, A. J., T. Samurkas, and J. M. Dealy, *Polym. Eng. Sci.* (1989), *29*, 499
2. Ostwald, W., *Kolloid-Z.* (1925), 36, 99
3. de Waale, A., *Oil and Color Chem. Assoc. Journal* (1923), *6*, 33
4. Laun, H. M., *Rheol. Acta* (1978), *17*, 1
5. Tadmor, Z., and R. B. Bird, *Polym. Eng. Sci.* (1974), *14*, 124
6. Agassant, J.-F., P. Avenas, J.-Ph. Sergent, and P. J. Carreau, *Polymer Processing: Principles and Modeling*, Hanser Publishers (1991), Munich
7. Vinogradov, G. V., A. Y., Malkin, Y. G. Yanovskii, E. K. Borisenkova, B. V. Yarlykov, and G. V. Berezhnaya, *J. Polym. Sci. Part A-2* (1972), *10*, 1061
8. Vlachopoulos, J., and M. Alam, *Polym. Eng. Sci.* (1972), *12*, 184
9. Osswald, T. A., and G. Menges, *Materials Science of Polymers for Engineers*, Hanser Publishers (2003), Munich
10. Spencer, R. S., and R. D. Dillon, *J. Colloid Inter. Sci.* (1947), *3*, 163
11. Denn, M. M., *Annu. Rev. Fluid Mech.* (1990), *22*, 13
12. Castro, J. M. and C. W. Macosko, *AIChe J.* (1982), *28*, 250
13. Castro, J. M., S. J. Perry and C. W. Macosko, *Polymer Comm.* (1984), *25*, 82
14. Guth, E., and R. Simha, Kolloid-Zeitschrift (1936), 74, 266
15. Keunings, R., *Simulation of Viscoelastic Fluid Flow*, in *Computer Modeling for Polymer Processing*, C. L. Tucker III (Ed.), Hanser Publishers (1989), Munich
16. Crocket, M. J., A. R., Davies, and K. Walters, *Numerical Simulation of Non-Newtonian Flow*, Elsevier (1984), Amsterdam
17. Debbaut, B., J. M. Marchal, and M. J. Crochet, *J. Non-Newtonian Fluid Mech.* (1988), *29*, 119
18. Dietsche, L., and J. Dooley, *SPE ANTEC* (1995), *53*, 188
19. Dooley, J., and K. Hughes, *SPE ANTEC* (1995), *53*, 69
20. Baaijens, J. P. W., *Evaluation of Constitutive Equations for Polymer Melts and Solutions in Complex Flows*, Ph.D. Thesis (1994), Eindhoven University of Technology, Eindhoven, The Netherlands
21. ASTM, 8.01, Plastics (I), ASTM (1994), Philadelphia
22. Hagen, G. H. L., *Annalen der Physik* (1839), *46*, 423
23. Poiseuille, L. J., *Comptes Rendus 11* (1840), 961
24. Dealy, J. M., *Rheometers for Molten Plastics*, Van Nostrand Reinhold Company (1982), New York
25. Bagley, E. B., *J. Appl. Phys.* (1957), *28*, 624
26. Rabinowitsch, B., *Z. Phys. Chem.* (1929), *145*, 1

Part II
Polymer Processes

4 Extrusion

During extrusion, a polymer melt is pumped through a shaping die and formed into a profile. This profile can be a plate, a film, a tube, or have any cross sectional shape. Ram-type extruders were first built by J. Bramah in 1797 to extrude seamless lead pipes. The first ram-type extruders for rubber were built by Henry Bewley and Richard Brooman in 1845. In 1846 a patent for cable coating was filed for trans-gutta-percha and cis-hevea rubber and the first insulated wire was laid across the Hudson River for the Morse Telegraph Company in 1849. The first screw extruder was patented by Mathew Gray in 1879 for the purpose of wire coating. However, the screw pump can be attributed to Archimedes, and the actual invention of the screw extruder in polymer processing by A. G. DeWolfe of the United States dates to the early 1860s. The first extrusion of thermoplastic polymers was done at the Paul Troester Maschinenfabrik in Hannover, Germany in 1935.

4.1 Pumping

Although ram and screw extruders are both used to pump highly viscous polymer melts through passages to generate specified profiles, they are based on different principles. The schematic in Fig. 4.1 shows the principles that rule the work of ram extruders, screw extruders, and other pumping systems.

The ram extruder is a positive displacement pump based on the pressure gradient term of the equation of motion. Here, as the volume is reduced, the fluid is displaced from one point to the other, resulting in a pressure rise. The gear pump, widely used in the polymer processing industry, also works on this principle. On the other hand, a screw extruder is a *viscosity pump* that works based on the pressure gradient term and the deformation of the fluid, represented as the divergence of the deviatoric stress tensor in Fig. 4.1. Figure 4.2 represents the simplest form of a viscosity pump. Here, the inner cylinder turns and drags the highly viscous fluid in the counter-clockwise direction, until it is pushed out of the system. Various situations can be described with this type of pump:

- (Case 1) Open discharge,
- (Case 2) Closed discharge, and
- (Case 3) Flow resisting die.

The *open discharge* case is where the pump is open to the atmosphere and consequently does not encounter a flow resistance. Here, the maximum flow rate is generated and no pressure is built up. In the *closed discharge* case the exit of the pump is blocked, leading to no flow rate and a maximum pressure build-up. In the flow restricting die case the fluid exiting the pump encounters a resistance, such as a die, where it is forced to flow through a narrow gap. Here, the pressure generated by the shear deformation is consumed by the die. For more detail on the underlying physics of the viscosity pump, please refer to Chapter 9 of this book.

Pumping in engineering

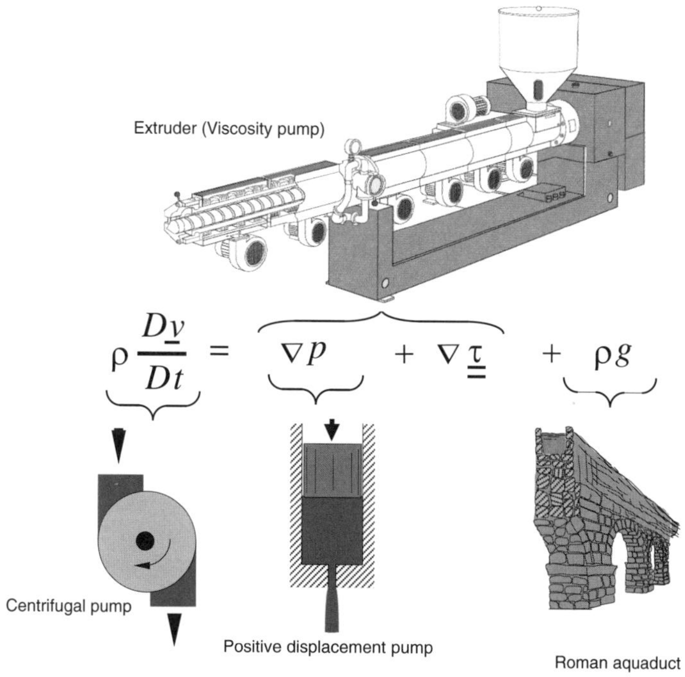

Extruder (Viscosity pump)
$\rho \frac{D\underline{v}}{Dt} = \nabla p + \nabla \underline{\underline{\tau}} + \rho g$
Centrifugal pump
Positive displacement pump
Roman aquaduct

Figure 4.1 Schematic of pumping principles

The viscosity pump principle

Case 1
Open discharge
Case 2
Closed discharge
Case 3
Die
n

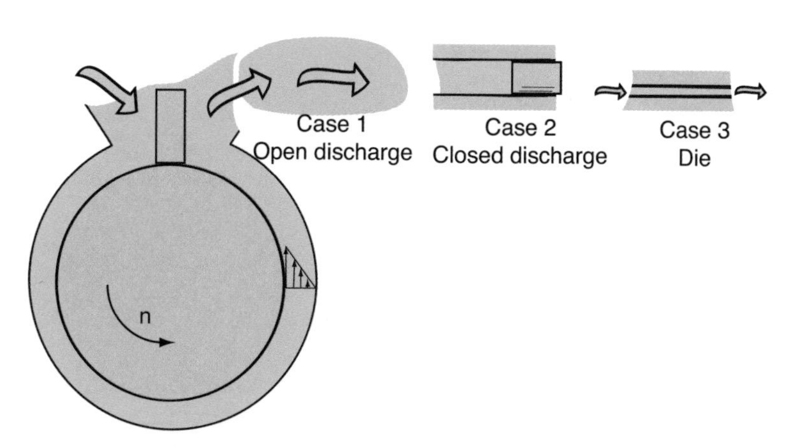

Figure 4.2 Schematic of a viscosity pump

The centrifugal pump, which works based on the fluid's inertia is also represented in the figure and is typical of low viscosity liquids. The Roman aqueduct, shown on the right of the figure, is driven by gravitational forces.

In today's polymer industry, the most commonly used extruder is the single screw extruder, schematically depicted in Fig. 4.3. The single screw extruder can either have a smooth inside barrel surface, called a *conventional single screw extruder*, or a grooved feed zone, called a *grooved feed extruder*. In some cases, an extruder can have a degassing zone, required to extract moisture, volatiles, and other gases that form during the extrusion process.

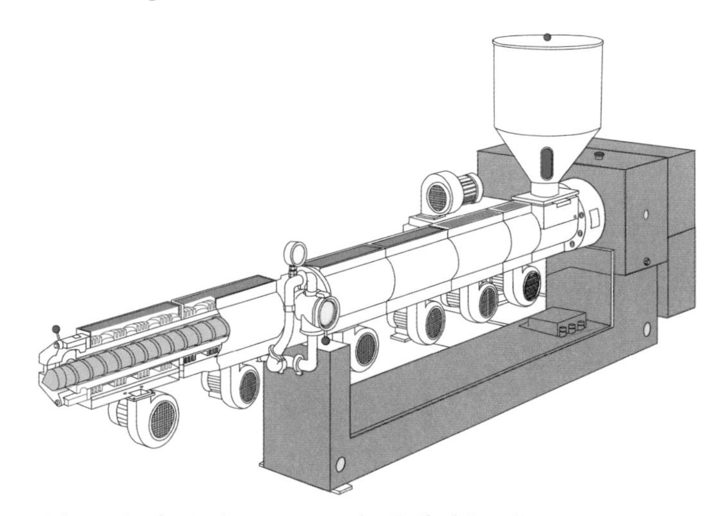

Figure 4.3 Schematic of a single screw extruder (Reifenhäuser)

Another important class of extruders are the twin screw extruders, schematically depicted in Fig. 4.4. Twin screw extruders can have co-rotating or counter-rotating screws, and the screws can be intermeshing or non-intermeshing. Twin screw extruders are primarily employed as mixing and compounding devices, as well as polymerization reactors. The mixing aspects of single and twin screw extruders are detailed in Chapter 5.

4.2 The Plasticating Extruder

The plasticating single screw extruder is the most common equipment in the polymer industry. It can be part of an injection molding unit and is found in numerous other extrusion processes, including blow molding, film blowing, and wire coating. A schematic of a plasticating, or three-zone, single screw extruder with its most important elements is shown in Fig. 4.5. Table 4.1 presents typical extruder dimensions and relationships common in single screw extruders.

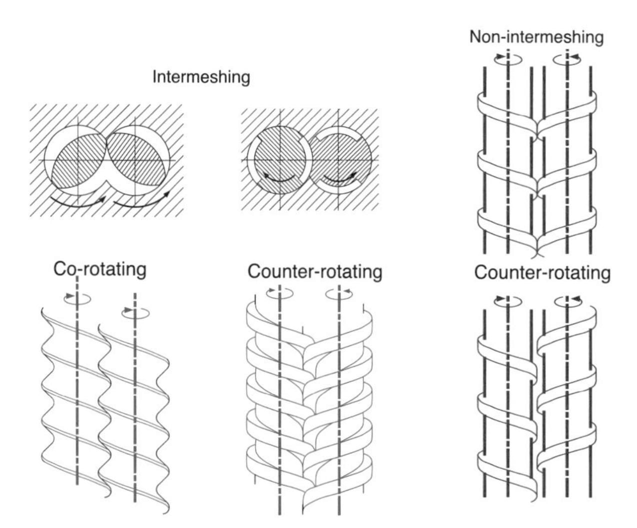

Figure 4.4 Schematic of different twin screw extruders

The three zone extruder — solids, melting and pumping zones

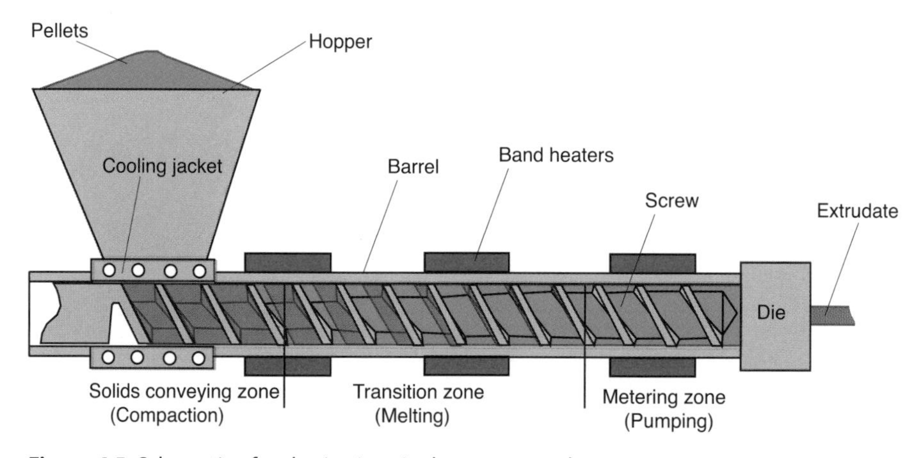

Figure 4.5 Schematic of a plasticating single screw extruder

The plasticating extruder can be divided into three main zones:

- the solids conveying zone
- the melting or transition zone
- the metering or pumping zone

The tasks of a plasticating extruder are to:

- transport the solid pellets or powder from the hopper to the screw channel
- compact the pellets and move them down the channel
- melt the pellets

Table 4.1 Typical extruder dimensions and relationships (the notation for Table 4.1 is defined in Fig. 4.6.)

L/D	Length to diameter ratio
	20 or less for feeding or melt extruders
	25 for blow molding, film blowing and injection molding
	30 or higher for vented extruders or high output extruders
D	Standard diameter
U.S. (in)	0.75, 1.0, 1.5, 2, 2.5, 3.5, 4.5, 6, 8, 10, 12, 14, 16, 18, 20, and 24
Europe (mm)	20, 25, 30, 35, 40, 50, 60, 90, 120, 150, 200, 250, 300, 350, 400, 450, 500, and 600
φ	Helix angle
	17.65° for a square pitch screw where $L_s = D$
	New trend: $0.8 < L_s/D < 1.2$
h	Channel depth in the metering section
	(0.05–0.07) D for $D < 30$ mm
	(0.02–0.05) D for $D > 30$ mm
β	Compression ratio: $h_{\text{feed}} = \beta h$
	2 to 4
δ	Clearance between the screw flight and the barrel
	0.1 mm for $D < 30$ mm
	0.15 mm for $D > 30$ mm
n	Screw speed
	1–2 rev/s (60–120 rpm) for large extruders
	1–5 rev/s (60–300 rpm) for small extruders
u_B	Barrel velocity (relative to screw speed) $= \pi D n$
	0.5 m/s for most polymers
	0.2 m/s for unplasticized PVC
	1.0 m/s for LDPE

Characteristic speed inside an extruder is 0.5 m/s for most polymers

- mix the polymer into a homogeneous melt
- pump the melt through the die

The pumping capability and characteristic of an extruder can be represented with sets of *die and screw characteristic curves.* Figure 4.7 presents such curves for a conventional (smooth barrel) single screw extruder.

The die characteristic curves are labeled K_1, K_2, K_3, and K_4 in ascending order of die restriction. Here, K_1 represents a low resistance die, such as for a thick plate, and K_4 represents a restrictive die, such as is used for film. The different screw characteristic curves represent different screw rotational speeds. In a screw characteristic curve the point of maximum throughput and no pressure build-up is called the point of *open discharge*. This occurs when there is no die. The point of maximum pressure build-

A square pitch screw — lead equals the diameter

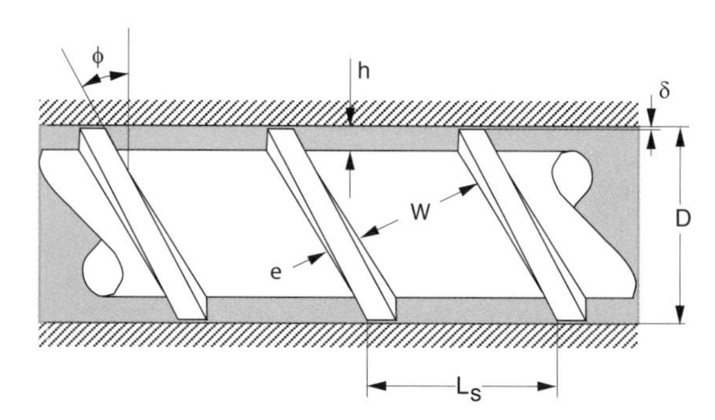

Figure 4.6 Schematic diagram of a screw section

Extruder characteristic curves

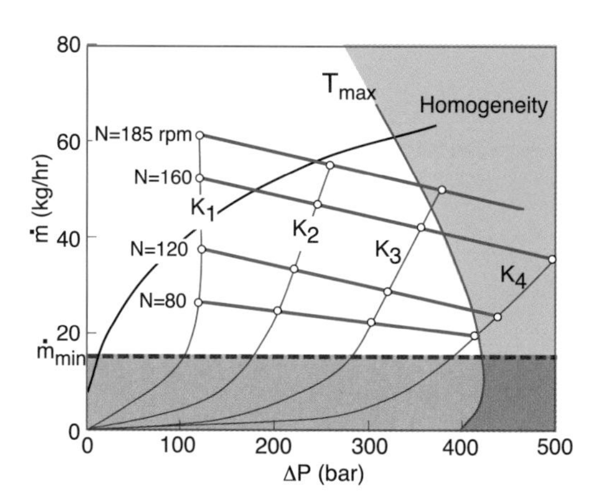

Figure 4.7 Screw and die characteristic curves for a 45 mm diameter extruder with an LDPE

up and no throughput is called the point of *closed discharge*. This occurs when the extruder is plugged.

The lines also shown in Fig. 4.7 represent critical aspects encountered during extrusion. The curve labeled T_{max} represents the conditions at which excessive temperatures are reached as a result of viscous heating. The feasibility line ($\dot{m}_{min}$) represents the throughput required to have an economically feasible system. The processing conditions to the right of the homogeneity line render a thermally and physically heterogeneous polymer melt.

4.2.1 The Solids Conveying Zone

The task of the solids conveying zone is to move the polymer pellets or powders from the hopper to the screw channel. Once the material is in the screw channel, it is compacted and transported down the channel. The process to compact the pellets and to move them can only be accomplished if the friction at the barrel surface exceeds the friction at the screw surface. This can be visualized if one assumes the material inside the screw channel to be a nut sitting on a screw. As we rotate the screw without applying outside friction, the nut (polymer pellets) rotates with the screw without moving in the axial direction. As we apply outside forces (barrel friction), the rotational speed of the nut is less than the speed of the screw, causing it to slide in the axial direction. Virtually, the solid polymer is then "unscrewed" from the screw.

The most complete analysis of the solids conveying zone in single screw extruders was performed by Darnell and Mol [1] and continued by Tadmor and Klein [2].

A useful limiting case is when the friction on the screw surface is negligible compared to the friction on the barrel, and when the pressure build-up in the screw channel is negligible (open discharge). These assumptions lead to a maximum mass throughput of

$$\dot{m} = \varrho_{\text{bulk}} \pi D_{\text{b}} N \tan\varphi \left(\frac{\pi}{4} \left(D_{\text{b}}^2 - D_{\text{s}}^2 \right) - \frac{e h_{\text{feed}}}{\sin\varphi} \right) \tag{4.1}$$

To maintain a high coefficient of friction between the barrel and the polymer, the feed section of the barrel must be cooled, usually with cold water cooling lines. The frictional forces also result in a pressure rise in the feed section. This pressure compresses the solids bed, which continues to travel down the channel as it melts in the transition zone. Figure 4.8 presents the pressure build-up in a conventional, smooth barrel extruder. In these extruders, most of the pressure required for pumping and mixing is generated in the metering section.

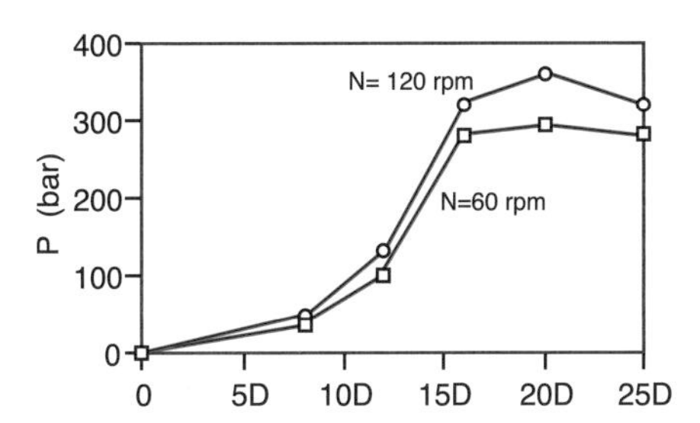

Figure 4.8 Conventional extruder pressure distribution

The simplest mechanism for ensuring high friction between the polymer and the barrel surface is grooving its surface in the axial direction [3, 4]. Extruders with a grooved feed section are called *grooved feed extruders*. To avoid excessive pressures that can lead to barrel or screw failure, the length of the grooved barrel section must not exceed 3.5D. A schematic diagram of the grooved section in a single screw extruder is presented in Fig. 4.9.

The key factors that propelled the development and refinement of the grooved feed extruder were the processing problems, excessive melt temperature, and reduced productivity posed by high viscosity and low coefficients of friction typical of high molecular weight polyethylenes and polypropylenes.

The friction on the barrel surface must be larger than the friction on the screw surface in order to start the motion of the polymer

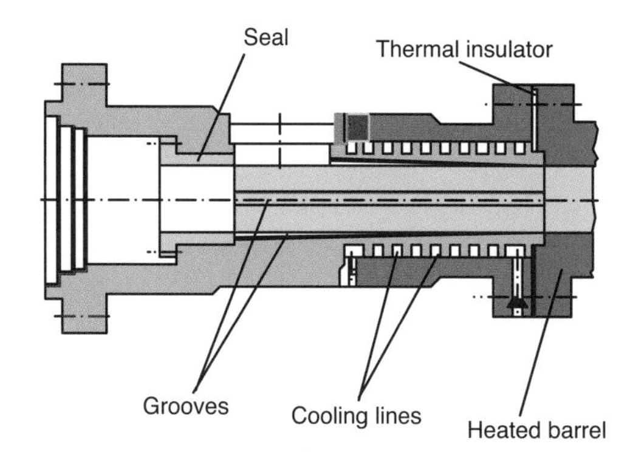

Figure 4.9 Schematic diagram of the grooved feed section of a single screw extruder

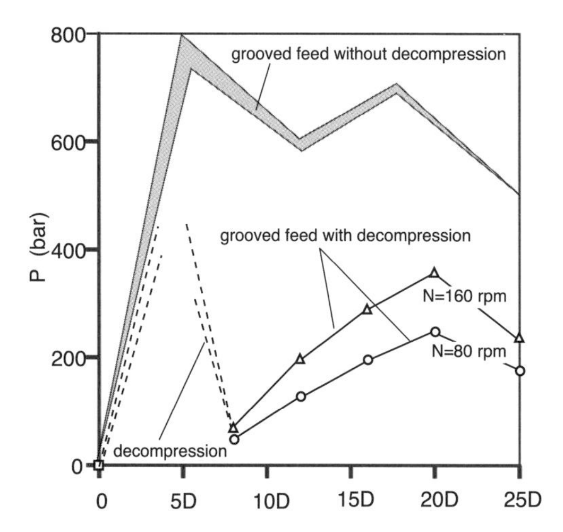

Figure 4.10 Grooved feed extruder pressure distribution

In a grooved feed extruder, the conveying and pressure build-up tasks are assigned to the feed section. Figure 4.10 presents the pressure build-up in a single screw extruder with a grooved feed section. The high pressures in the feed section lead to the main advantages over conventional systems. With grooved feed systems come higher productivity, higher melt flow stability and pressure invariance. This is demonstrated with the screw characteristic curves in Fig. 4.11, which presents screw characteristic curves for a 45-mm diameter grooved feed extruder with comparable mixing sections and die openings, as shown in Fig. 4.7.

The behavior of the two extruders in Figs. 4.7 and 4.11 are best compared if the throughput and the pressure build-up are non-dimensionalized. The dimensionless throughput is

$$\hat{m} = \frac{\dot{m}}{\varrho N D^3} \tag{4.2}$$

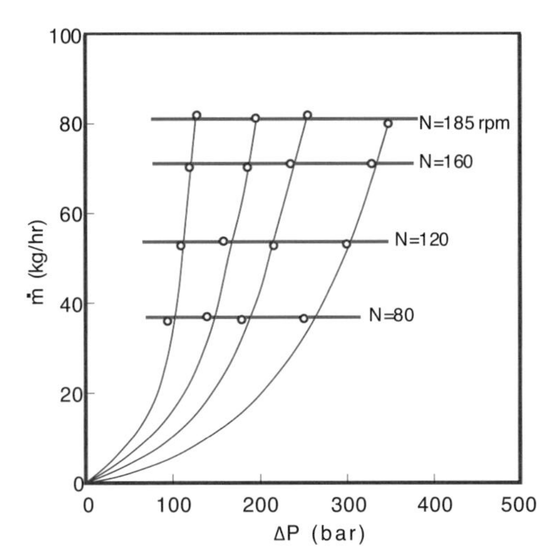

Figure 4.11 Screw and die characteristic curves for a grooved feed 45-mm diameter extruder with an LDPE

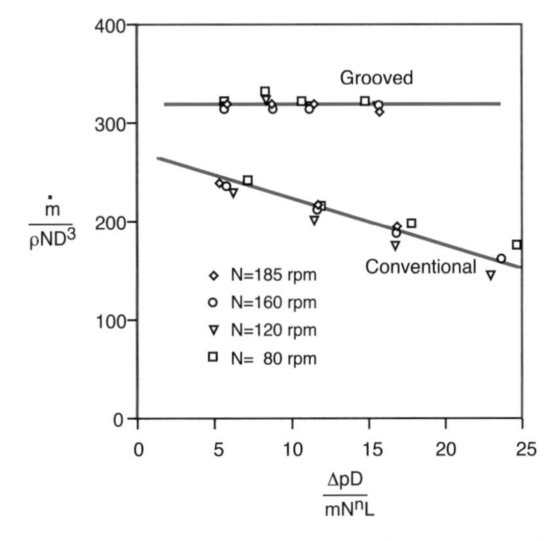

Figure 4.12 Dimensionless screw characteristic curves for conventional and grooved feed extruders

and the dimensionless pressure build-up is

$$\Delta\hat{p} = \frac{\Delta p D}{m N^n L} \tag{4.3}$$

where L represents the total channel length and for a 25 L/D extruder is

$$L = \frac{25D}{\sin(\varphi)} \tag{4.4}$$

where φ is assumed to be 17.65° (square pitch). Figure 4.12 presents the results shown in Figs. 4.7 and 4.11 after having been non-dimensionalized using Eqs. 4.2 and 4.3. The figure clearly shows the higher productivity of the grooved feed extruder, where the throughput is at least 50 % more than that observed with the conventional system for a comparable application. Used with care, Fig. 4.12 can also be used for scale-up.

4.2.2 The Melting Zone

The melting or transition zone is the portion of the extruder where the material melts. The length of this zone is a function of the material properties, screw geometry, and processing conditions. During melting, the size of the *solid bed* shrinks as a *melt pool* forms at its side, as depicted in Fig. 4.13(a), which shows the polymer unwrapped from the screw channel.

Figure 4.13b presents a cross section of the screw channel in the melting zone. The solid bed is pushed against the leading flight of the screw as freshly molten polymer is wiped from the *melt film* into the melt pool by the relative motion between the solids bed and the barrel surface.

Knowing where the melt starts and ends is important when designing a screw for a specific application. The most widely used model to predict melting in a plasticating

The solid bed profile in a single screw extruder

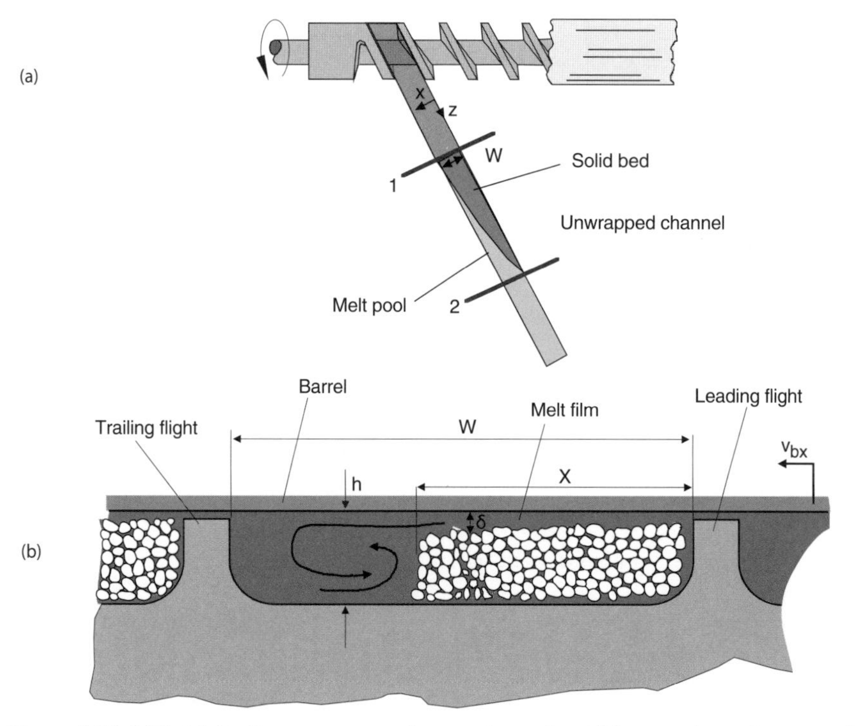

Figure 4.13 (a) Solids bed in an unwrapped screw channel and (b) screw channel cross section

Table 4.2 Extruder parameters, processing conditions, and material properties for the solids bed profile results in Fig. 4.14

Extruder Geometry:
Square pitch screw, $D = 63.5$ mm, $L/D = 26.5$, $W = 54.16$ mm
Feed zone — 12.5 turns $h_1 = 9.4$ mm
Transition zone — 9.5 turns $h_1 = 9.4$ mm $h_2 = 3.23$ mm
Metering zone — 4.5 turns $h_2 = 3.23$ mm
Processing Conditions:
$T_0 = 24$ °C $T_b = 149$ °C $N = 60$ rpm $\Delta p = 204$ bar $\dot{m} = 61.8$ kg/hr
Material Properties (LDPE):
Viscosity: $n = 0.345$ $a = 0.01$ °C^{-1} $m_0 = 5.6 \times 10^4$ Pa·s^n
$T_m = 110$ °C
Thermal: $k_m = 0.1817$ W/m°C $C_m = 2.596$ kJ/kg°C $C_s = 2.763$ kJ/kg°C
$\varrho_{bulk} = 595$ kg/m^3 $\varrho_s = 915.1$ kg/m^3 $\varrho_m = 852.7 + 5.018 \times 10^{-7} p - 0.4756 T$
$\lambda = 129.8$ kJ/kg

single screw extruder is the well known Tadmor Model [5]. Using the Tadmor Model, one can accurately predict the solid bed profile in the single screw extruder. Figure 4.14 presents the experimental and predicted solids bed profile of an LDPE in a single screw extruder. The material properties and processing conditions used in the calculations are given in Table 4.2. Chapter 9 discussed the Tadmor model in detail.

From experiment to experiment there are always large variations in the experimental solids bed profiles. The variations in this section of the extruder are caused by slight variations in processing conditions and by the uncontrolled solids bed break-up towards the end of melting. This effect can be eliminated by introducing a screw with a barrier flight that separates the solids bed from the melt pool. The Maillefer screw and barrier screw in Fig. 4.15 are commonly used to ensure high quality and reproducibility. The Maillefer screw maintains a constant solids bed width, maximizing the contact area between the solid bed and the heated barrel surface. On the other hand,

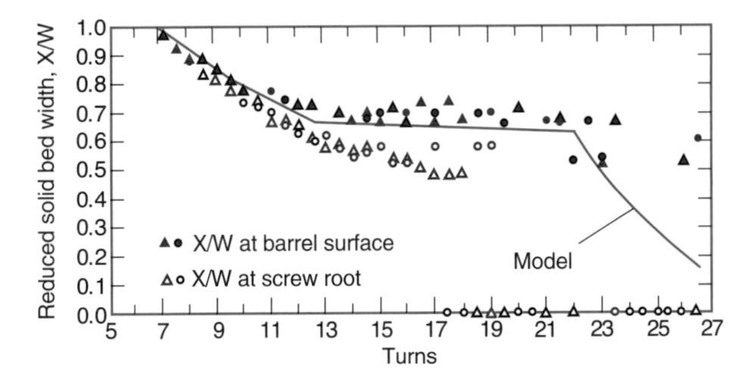

Figure 4.14 Predicted (Tadmor Model) and experimental solids bed profile

Barrier screws assure process repeatability

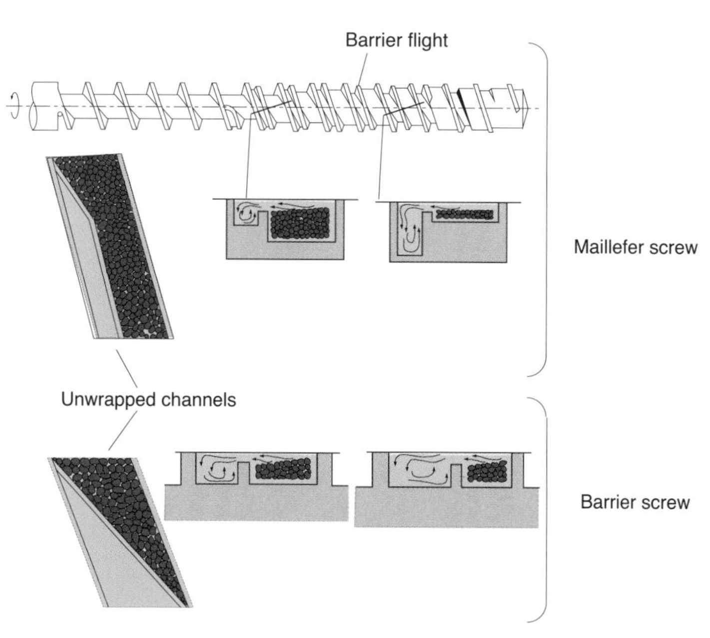

Figure 4.15 Schematic diagram of screws with different barrier flights

the barrier screw uses a constant channel depth with a gradually decreasing solids bed width.

4.2.3 The Metering Zone

The metering zone is the most important section in melt extruders and conventional single screw extruders that rely on it to generate pressures sufficient for pumping. The pumping capabilities in the metering section of a single screw extruder can be estimated by solving the equation of motion with appropriate constitutive laws. For a Newtonian fluid in an extruder with a constant channel depth, the screw and die characteristic curves for different cases are represented in Fig. 4.16. The figure shows the influence of the channel depth on the screw characteristic curves. A restrictive extrusion die would clearly work best with a shallow channel screw, and a less restrictive die would render the highest productivity with a deep channel screw.

In both the grooved barrel and the conventional extruder, the diameter of the screw determines the metering or pumping capacity of the extruder. Figure 4.17 presents typical normalized mass throughput as a function of screw diameter for both systems.

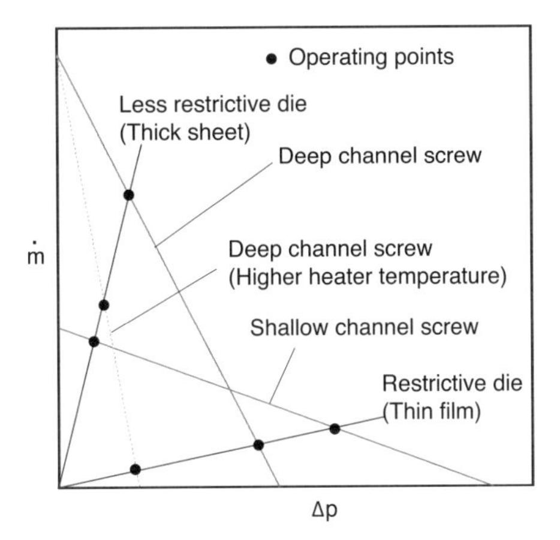

Figure 4.16 Screw characteristic curves (Newtonian)

Restrictive versus non-restrictive dies

Shallow versus deep screw channels

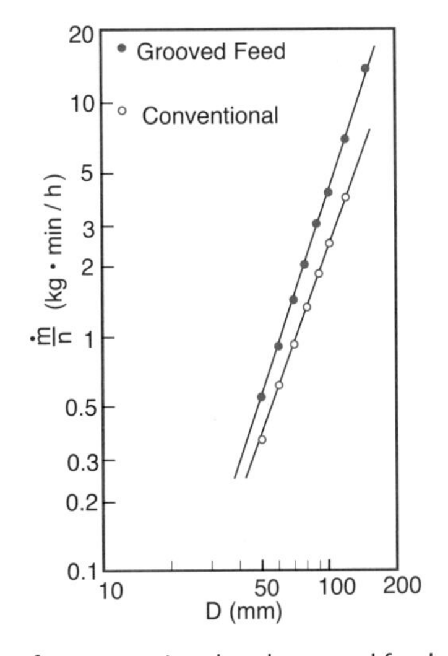

Figure 4.17 Throughput for conventional and grooved feed extruders

Grooved versus smooth barrel zone

4.3 Extrusion Dies

The extrusion die shapes the polymer melt into its final profile. The extrusion die is located at the end of the extruder and it used to extrude:

- flat films and sheets
- pipes and tubular films for bags
- filaments and strands

- hollow profiles for window frames
- open profiles

As shown in Fig. 4.18, depending on the functional needs of the product, several rules of thumb can be followed when designing an extruded plastic profile. These are:

- Avoid thick sections. They add to the material cost and increase sink marks caused by shrinkage.
- Minimize the number of hollow sections. hollow sections add to die cost and make the die more difficult to clean.
- Generate profiles with constant wall thickness. Constant wall thickness in a profile makes it easier to control the thickness of the final profile and results in a more even crystallinity distribution in semi-crystalline profiles.

Die design rules

- Avoid thickness variations
- Avoid thick sections
- Minimize hollow sections

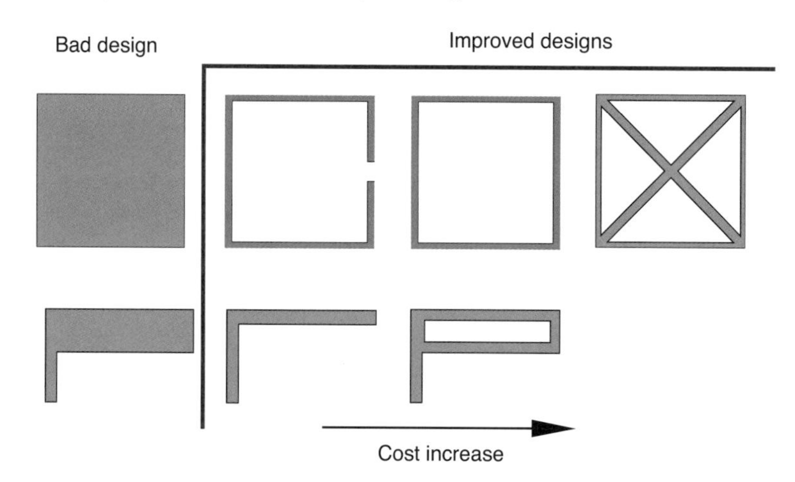

Figure 4.18 Extrusion profile designs

4.3.1 Sheeting Dies

One of the most widely used extrusion dies is the *coat-hanger sheeting die*. A sheeting die, such as depicted in Fig. 4.19, is formed by the following elements:

- *Manifold:* evenly distributes the melt to the approach or land region
- *Approach or land:* carries the melt from the manifold to the die lips
- *Die lips:* perform the final shaping of the melt
- *Flex lips:* for fine tuning when generating a uniform profile

To generate a uniform extrudate geometry at the die lips, the geometry of the manifold must be specified appropriately. Figure 4.20 presents the schematic of a coat-hanger die with a pressure distribution that corresponds to a die that renders a uniform extrudate.

The goal in sheeting die design is to have a uniform product

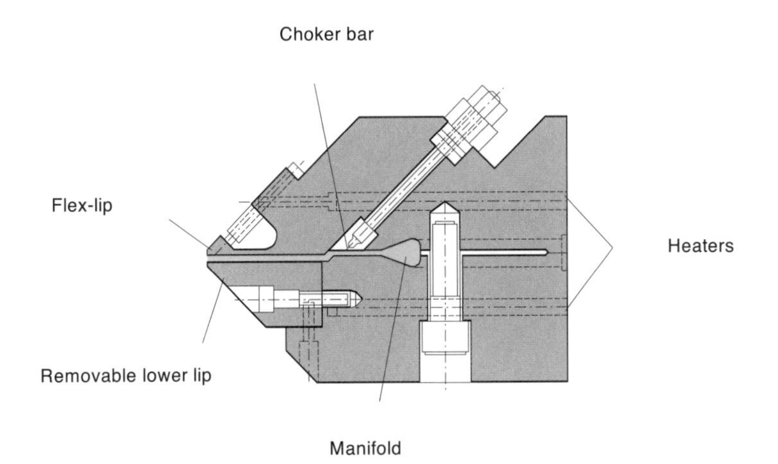

Figure 4.19 Cross section of a coat-hanger die

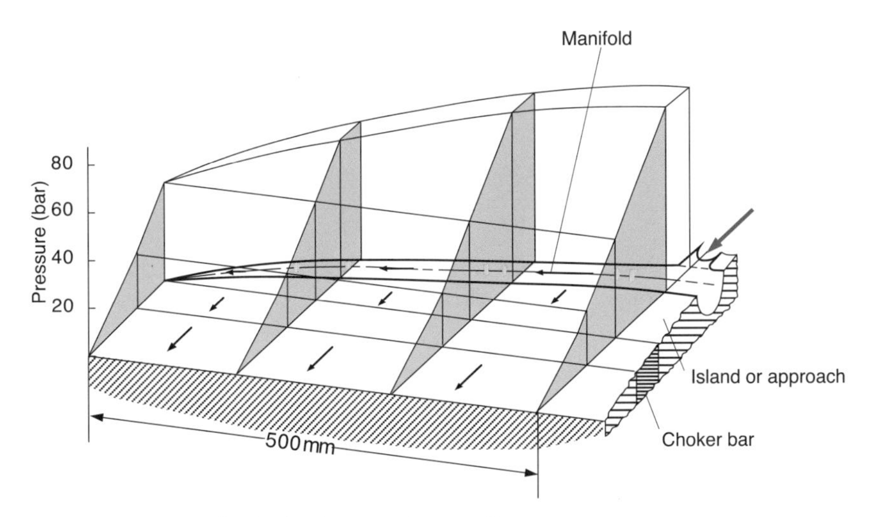

Figure 4.20 Pressure distribution in a coat-hanger die

It is important to mention that the flow through the manifold and the approach zone depend on the non-Newtonian properties of the polymer extruded. So the design of the die depends on the shear thinning behavior of the polymer. Hence, a die designed for one material does not necessarily work for another. Die design is discussed in more detail in Chapter 9 of this book.

4.3.2 Tubular Dies

In a tubular die, the polymer melt exits through an annulus. These dies are used to extrude plastic pipes and tubular film. The film blowing operation is discussed in more detail in Chapter 7.

The simplest tubing die is the spider die, depicted in Fig. 4.21. Here, a symmetric mandrel is attached to the body of the die by several legs. The polymer must flow around the spider legs, causing weld lines along the pipe or film. These weld lines, visible streaks along the extruded tube, are weaker regions.

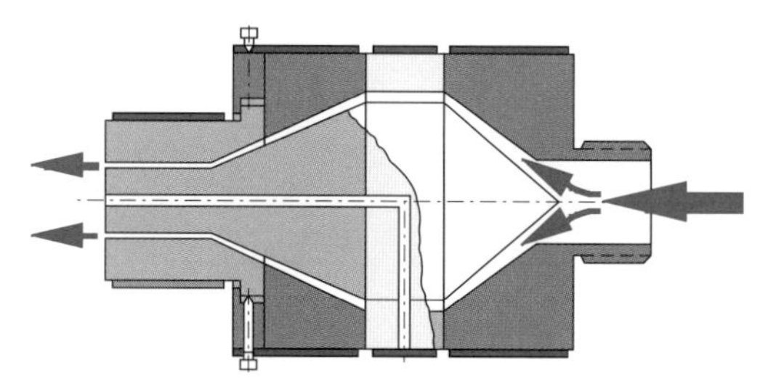

Figure 4.21 Schematic of a spider leg tubing die

To overcome weld line problems, the cross-head tubing die is often used. Here, the die design is similar to that of the coat-hanger die, but wrapped around a cylinder. This die is depicted in Fig. 4.22. Since the polymer melt must flow around the mandrel, the extruded tube exhibits one weld line. In addition, although the eccentricity of a mandrel can be controlled using adjustment screws, there is no flexibility to perform

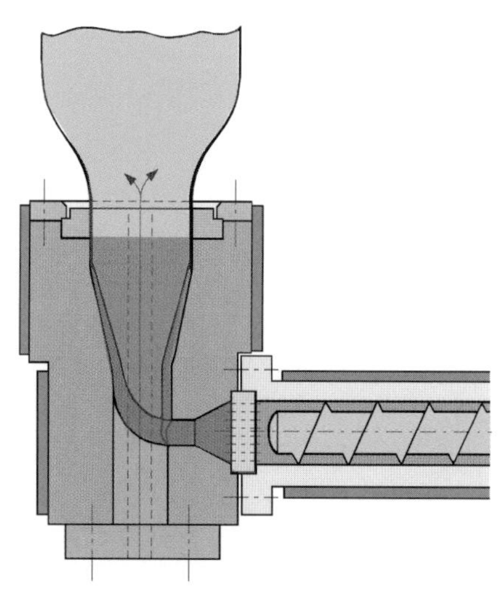

Figure 4.22 Schematic of a cross-head tubing die used in film blowing

fine-tuning such as in the coat-hanger die. This can result in tubes with uneven thickness distributions.

The spiral die, commonly used to extrude tubular blown films, eliminates weld line effects and produces a thermally and geometrically homogeneous extrudate. The polymer melt in a spiral die flows through several feed ports into independent spiral channels wrapped around the circumference of the mandrel. This type of die is schematically depicted in Fig. 4.23.

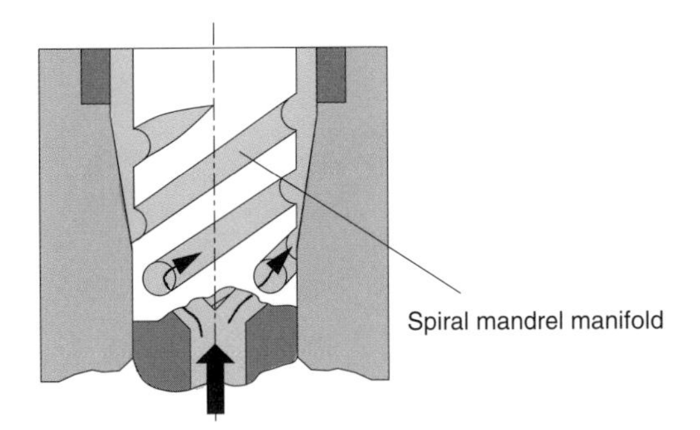

Figure 4.23 Schematic of a spiral die

References

1. Darnell, W.H., and E.A.J. Mol, *Soc. Plastics Eng. J.* (1956), *12*, 20
2. Tadmor, Z., and I. Klein, *Engineering Principles of Plasticating Extrusion*, Van Nostrand Reinhold Company (1970), New York
3. Menges, G., W. Predöhl, R. Hegele, U. Kosel, and W. Elbe, Plastverarbeiter (1969), (20) 79 and 188
4. Menges, G., U. Kosel, R. Hegele, and W. Elbe, SPE ANTEC (1972), 784
5. Z. Tadmor, and C.G. Gogos, *Principles of Polymer Processing*, 2nd Edition, John Wiley & Sons (2006), New York

5 Mixing

The quality of the finished product in almost all polymer processes depends in part on how well the material was mixed. Mixing can occur in internal mixers or during processing (e.g., in single and twin screw extruders). Both the material properties and the formability of the compound into shaped parts are highly influenced by the mixing quality. Hence, a better understanding of the mixing process helps one optimize processing conditions and increase part quality.

The process of polymer blending or mixing is accomplished by distributing or dispersing a minor or secondary component within a major component serving as a matrix. The major component can be thought of as the continuous phase, while the minor components can be interpreted as distributed or dispersed phases in the form of droplets, filaments, or agglomerates.

When creating a polymer blend, one must always keep in mind that the blend will probably be re-melted in subsequent processing or shaping processes. For example, a rapidly cooled system, frozen as a homogenous mixture, may separate into phases because of coalescence when re-heated. For all practical purposes, such a blend is not processable. To avoid this, special macromolecules are commonly used to compatibilize the boundary layers between phases [1].

The mixing can be *distributive* or *dispersive*. The morphology development of polymer blends is determined by three competing mechanisms: distributive mixing, dispersive mixing, and coalescence. Figure 5.1 presents a model, proposed by Macosko and co-workers [1, 2], that helps us visualize the mechanisms governing morphology development in polymer blends.

5.1 Distributive Mixing

Distributive mixing or laminar mixing of compatible liquids is usually characterized by the distribution of the droplet or secondary phase within the matrix. This distribution is achieved by imposing large strains on the system such that the interfacial area between phases increases and the local dimensions, or striation thicknesses, of the secondary phases decrease. This concept is shown in Fig. 5.2 [3]. Here we have a Couette flow device with the secondary component having an initial striation thickness of δ_0. As the inner cylinder rotates, the secondary component is distributed through the systems with constantly decreasing striation thickness.

Distributive mixing is controlled by deformation and orientation

Imposing large strains on the system may not homogenize the mixture sufficiently. The type of mixing device, initial orientation, and position of the fluid components play a significant role in the quality of mixing. For example, the mixing mechanism shown in Fig. 5.2 homogeneously distributes the melt within the region contained by

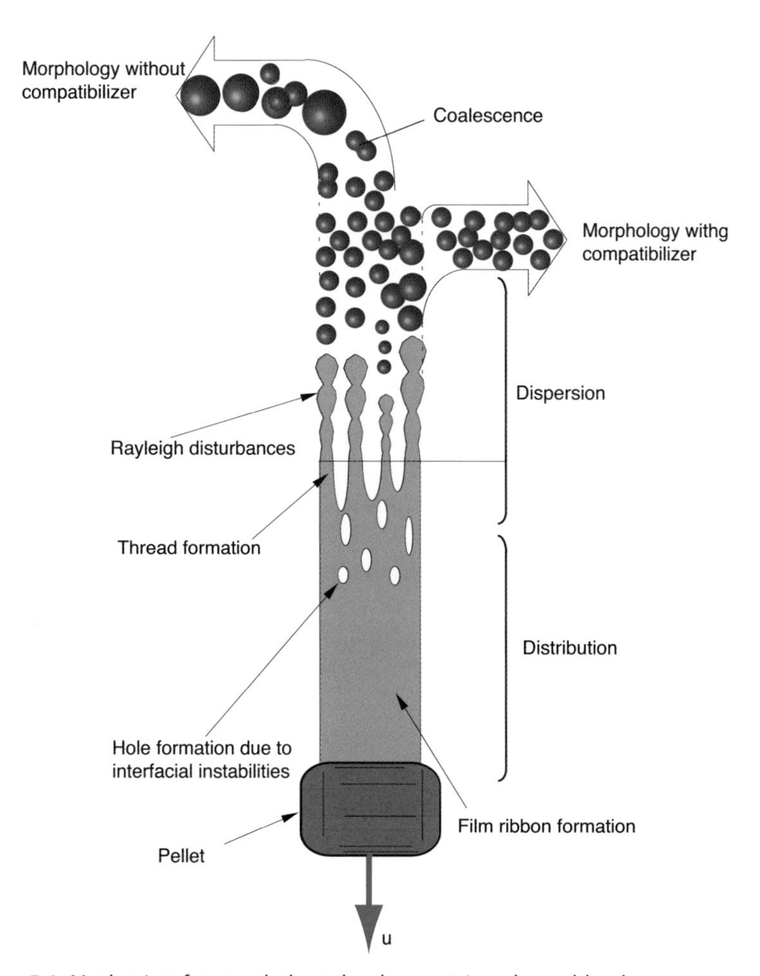

Figure 5.1 Mechanism for morphology development in polymer blends

the streamlines cut across by the initial secondary component. Figure 5.3 [4] shows another variation of initial orientation and arrangement of the secondary component. Here, the secondary phase cuts across all streamlines, which can lead to a homogeneous mixture throughout.

Distributive mixing can be quantified by the reduction in striation thickness. For a sphere deformed into an ellipsoid, the striation thickness can be related to the total strain using

$$\delta = 2R(1+\gamma^2)^{-1/4} \tag{5.1}$$

Another common way of quantifying mixing is by following the growth of the interface between the primary and secondary fluids. In simple shear, a simple relation exists between the interfacial area, the strain and the orientation of the area of the secondary

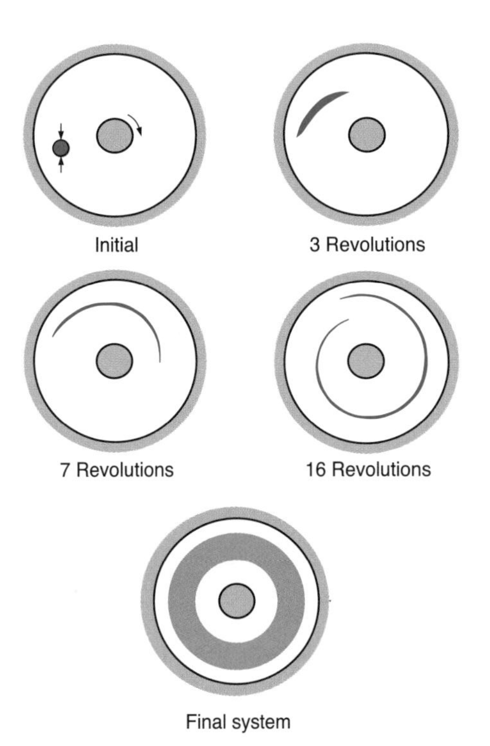

Figure 5.2 Experimental results of distributive mixing in Couette flow, and schematic of the final mixed system

fluid with respect to the flow direction [5]:

$$\frac{A}{A_0} = \gamma \cos \alpha \tag{5.2}$$

where A_0 is the initial interfacial area, A is the final interfacial area, γ is the total strain and α the angle between the surface normal vector and the flow direction. Figure 5.4 [6] demonstrates this.

In Fig. 5.4, both cases (a) and (b) start-up with equal initial areas, A_0, and undergo the same amount of strain, $\gamma = 10$. The circular secondary component in (a) has a surface that is randomly oriented, between 0 and 2π, whereas most of the surface of the elongated secondary component in case (b) is oriented at $\frac{p}{2}$ so that of the interfacial area hardly changes. An ideal case would have been a long slender secondary component with a surface oriented in the flow direction or vertically between the parallel plates. Hence, the maximum interface growth inside a simple shear mixer can be achieved if the direction of the interface is maintained in an optimal orientation ($\cos \alpha = 1$). In a simple shear flow, this would require a special stirring mechanism that would maintain the interface between the primary and secondary fluid components in a vertical position. Using this concept, Erwin [7] demonstrated that the upper bound for the ideal mixer is found in a mixer that applies a plane strain extensional flow or

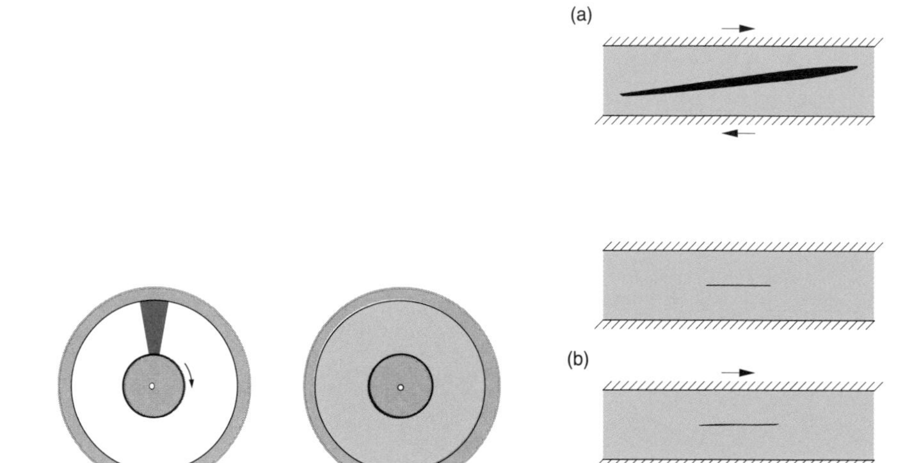

Figure 5.3 Distributive mixing in Couette flow

Figure 5.4 Effect of initial surface orientation on distributive mixing

pure shear flow to the fluid. In such a system the growth of the interfacial areas follows the relation given by

A system where the surfaces are always ideally oriented results in Erwin's ideal mixer

$$\frac{A}{A_0} = e^{\gamma/2} \tag{5.3}$$

In Erwin's ideal mixer, the amount of mixing increases exponentially compared to a linear increase if the orientation of the fluids' interfaces remain undisturbed.

5.2 Dispersive Mixing

Dispersive mixing is controlled by stresses and type of flow

Dispersive mixing in polymer processing involves breaking a secondary immiscible fluid or an agglomerate of solid particles and dispersing them throughout the matrix. Here, the imposed strain is not as important as the imposed stress that causes the system to break up. Hence, the type of flow inside a mixer plays a significant role in the break up of solid particle clumps or fluid droplets in dispersion.

5.2.1 Break-Up of Particulate Agglomerates

The most common example of dispersive mixing of particulate solid agglomerates is the dispersion and mixing of carbon black into a rubber compound. The dispersion of such a system is schematically represented in Fig. 5.5. However, the break-up of particulate agglomerates is best explained using an ideal system of two small spherical particles that must be separated and dispersed during mixing.

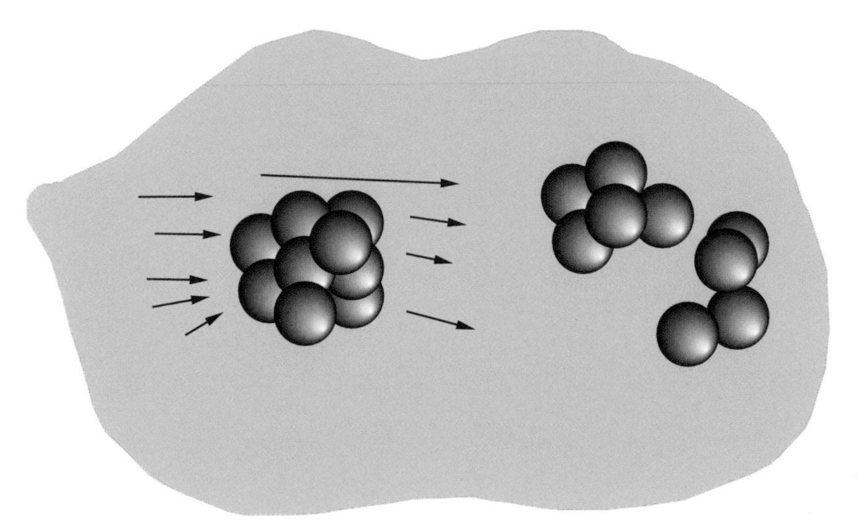

Figure 5.5 Break-up of particulate agglomerates during flow

If the mixing device generates a simple shear flow, as shown in Fig. 5.6, the maximum separation forces on the particles occur when they are oriented in a 45° position as they continuously rotate during flow. The magnitude of the force trying to separate the "agglomerate" is given by [8]

$$F_{\text{shear}} = 3\pi\eta\dot{\gamma}r^2 \tag{5.4}$$

where η is the viscosity of the carrier fluid, $\dot{\gamma}$ the magnitude of the strain rate tensor, and r the radii of the particles.

However, if the flow field generated by the mixing device is purely elongational, such as shown in Fig. 5.7, the particles are always oriented at 0°, the position of maximum

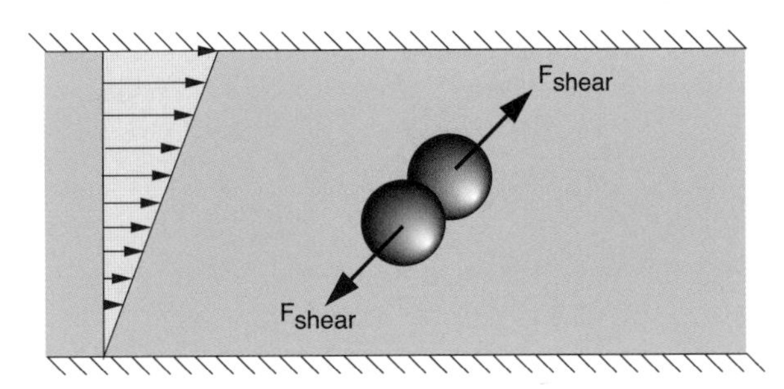

Figure 5.6 Force applied to a two-particle agglomerate in simple shear

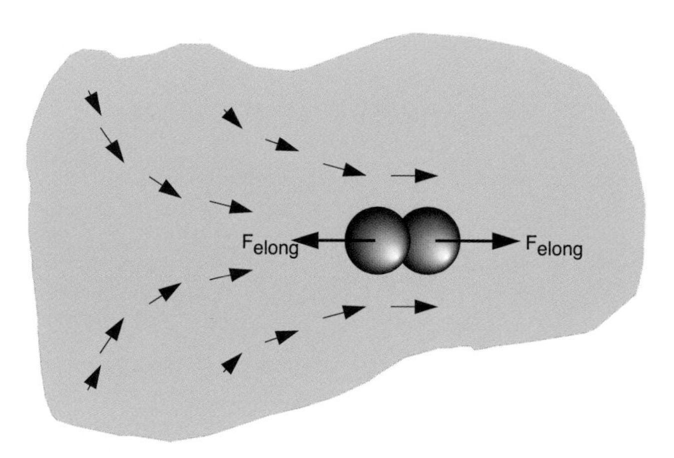

Figure 5.7 Force applied to a two-particle agglomerate in an elongational flow

force. The magnitude of the force for this system is given by

$$F_{\text{elong}} = 6\pi\eta\dot{\gamma}r^2 \tag{5.5}$$

which is *twice* the maximum tangential force producing simple shear. In addition, in elongational flow, the agglomerate is always oriented in the direction of maximum force generation, whereas in simple shear flow, the agglomerate tumbles quickly through the position of maximum force.

The above analysis makes it clear that for mixing processes requiring break-up and dispersion of agglomerates, elongation is the preferred mode of deformation. This is only valid if the magnitude of the rate of deformation tensor can be kept the same in elongation as in shear. Hence, when optimizing mixing devices, it is important to know which mode of deformation dominates, which can be accomplished by computing a *flow number or Manas-Zloczower number (Mz)* [9], defined by

The flow number, also known as Manas-Zloczower number, reveals the type of flow that exists in a process

$$\lambda = \frac{\dot{\gamma}}{\dot{\gamma} + \omega} \tag{5.6}$$

where $\dot{\gamma}$ is the magnitude of the rate of deformation tensor and ω the magnitude of the vorticity tensor. A flow number of 0 implies pure rotational flow, a value of 0.5 represents simple shear flow, and pure elongational flow is implied when λ equals 1.

5.2.2 Break-Up of Fluid Droplets

Droplets in an incompatible matrix tend to be spherical as a result of their natural tendency to maintain the lowest possible surface to volume ratio. However, a flow field within the mixer applies a stress on the droplets, causing them to deform. If this stress is high enough, it eventually causes the drops to disperse. The droplets disperse when the surface tension can no longer maintain their shape in the flow field and the filaments break up into smaller droplets. This phenomenon of dispersion and distribution continues to repeat itself until the stresses of the flow field can no longer overcome the surface tension of the new droplets formed.

As can be seen, the mechanisms of fluid agglomerate break-up are similar in nature to solid agglomerate break-up in the sense that both rely on forces to disperse the agglomerates. Hence, elongation is also the preferred mode of deformation when breaking up fluid droplets and threads, making the flow number, λ, indispensable when quantifying mixing processes.

A parameter commonly used to determine whether a droplet disperses is the capillary number defined by

$$Ca = \frac{\tau R}{\sigma_s} \tag{5.7}$$

where τ is the flow induced or deviatoric stress, R the characteristic dimension of the droplet and σ_s the surface tension acting on the drop. The capillary number is the ratio of flow stresses to droplet surface stresses. Droplet break-up occurs when a critical capillary number, Ca_{crit}, is reached. This break-up is clearly shown in Fig. 5.8 [10], which depicts the disintegration of a Newtonian thread in a Newtonian matrix.

A fluid thread breaks up when the critical capillary number is reached

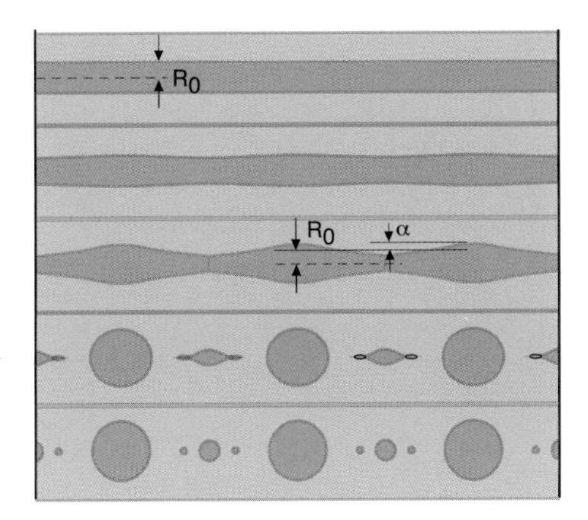

Figure 5.8 Disintegration of a Newtonian 0.35 mm diameter castor oil thread in a Newtonian silicon oil matrix. Redrawn from photographs taken every second

Because of the continuously decreasing thread radius, the critical capillary number is reached at some specific point in time. As a result of the competing deviatoric stresses and surface forces, the cylindrical shape becomes unstable and small disturbances at the surface lead to a growth of capillary waves, commonly called *Rayleigh disturbances*. Disturbances with various wavelengths form on the cylinder surface, but only those with a wavelength exceeding the circumference ($2\pi R_0$) of the thread lead to a monotonic decrease of the interfacial area.

The critical capillary number is reached when the surface tension stresses are overcome by deformation stresses

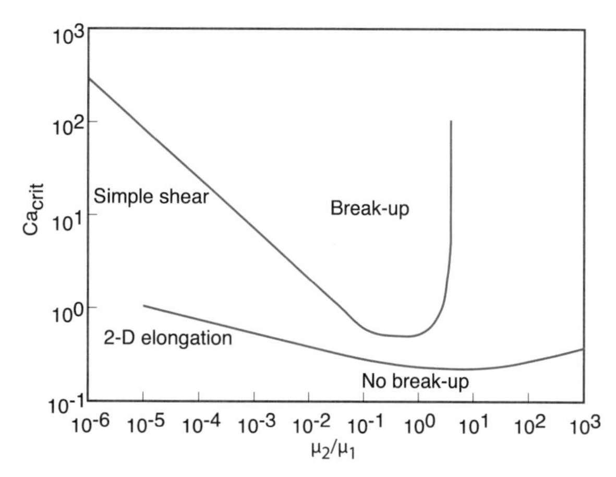

Figure 5.9 Critical capillary number for drop break-up as a function of viscosity ratio in a simple shear and a 2-D elongational flow

Figure 5.9 [11] shows the critical capillary number as a function of viscosity ratio, φ, and flow type, described by the mixing parameter, λ. For a viscosity ratio of 1 the critical capillary number is of order 1 [12]. Distributive mixing is implied when Ca is much greater than Ca_{crit} because the interfacial stress is much smaller than shear stresses. For such a case, the capillary waves, which cause droplet break-up, would not develop. Dispersive mixing is implied when Ca approaches its critical value or when interfacial stresses nearly equal the deviatoric stresses, causing droplet break-up. In addition, break-up can occur only if enough time elapses.

The time required for break-up, t_{b}, can be computed using

Break-up time of a thread depends on the growth rate of the Rayleigh disturbances

$$t_{\text{b}} = \frac{5.61}{q} \tag{5.8}$$

where q is given by

$$q = \frac{\sigma_{\text{s}} \Omega}{2\eta_1 R_0} \tag{5.9}$$

and Ω is a dimensionless growth rate represented in Fig. 5.10. The break-up time decreases as the critical capillary number is exceeded. The reduced break-up time t_{b}^* can be approximated using [11]

The break-up time reduces as the capillary number increases from its critical value

$$t_{\text{b}}^* = t_{\text{b}} \left(\frac{Ca}{Ca_{\text{crit}}} \right)^{-0.63} \tag{5.10}$$

As mentioned before, surface tension plays a large role in mixing, especially when dealing with dispersive mixing, when the capillary number approaches its critical value. Because of the stretching of the interfacial area caused by distributive mixing, the local radii of the suspended components decrease as surface tension starts to play a role.

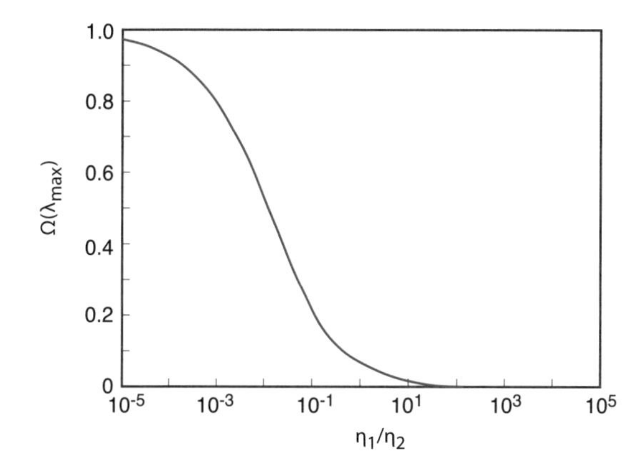

Figure 5.10 Dominant growth rate of interfacial disturbances as a function of viscosity ratio

Once the capillary number goes below its critical value, only slight deformations occur and internal circulation maintains an equilibrium elliptical droplet in the flow. The mixing process then reduces to the distribution of the dispersed droplets.

Example 5.1 Dispersive mixing of a polymer blend

You are to use a 45-mm diameter single screw extruder to create a polymer polycarbonate/ polypropylene polymer blend. The maximum screw rotation is 160 rpm and the screw channel depth is 4 mm. Assuming a barrel temperature of 300 °C, a surface tension, σ_s between the two polymers of 8×10^{-3} N/m, and using the viscosity curves given in Figs. 5.11 and 5.12, determine:

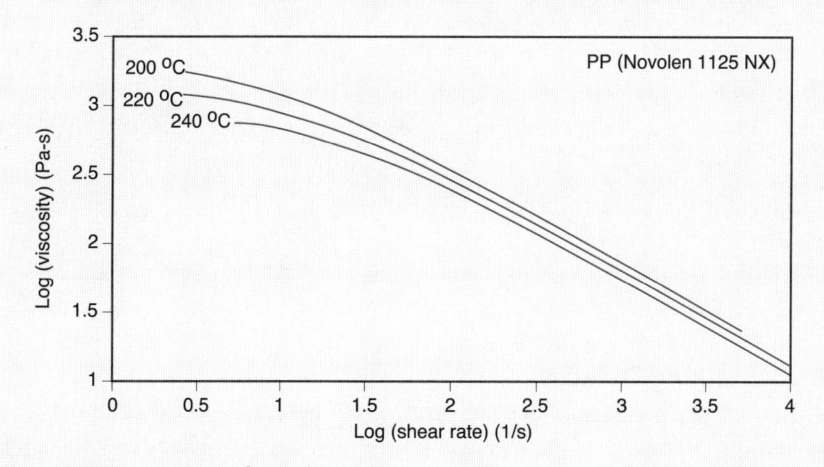

Figure 5.11 Viscosity curves for a polypropylene

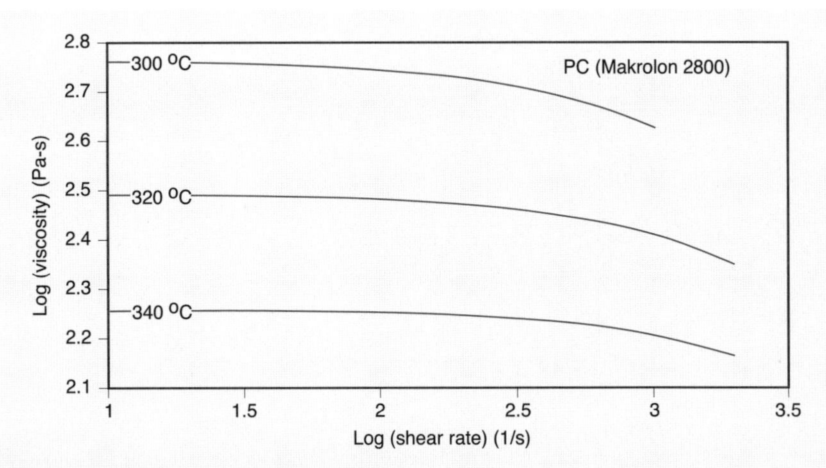

Figure 5.12 Viscosity curves for a polycarbonate

- Whether one can disperse 20 % PC into 80 % PP
- Whether one can disperse 20 % PP into 80 % PC
- The minimum size of the dispersed phase

We start this problem by first calculating the average speed in the extruder using

$$v_0 = \pi D n = \pi(45)(160)(1/60) = 377 \text{ mm/s}$$

which results in an average rate of deformation of

$$\dot{\gamma} = \frac{v_0}{h} = \frac{377}{4} = 94\frac{1}{s}$$

In shear the viscosity of the droplet must be less than 3.8 times the viscosity of the matrix

From the viscosity curves we get $\eta_{PC} \approx 600$ Pa·s and $\eta_{PP} \approx 150$ Pa·s. Using Fig. 5.9 we can deduce that one cannot disperse polycarbonate into polypropylene using a single screw extruder that only induces shear deformation, because $\eta_{PC}/\eta_{PP} > 4$. On the other hand, one can disperse polypropylene into polycarbonate using the same single screw extruder. Using Fig. 5.9 we can see that dispersive mixing for a $\eta_{PP}/\eta_{PC} = 0.25$ will occur at $Ca_{crit} \approx 0.7$. Hence, neglecting the effects of coalescence we can calculate the minimum size of the dispersed phase using

$$Ca_{crit} = 0.7 = \frac{\tau R}{\sigma_s} = \frac{600(94)R}{8 \times 10^{-3}} \rightarrow D = 2R = 0.2\ \mu\text{m}$$

To achieve this dispersion we must maintain the stresses for an extended period. ■

5.3 Mixing Devices

The final properties of a polymer component are heavily influenced by blending or mixing during processing or as a separate step in the manufacturing process. As mentioned earlier, when measuring the quality of mixing, one must also evaluate the mixing efficiency. For example, the amount of power required to achieve the highest mixing quality for a blend may be unachievable. This section presents mixing devices commonly encountered in polymer processing.

In general, mixers can be classified into two categories: internal batch mixers and continuous mixers. Internal batch mixers, such as the Banbury mixer, are the oldest mixing devices in polymer processing but are slowly being replaced by continuous mixers. This is because most continuous polymer processes involve mixing in addition to their normal processing tasks. Typical examples are single and twin screw extruders that often have mixing heads or kneading blocks incorporated into their systems.

5.3.1 Banbury Mixer

The Banbury mixer, shown in Fig. 5.13, is perhaps the most commonly used internal batch mixer. Internal batch mixers are high intensity mixers that generate complex shearing and elongational flows and work especially well in the dispersion of solid particle agglomerates within polymer matrices. One of the most common applications for high intensity internal batch mixing is the break-up of carbon black agglomerates into rubber compounds. The dispersion of agglomerates depends strongly on mixing time, rotor speed, temperature, and rotor blade geometry [13]. Figure 5.14 [14, 15] shows the fraction of undispersed carbon black versus time in a Banbury mixer at 77 rpm and 100 °C. The broken line in the figure represents the fraction of particles smaller than 500 nm.

The internal batch mixer is used extensively in the rubber industry

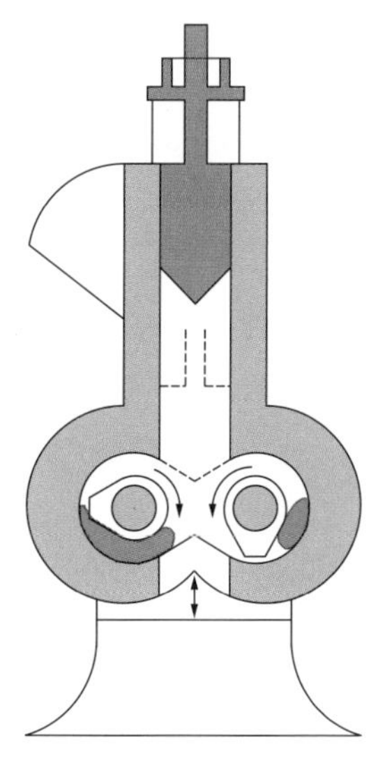

Figure 5.13 Schematic diagram of a Banbury mixer

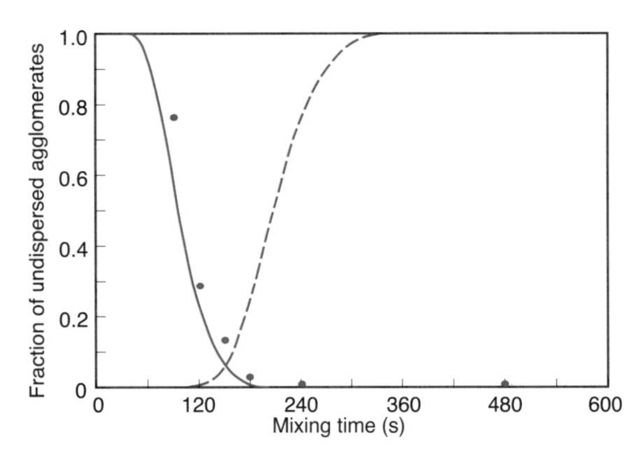

Figure 5.14 Fraction of undispersed carbon black, of size above 9 µm, as a function of mixing time inside a Banbury mixer; (O) denotes experimental results and solid line theoretical predictions; the broken line denotes the fraction of aggregates below 500 nm in size

5.3.2 Mixing in Single Screw Extruders

One task of a single screw extruder is to mix polymers with additives and to homogenize the melt before pumping it through the die. As discussed in the previous chapter, even without a mixing section, there is a cross flow component in the polymer melt traveling down the channel in a single screw extruder. This cross flow component acts as a stirring mechanism that causes mixing. By circulating the fluid from the top of the channel to the bottom, and vice versa, the cross-channel component re-orients the interfaces between the primary and secondary fluids, enhancing mixing during extrusion.

Pins enhance distributive mixing by re-orienting surfaces

Mixing caused by the cross-channel flow component can be further enhanced by introducing pins in the flow channel. These pins can either sit on the screw, as shown in Fig. 5.15 [16], or on the barrel, as shown in Fig. 5.16 [17]. The extruder with the adjustable pins on the barrel is called a QSM-extruder.[1)] In both cases the pins disturb the flow by re-orienting the surfaces between fluids and by creating new surfaces by splitting the flow. The pin type extruder is especially useful for mixing high viscosity materials such as rubber compounds; thus, it is often used in *cold feed rubber extruders*, a machine widely used to produce rubber profiles.

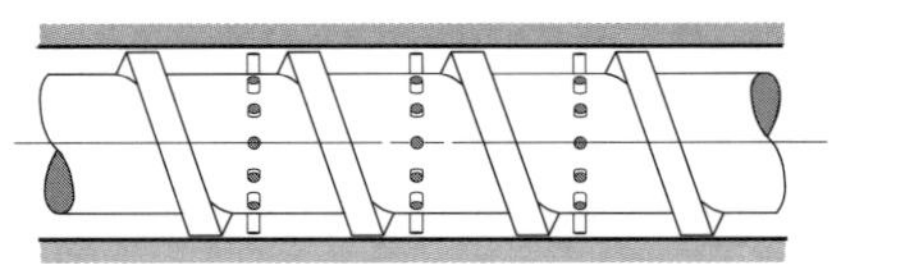

Figure 5.15 Pin mixing section on the screw of a single screw extruder

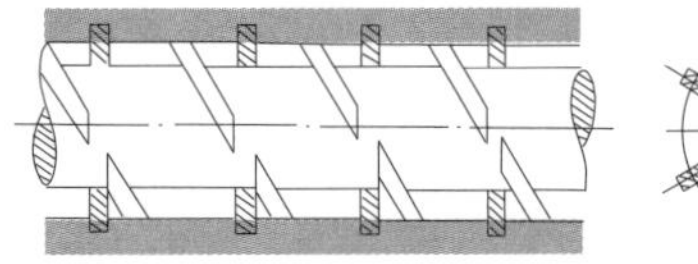

Figure 5.16 Pin barrel extruder (QSM)

1) QSM comes from the German *Quer Strom Misch* which translates as "cross-flow mixing".

For lower viscosity fluids, such as thermoplastic polymer melts, often the mixing action caused by the cross-flow is insufficient to re-orient, distribute, and disperse the mixture, so one must use special mixing sections. Re-orientation of the interfaces between primary and secondary fluids and distributive mixing can be induced by any disruption in the flow channel. Figure 5.17 [16] presents commonly used distributive mixing heads for single screw extruders. These mixing heads introduce several disruptions in the flow field that improve mixing.

The rhomboidal mixer is a very common distributive mixer in single screw extrusion

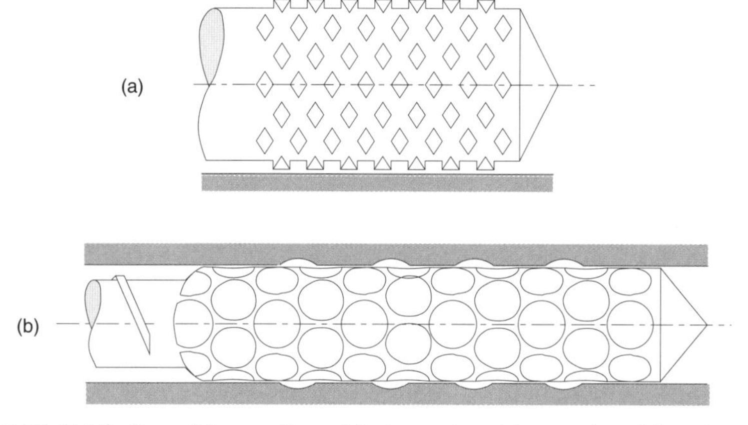

Figure 5.17 Distributive mixing sections: (a) pineapple mixing section, (b) cavity transfer mixing section

As mentioned earlier, dispersive mixing is required when breaking up particle agglomerates or when surface tension effects exist between primary and secondary fluids. To disperse such systems, the mixture must be subjected to large stresses. Barrier-type screws are often sufficient to apply high stresses; however, more intensive mixing results from a mixing head. When using barrier screws or a mixing head, as shown in Fig. 5.18 [16], the mixture is forced through narrow gaps, causing high stresses in the melt. Most dispersive and distributive mixing heads result in flow resistance, causing viscous heating and pressure losses during extrusion.

Current dispersive mixing heads have two important drawbacks. One, they rely mostly on shear rather than elongational stresses, and two, the material passes over the high

The Maddock mixer is a very common dispersive mixer in single screw extrusion

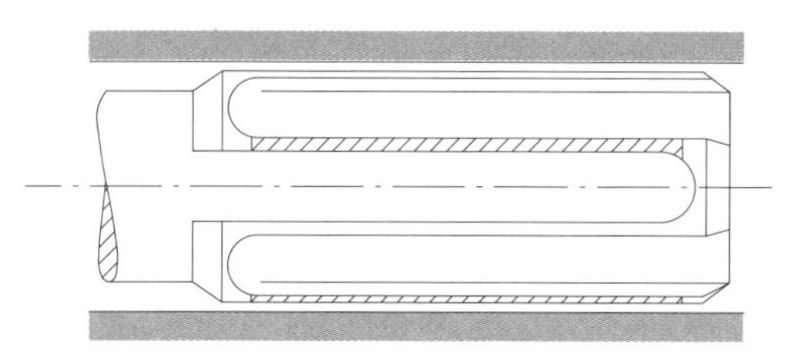

Figure 5.18 Maddock or Union Carbide mixing section

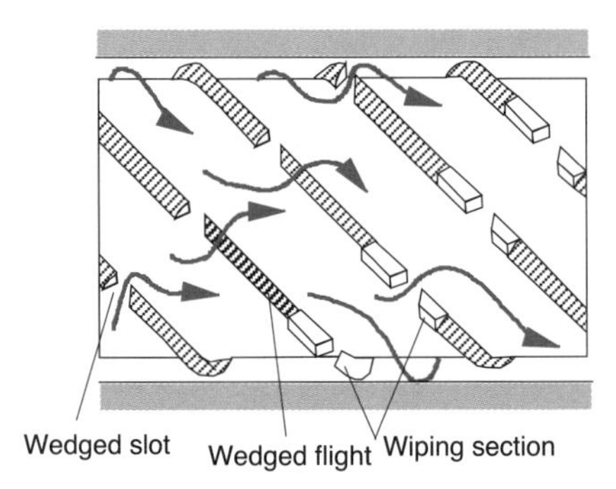

Figure 5.19 CRD mixing section

stress region only once. As already discussed earlier, elongational deformation achieves dispersion more effectively than shear, but also, in terms of energy and power consumption, simple shear flows are significantly inferior to any extensional flow. The disadvantages of existing single screw extruder mixing heads can be overcome by adding wedged-shaped pushing flights with a large gap, and wedged-shaped slots within the flights, as shown in Fig. 5.19 [17]. The wedge shapes generate strong elongational deformations with little viscous dissipation. Experiments performed on PS/HDPE blends show that such a mixing head can break down the dispersed PS phase to an average of 2 μm droplets, with negligible pressure losses and viscous dissipation. The size of the dispersed phase is comparable to co- and counter-rotating twin screw extruders, as discussed in Section 5.3.5.

5.3.3 Static Mixers

Static mixers or motionless mixers are pressure-driven, continuous mixing devices through which the melt is pumped, rotated, and divided, leading to effective mixing without movable parts and mixing heads. One of the most commonly used static mixers is the twisted tape static mixer schematically shown in Fig. 5.20. As the fluid is rotated by the dividing wall, the interfaces between the fluids increase. The interfaces are then re-oriented by 90° once the material enters a new section. The stretching and re-orientation sequence is repeated until the number of striations is so high that a

The static mixer uses Erwin's distributive mixing scheme of deformation and re-orientation

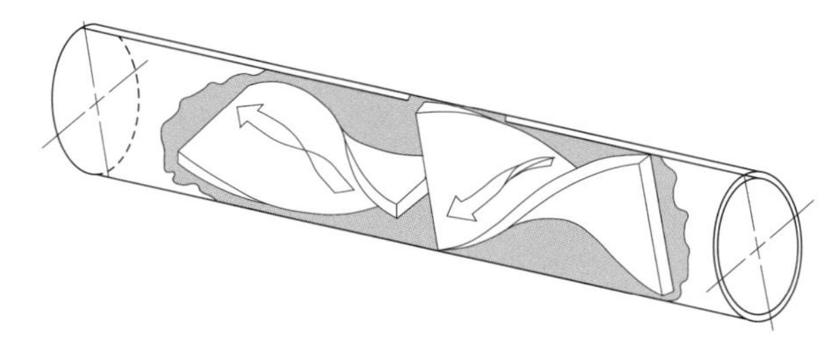

Figure 5.20 Static mixer

Figure 5.21 Experimental progression of the layering of colored resins in a Kenics static mixer (Courtesy of Chemineer, Inc.)

nearly homogeneous mixture is achieved. Figure 5.21 shows a sequence of cuts down a Kenics static mixer.

The number of striations increase from section to section by 2, 4, 8, 16, 32, ... or simply by $N = 2^n$, where N is the number of striations and n the number of sections in the mixer.

Similar to dispersive mixers in single screw extrusion, a major drawback of existing static mixers is that their main mode of deformation is shear rather than elongation. A dispersive static mixer has also been developed to generate elongational deformations, which lead to higher dispersive mixing [18]. Figure 5.22 presents a schematic of the new dispersive static mixer.

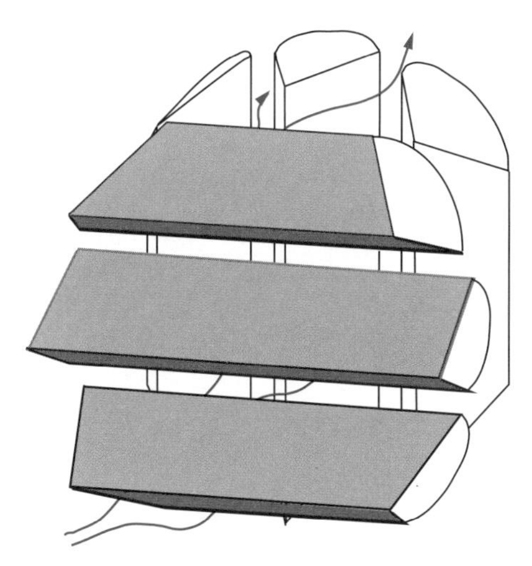

Figure 5.22 Schematic diagram of a dispersive static mixer

5.3.4 Cokneader

The cokneader is a single screw extruder with pins on the barrel and a screw that oscillates in the axial direction. Figure 5.23 shows a schematic diagram of a cokneader. The pins on the barrel wipe the entire surface of the screw, making it the only self-cleaning single-screw extruder. This reduces residence time, which makes it appropriate for processing thermally sensitive materials. The pins on the barrel also disrupt the solids bed creating a *dispersed melting* [19], which improves the overall melting rate and reduces the material temperature.

A simplified analysis of a cokneader gives the following number of striations per L/D [20]

$$N_s = 2^{12} \tag{5.11}$$

so that over a section of $4\,D$ the number of striations is $2^{12(4)} = 2.8 \times 10^{14}$. A detailed discussion on the cokneader is given by Rauwendaal [20] and Elemans [21].

5.3.5 Twin Screw Extruders

In the past two decades, twin screw extruders have developed into the most popular continuous mixing devices. In general, they can be classified into intermeshing or non-intermeshing and co-rotating or counter-rotating twin screw extruders. The intermeshing twin screw extruders are *self-cleaning*, which reduces the polymer residence time. The main characteristic of this configuration is that the screw surfaces slide past each other, constantly removing polymer stuck to the screw.

In the last two decades, the co-rotating twin screw extruder systems have established themselves as efficient continuous mixers for most processes, including reactive extrusion. In essence, the co-rotating systems have a high pumping efficiency caused by the double transport action of the two screws. The counter-rotating systems generate high temperature pulses, making them inappropriate for reactive extrusion, but they

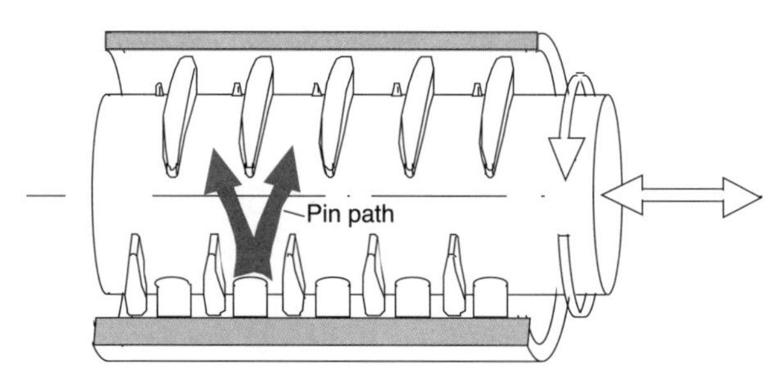

Figure 5.23 Schematic diagram of a cokneader

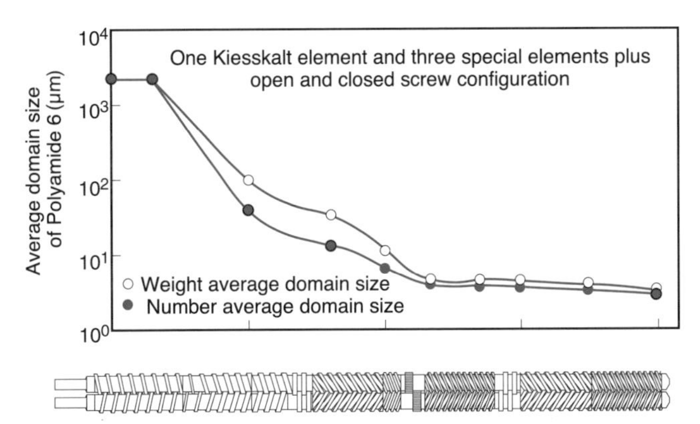

Figure 5.24 Number and weight average of polyamide 6 domain sizes along the screws for a counter-rotating twin screw extruder with special mixing elements

Counter-rotating twin screw extruders are excellent dispersive mixing devices – superior to co-rotating systems

generate high stresses because of the calendering action between screws, making them efficient machines to disperse pigments and lubricants.[2)]

Several studies evaluate the mixing capabilities of twin screw extruders. Two studies by Lim and White [22, 23] evaluated morphology development in a 30.7-mm screw diameter co-rotating [24] and a 34-mm screw diameter counter-rotating [13] intermeshing twin screw extruder. In both studies, they dry-mixed a 75/25 blend of polyethylene and polyamide 6 pellets and then fed the mix into the hopper at 15 kg/hour. Small samples were taken along the axis of the extruder and evaluated using optical and electron microscopy.

The blend dispersion is manifested by the reduction of the characteristic size of the polyamide 6 phase. Figure 5.24 is a plot of the weight average and number average domain size of the polyamide 6 phase along the screw axis with one kneading-pump element and three special mixing elements.

Using a co-rotating twin screw extruder with three kneading disk blocks, a final morphology with polyamide 6 weight average phase sizes of 2.6 μm was achieved. Figure 5.25 shows the morphology development along the screw axes. When comparing counter-rotating (Fig. 5.24) with co-rotating (Fig. 5.25), both achieve similar final mixing quality. However, the counter-rotating extruder achieved the final morphology much earlier in the screw than the co-rotating twin screw extruder. A possible explanation for this is that the blend travelling through the counter-rotating configuration melted earlier than in the co-rotating geometry. In addition, the phase size was slightly smaller, possibly as a result of the calendering effect between the screws in the counter-rotating system.

2) There seems to be considerable disagreement about co- versus counter-rotating twin screw extruders between different groups in the polymer processing industry and academic community.

Co-rotating twin screw extruders are the most popular industrial compounding devices when making polymer blends

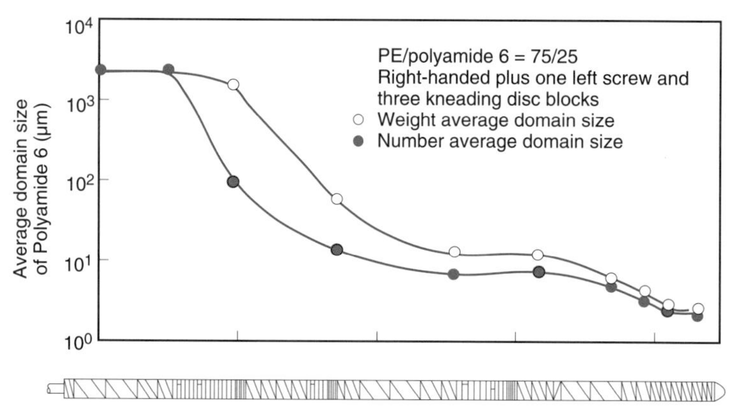

Figure 5.25 Number and weight average of polyamide 6 domain sizes along the screws for a co-rotating twin screw extruder with special mixing elements

References

1. Sundararaj, U., and C.W. Macosko, *Macromolecules* (1995), *28*, 2647
2. Scott, C.E., and C.W. Macosko, *Polymer Bulletin* (1991), *26*, 341
3. Gramann, P.J., L. Stradins, and T.A. Osswald, *Intern. Polymer Processing* (1993), 8, 287
4. Tadmor, Z., and C.G. Gogos, *Principles of Polymer Processing*, 2nd Edition, John Wiley & Sons (2006), New, York
5. Erwin, L., *Polym. Eng. & Sci.* (1978), *18*, 572
6. Rauwendaal, C., *Mixing in Polymer Processing*, Marcel Dekker, Inc. (1991), New York
7. Erwin, L., *Polym. Eng. & Sci.* (1978), *18*, 738
8. Tadmor, Z., *Ind. Eng. Fundam.* (1976), *15*, 346
9. Cheng, J., and I. Manas-Zloczower, *Internat. Polym. Proc.* (1990), 5, 178
10. Sundararaj, U., and C.W. Macosko, *Macromolecules* (1995), *28*, 2647
11. Grace, H.P., *Chem. Eng. Commun.* (1982), *14*, 225
12. Janssen, J.M.H., Ph.D. Thesis (1993), Eindhoven University of Technology, The Netherlands
13. Biswas, A., and T.A. Osswald (1994), unpublished research
14. Cox, R.G., *J. Fluid Mech.* (1969), 37, 3, 601–623
15. Boonstra, B.B., and A.I. Medalia, *Rubber Age* (1963), 3, 4
16. Rauwendaal, C., *Polymer Extrusion*, 4th Edition, Hanser Publishers (2001), Munich
17. Rauwendaal, C., T.A. Osswald, P.J. Gramann, and B.A. Davis, *SPE ANTEC Tech. Pap.* (1998), 56
18. Gramann, P.J., B.A. Davis, T.A. Osswald and C. Rauwendaal (2000), US Patent number 6,136,246
19. Rauwendaal, C., *SPE ANTEC Tech. Pap.* (1993), *39*, 2232

20. Rauwendaal, C., *Mixing in Reciprocating Extruders*, A chapter in *Mixing and Compounding of Polymers*, 2nd Edition, I. Manas-Zloczower (Ed.), Hanser Publishers (2009), Munich

21. Elemans, P.H.M., Modeling of the Cokneater, A chapter in *Mixing and Compounding of Polymers*, I. Manas-Zloczower (Ed.), Hanser Publishers (2009), Munich

22. Lim, S. and J.L. White, *Intern. Polymer Processing* (1993), *8*, 119

23. Lim, S. and J.L. White, *Intern. Polymer Processing* (1994), *9*, 33

24. Stone, H.A. and L.G. Leal, *J. Fluid Mech.* (1989), 198, 399–427

6 Injection Molding

Injection molding is the most important process used to manufacture plastic products. Today, more than one-third of all thermoplastic materials are injection molded and more than half of all polymer processing equipment is for injection molding. The injection molding process is ideally suited to manufacture mass produced parts of complex shapes requiring precise dimensions. The process goes back to 1872, when the Hyatt brothers patented their stuffing machine to inject cellulose into molds. However, today's injection molding machines are mainly related to the reciprocating screw injection molding machine patented in 1956. A modern injection molding machine with its most important elements is shown in Fig. 6.1. The components of the injection molding machine are the plasticating unit, clamping unit, and the mold.

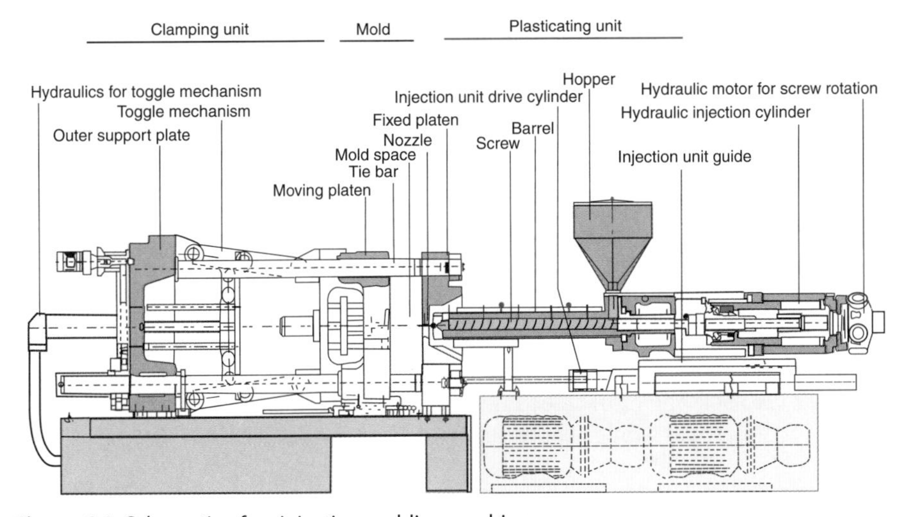

Figure 6.1 Schematic of an injection molding machine

Today, injection molding machines are classified by the following international convention[1)]

Manufacturer type T/P

International designation of injection molding machines

where T is the clamping force in metric tons and P is defined as

$$P = \frac{v_{\max} p_{\max}}{1000} \tag{6.1}$$

1) The old US convention uses MANUFACTURER T-v where T is the clamping force in British tons and v the shot size in ounces of polystyrene.

where v_{max} is the maximum shot size in cm^3 and p_{max} is the maximum injection pressure in bar. The clamping forced T can be as low as 1 metric ton for small machines, and as high as 11 000 tons.

6.1 The Injection Molding Cycle

The sequence of events during the injection molding of a plastic part, as shown in Fig. 6.2, is called the injection molding cycle. The cycle begins when the mold closes, followed by the injection of the polymer into the mold cavity. Once the cavity is filled, a holding pressure is maintained to compensate for material shrinkage. In the next step, the screw turns, feeding the next shot to the front of the screw. This causes the screw to retract as the next shot is prepared. Once the part is sufficiently cool, the mold opens and the part is ejected. Fig. 6.3 presents the sequence of events during the injection molding cycle. The figure shows that the cycle time is dominated by the cooling of the part inside the mold cavity. The total cycle time can be calculated using

$$t_{cycle} = t_{closing} + t_{cooling} + t_{ejection} \tag{6.2}$$

Typical cycle times are between 15 seconds and 1 minute

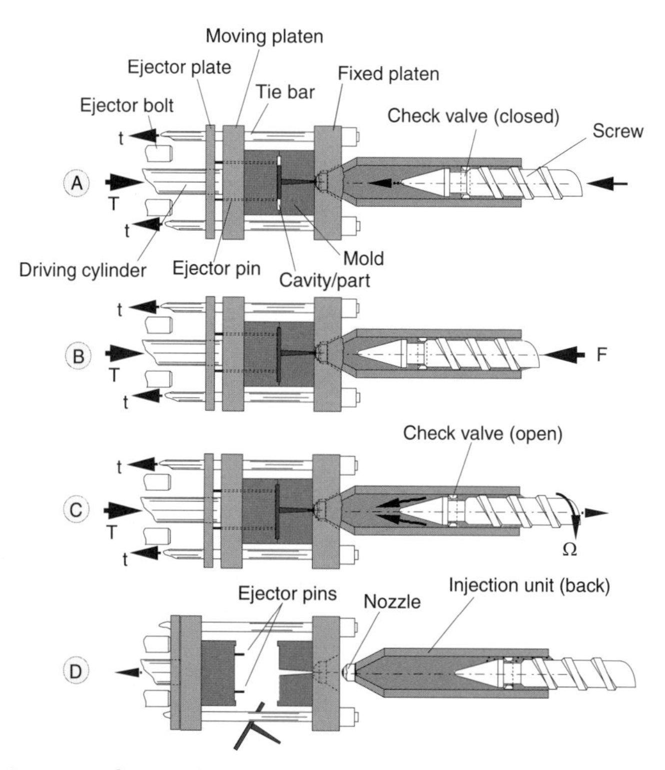

Figure 6.2 Sequence of events during an injection molding cycle

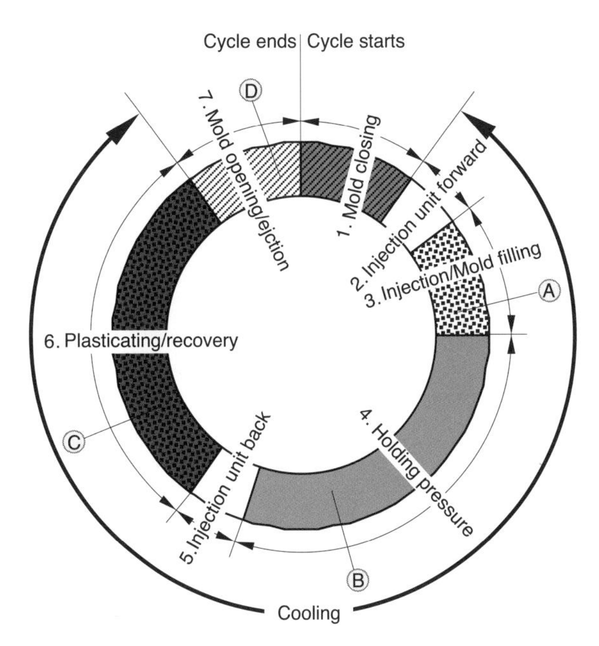

Figure 6.3 Injection molding cycle

where the closing and ejection times, $t_{closing}$ and $t_{ejection}$, can last from a fraction of a second to a few seconds, depending on the size of the mold and machine.

The cooling time for a plate-like part of thickness h can be estimated using [1]

$$t_{cooling} = \frac{h^2}{\pi\alpha} \ln\left(\frac{8}{\pi^2}\frac{T_M - T_W}{T_D - T_W}\right) \tag{6.3}$$

Cycle time increases by the square of the part thickness

and for cylindrical geometry of diameter D using

$$t_{cooling} = \frac{D^2}{23.14\,\alpha} \ln\left(0.692\frac{T_M - T_W}{T_D - T_W}\right) \tag{6.4}$$

In Eqs. 6.3 and 6.4, α represents thermal diffusivity, T_M represents the melt temperature, T_W the mold temperature, and T_D the average part temperature at ejection.

Using the average part temperature history and the cavity pressure history, the process can be followed and assessed using the PvT-diagram, as depicted in Fig. 6.4. [2–3]. To follow the process on the PvT-diagram, we must transfer both the temperature and the pressure at matching times. The diagram reveals four basic processes: an isothermal injection (0–1) with pressure rising to the holding pressure (1–2), an isobaric cooling process during the holding cycle (2–3), an isochoric cooling after the gate freezes with a pressure drop to atmospheric (3–4), and then isobaric cooling to room temperature (4–5).

The point on the PvT-diagram where the final isobaric cooling begins (4), controls the total part shrinkage, Δv. This point is influenced by the two main processing

The PvT diagram is used to predict thickness shrinkage in injection molded parts

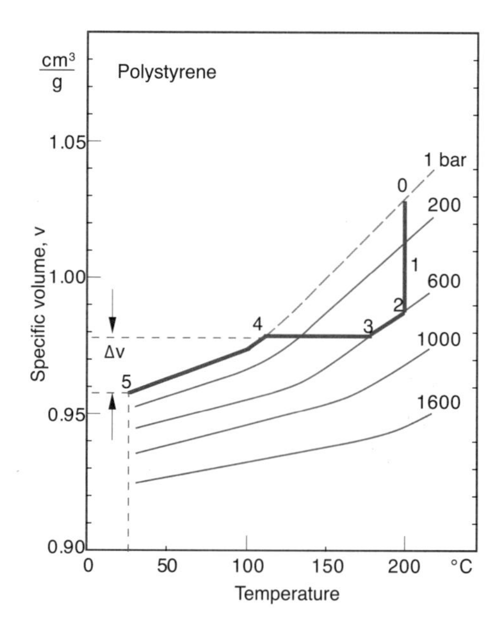

Figure 6.4 Trace of an injection molding cycle in a PvT-diagram

An increase in pack pressure reduces the thickness shrinkage

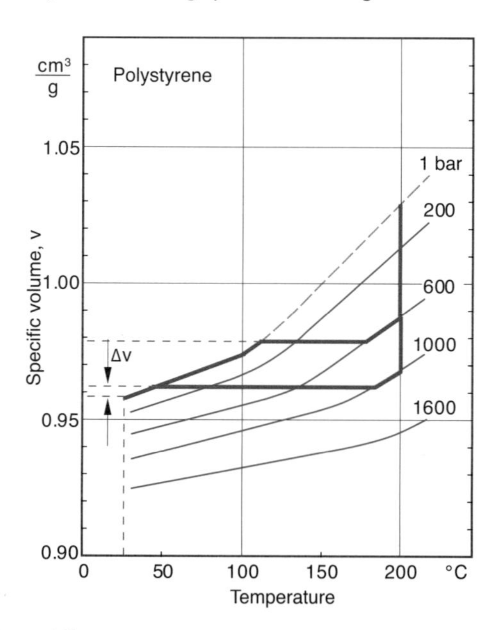

Figure 6.5 Trace of two different injection molding cycles in a PvT-diagram

conditions — the melt temperature, T_M, and the holding pressure, P_H, as depicted in Fig. 6.5. Here the process in Fig. 6.4 is compared to one with a higher holding pressure. Of course, there is an infinite combination of conditions that render acceptable parts,

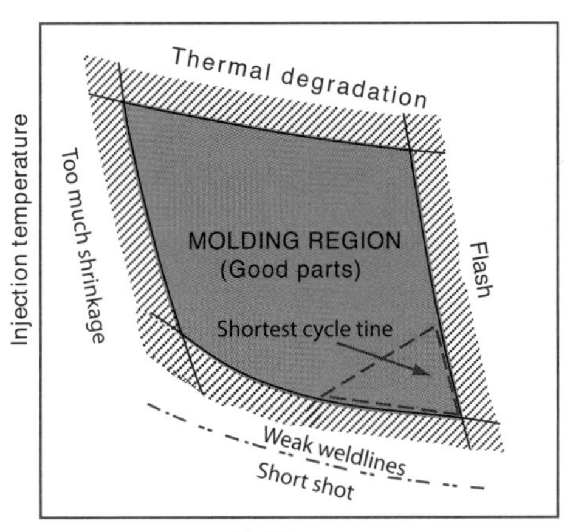

Figure 6.6 The molding diagram

Shortest cycle time at high pack pressures and lowest allowed injection temperatures

bound by minimum and maximum temperatures and pressures. Figure 6.6 presents *the molding diagram* with all limiting conditions. The melt temperature is bound by a low temperature that results in a *short shot* or unfilled cavity, and a high temperature that leads to material degradation. The hold pressure is bound by a low pressure that leads to excessive shrinkage or low part weight, and a high pressure that results in *flash*. Flash results when the cavity pressure force exceeds the machine clamping force, leading to melt flowing across the mold parting line. The holding pressure determines the corresponding clamping force required to size the injection molding machine. An experienced polymer processing engineer can usually determine which injection molding machine is appropriate for a specific application. For the untrained polymer processing engineer, finding this appropriate holding pressure and its corresponding mold clamping force can be difficult.

With difficulty, one can control and predict the component's shape and residual stresses at room temperature. For example, *sink marks* in the final product are caused by material shrinkage during cooling, and residual stresses can lead to environmental stress cracking under certain conditions [4].

Warpage in the final product is often caused by processing conditions that lead to asymmetric residual stress distributions through the part thickness. The formation of residual stresses in injection molded parts is attributed to two major coupled factors: cooling and flow stresses. The first and most important is the residual stress formed as a result of rapid cooling.

The parabolic temperature distribution through the thickness of a solidified injection molded part leads to a parabolic residual stress distribution, compressive in the outer surfaces of the component and tensile in the core. Assuming no residual stress build-up during phase change, a simple function based on the parabolic temperature distribution, can be used to approximate the residual stress distribution in thin sections [5]:

$$\sigma = \frac{2}{3}\alpha E\left(T_s - T_f\right)\left(\frac{6z^2}{4L^2} - \frac{1}{2}\right) \tag{6.5}$$

Here, T_f is the final temperature of the part, E is the modulus, α the thermal expansion coefficient, L the half thickness and T_s denotes the solidification temperature: glass transition temperature for amorphous thermoplastics or the melting temperature for semi-crystalline polymers. Figure 6.7 [6] compares the compressive stresses measured on the surface of PMMA moldings to those predicted by Eq. 6.10.

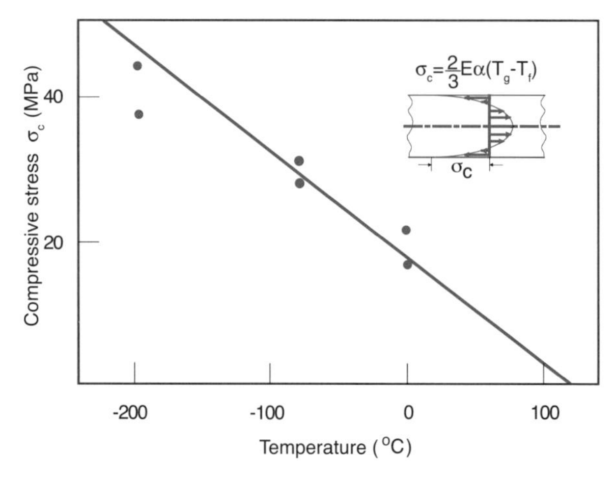

Figure 6.7 Comparison between computed, Eq. 6.10, and measured compressive stresses on the surface of injection molded PMMA plates

6.2 The Injection Molding Machine

6.2.1 The Plasticating and Injection Unit

A plasticating and injection unit is shown in Fig. 6.8. The major tasks of the plasticating unit are to melt the polymer, to accumulate the melt in the screw chamber, to inject the melt into the cavity, and to maintain the holding pressure during cooling.

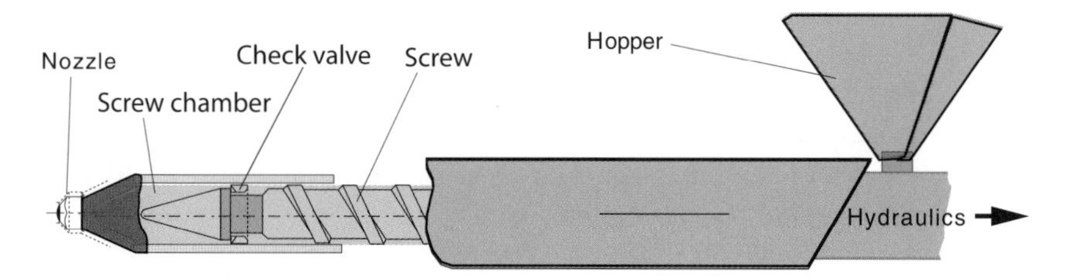

Figure 6.8 Schematic of the plasticating unit

The main elements of the plasticating unit are:

- hopper
- screw
- heater bands
- check valve
- nozzle

The plasticating unit in an injection molding machine is an extruder with a reciprocating screw

The hopper, heating bands, and the screw are similar to a plasticating single screw extruder, except that the screw in an injection molding machine can slide back and forth to allow for melt accumulation and injection. This characteristic gives it the name *reciprocating screw*. The maximum stroke in a reciprocating screw is $3D$.

Although the most common screw used in injection molding machines is the three-zone plasticating screw, two-stage vented screws are often used to extract moisture and monomer gases just after the melting stage.

The check valve, or non-return valve, is at the end of the screw and enables it to work as a plunger during injection and packing, without allowing polymer melt back flow into the screw channel. A check valve and its function during operation is depicted in Figs. 6.2 and 6.8. A high quality check valve allows less then 5 % of the melt back into the screw channel during injection and packing.

The nozzle is at the end of the plasticating unit and fits tightly against the sprue bushing during injection. The nozzle type is either open or shut-off. The open nozzle is the simplest, rendering the lowest pressure consumption.

6.2.2 The Clamping Unit

The job of a clamping unit in an injection molding machine is to open and close the mold, and to close the mold tightly to avoid flash during the filling and holding. Modern injection molding machines have two predominant clamping types: mechanical and hydraulic.

Figure 6.9 presents a toggle mechanism in the open and closed mold positions. Although the toggle is essentially a mechanical device, it is actuated by a hydraulic cylinder. The advantage of using a toggle mechanism is that, as the mold approaches closure, the available closing force increases and the closing decelerates significantly. However, the toggle mechanism only transmits its maximum closing force when the system is fully extended.

Figure 6.10 presents a schematic of a hydraulic clamping unit in the open and closed positions. The advantages of the hydraulic system is that a maximum clamping force is attained at any mold closing position and that the system can take different mold sizes without major system adjustments.

The toggle clamping mechanism

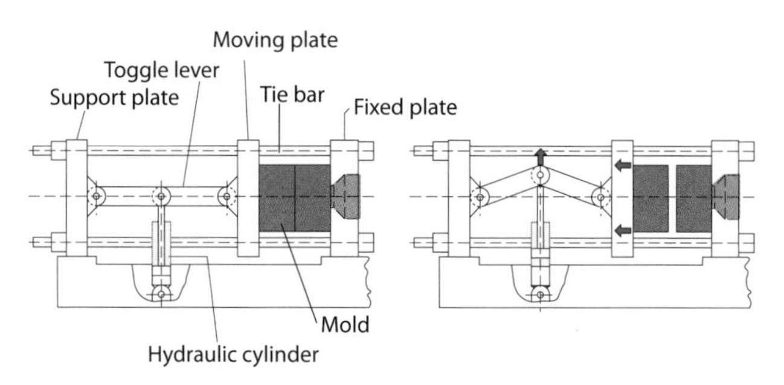

Figure 6.9 Clamping unit with a toggle mechanism

The hydraulic clamping mechanism

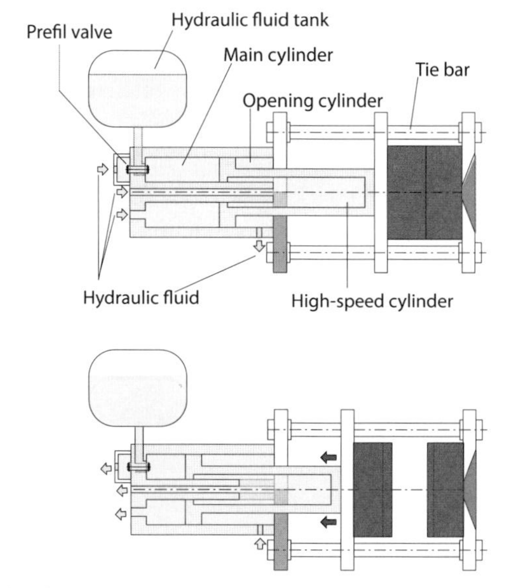

Figure 6.10 Hydraulic clamping unit

6.2.3 The Mold Cavity

The central point in an injection molding machine is the mold. The mold distributes polymer melt into and throughout the cavities, shapes the part, cools the melt, and ejects the finished product. As depicted in Fig. 6.11, the mold is custom-made and consists of the following elements:

- sprue and runner system
- gate
- mold cavity
- cooling system (thermoplastics)
- ejector system

During mold filling, the melt flows through the sprue and is distributed into the cavities by the runners, as shown in Fig. 6.12.

Flow path:
↓ Screw chamber
↓ Nozzle
↓ Sprue
↓ Runner
↓ Gate
↓ Cavity

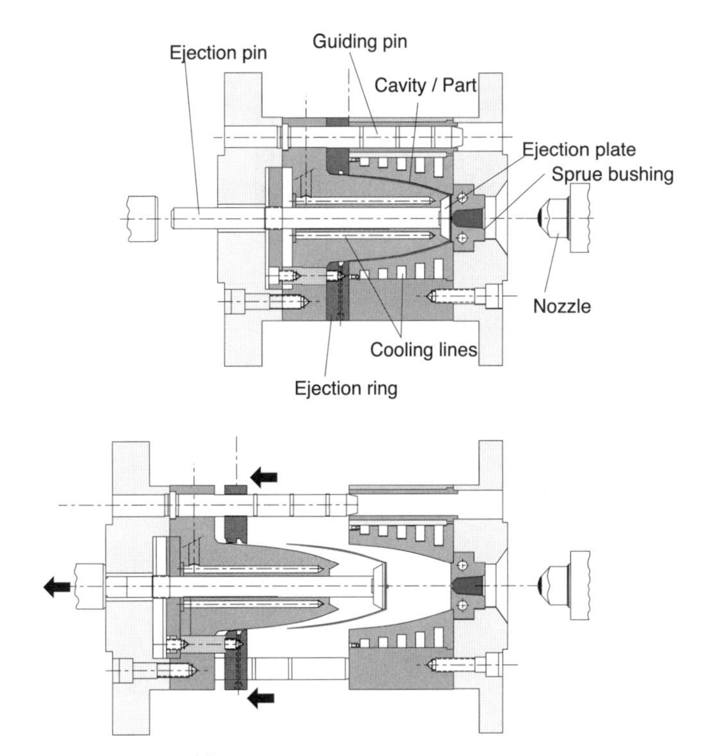

Figure 6.11 An injection mold

Cold runners reduce mold cost

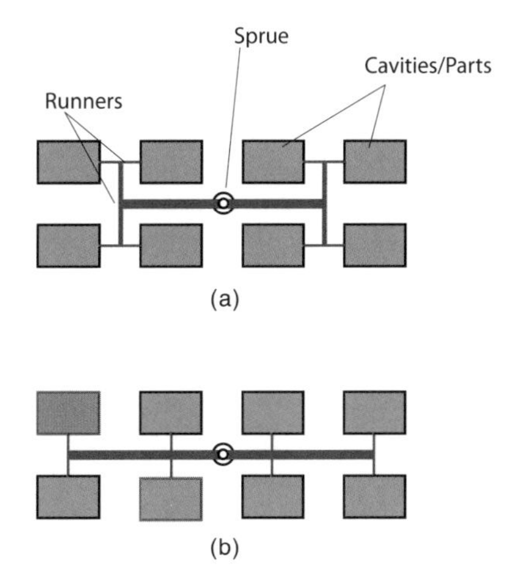

Figure 6.12 Schematic of different runner system arrangements

Hot runners are used to eliminate waste

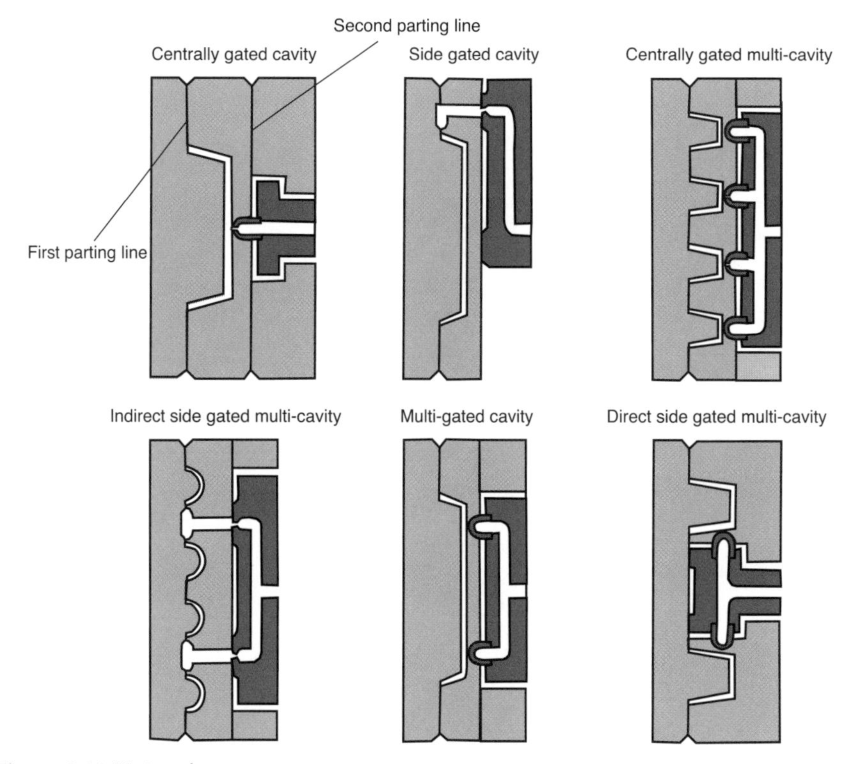

Figure 6.13 Various hot runner system arrangements

The runner system in Fig. 6.12(a) is symmetric, where all cavities fill at the same time, causing the polymer to fill all cavities in the same way. The disadvantage of this balanced runner system is that the flow paths are long, leading to high material and pressure consumption. On the other hand, the asymmetric runner system shown in Fig. 6.12(b) leads to parts of different quality. Equal filling of the mold cavities can also be achieved by varying runner diameters. There are two types of runner systems — cold and hot. Cold runners are ejected with the part and are trimmed after mold removal. The advantage of the cold runner is lower mold cost. The hot runner keeps the polymer at its melt temperature. The material stays in the runners system after ejection, and is injected into the cavity in the following cycle. There are two types of hot runner system: externally and internally heated. The externally heated runners have a heating element surrounding the runner that keeps the polymer isothermal. The internally heated runners have a heating element running along the center of the runner, maintaining a polymer melt that is warmer at its center and possibly solidified along the outer runner surface. Although a hot runner system considerably increases mold cost, its advantages include elimination of trim and lower pressures for injection. Various arrangements of hot runners are schematically depicted in Fig. 6.13. It should be noted

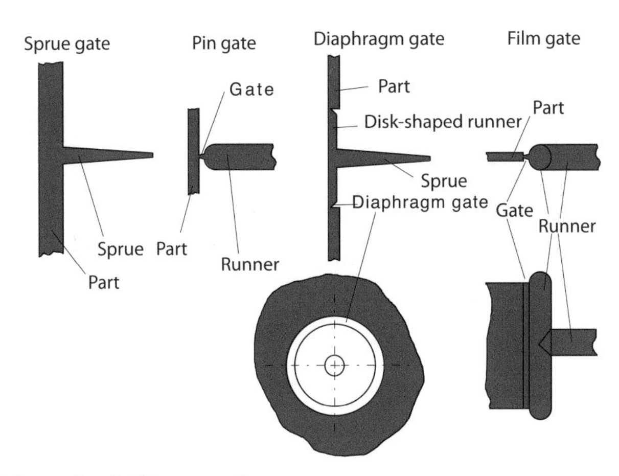

Figure 6.14 Schematic of different gating systems

that there are two parting lines in a hot runner cavity system, and that the second parting line is only opened during maintenance of the molds.

When large parts are injection molded, the sprue sometimes serves as the gate, as shown in Fig. 6.14. The sprue must be subsequently trimmed, often requiring further surface finishing. On the other hand, a pin-type gate (Fig. 6.14) is a small orifice that connects the sprue or the runners to the mold cavity. The part is easily broken off from such a gate, leaving only a small mark that usually does not require finishing. Other types of gates, also shown in Fig. 6.14, are film gates, used to eliminate orientation, and disk or diaphragm gates for symmetric parts such as compact discs.

6.3 Special Injection Molding Processes

There are numerous variations of injection molding processes, many of which are still under development. Furthermore, due to the diversified nature of these special injection molding processes, there is no unique method to categorize them. Figure 6.15 attempts to schematically categorize special injection molding processes for thermoplastics. The most common special injection molding processes are multi-component injection molding, co-injection molding, gas assisted injection molding, injection-compression molding, reaction injection molding, and injection molding of liquid silicone rubber.

6.3.1 Multi-Component Injection Molding

Multi-component (or multi-color) injection molding occurs when two or more components are injected through different runner and gate systems at different stages during the molding process. Each component is injected using its own plasticating

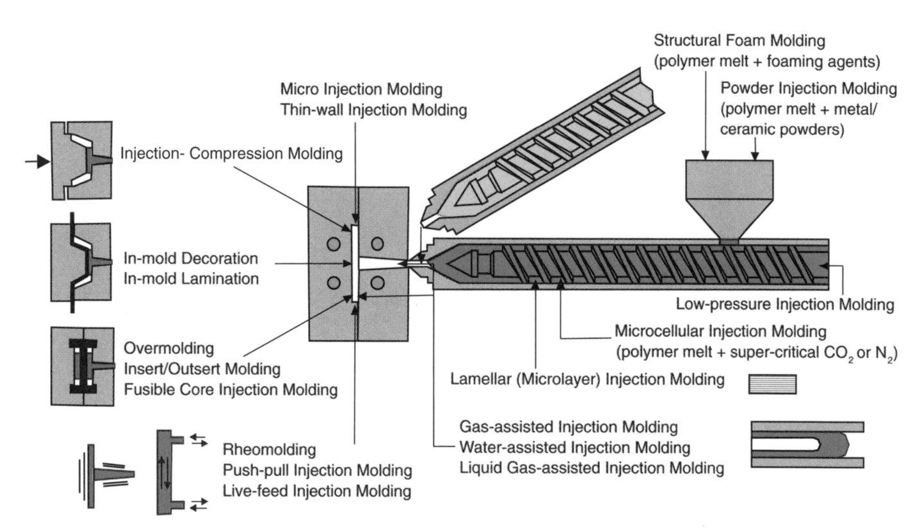

Figure 6.15 Schematic classification of special injection molding processes for thermoplastics

Whole parts can be assembled using multi-component injection molding processes

unit. The molds are often located on a turntable. Multi-color automotive stop lights are molded this way. In multi-component processes, often two incompatible materials are molded or one component is cooled sufficiently so that the two components do not adhere to each other. For example, to mold a ball and socket system, either the ball or the socket of the linkage is molded first. The component that is injected first is allowed to cool somewhat before the second component is molded in. This results in a perfectly movable system; if the socket is injected first, the assembly will be loose and if the ball is molded first, the assembly will be tight, as the socket shrinks over the ball. This type of injection molding process is used to replace tedious assembling tasks and is becoming popular in countries where labor costs are high. Hence, this type of process is referred to as *assembly injection molding.* A commonly used method of multi-component injection molding employs a rotating mold and multiple injection units, as shown in Fig. 6.16. Once the insert is molded, a hydraulic or electric servo drive rotates the core and the part by 180 degrees (or 120 degrees for a three-shot part), allowing alternating polymers to be injected. This is the fastest and most common method because two or more parts can be molded every cycle. Another variation of multi-component injection molding involves automatically expanding the original cavity geometry using retractable (movable) cores or slides while the insert is still in the mold. This process is called core-pull or core-back, as shown in Fig. 6.17. To be specific, the core retracts after the insert has solidified to create open volume to be filled by the second material within the same mold.

6.3.2 Co-Injection Molding

In contrast to multi-color or multi-component injection molding, co-injection molding uses the same gate and runner system. Here, the component that ends as the

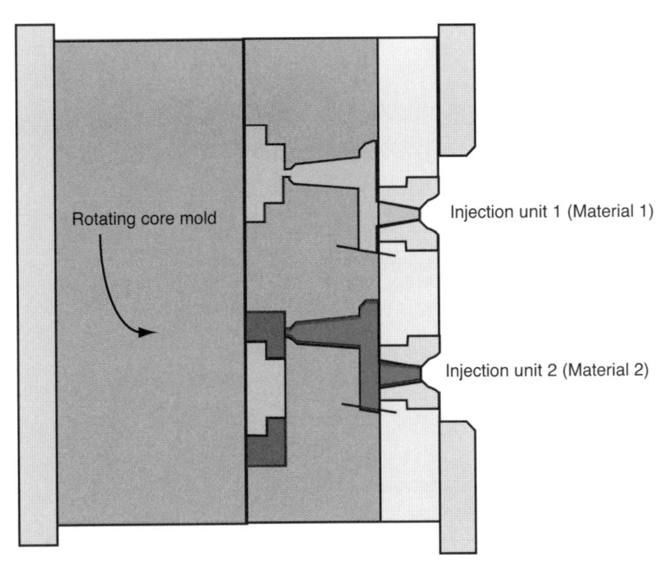

Figure 6.16 Schematic diagram of a rotating mold used to produce multi-component injection molded parts

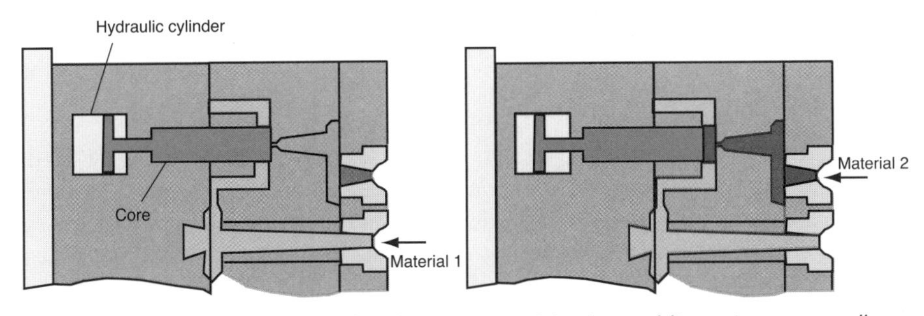

Figure 6.17 Schematic diagram of multi-component injection molding using a *core pull* or a *core back* technique

outer skin of the part is injected first, followed by the core component. Figure 6.18 illustrates the typical sequences of the co-injection molding process using the *one-channel technique* and the resulting flow of skin and core materials inside the cavity. This is accomplished with the use of a machine that has two separate, individually controllable injection units and a common injection nozzle block with a switching head. Due to the flow behavior of the polymer melts and the solidification of skin material, a frozen layer of polymer starts to grow from the colder mold walls. The polymer flowing in the center of the cavity remains molten. As the core material is injected, it flows within the frozen skin layers, pushing the molten skin material at the hot core to the extremities of the cavity. Because of the fountain-flow effect at the advancing melt front, the skin material at the melt front will show up at the region

Co-injection molding can be used to make parts with different skin and core materials

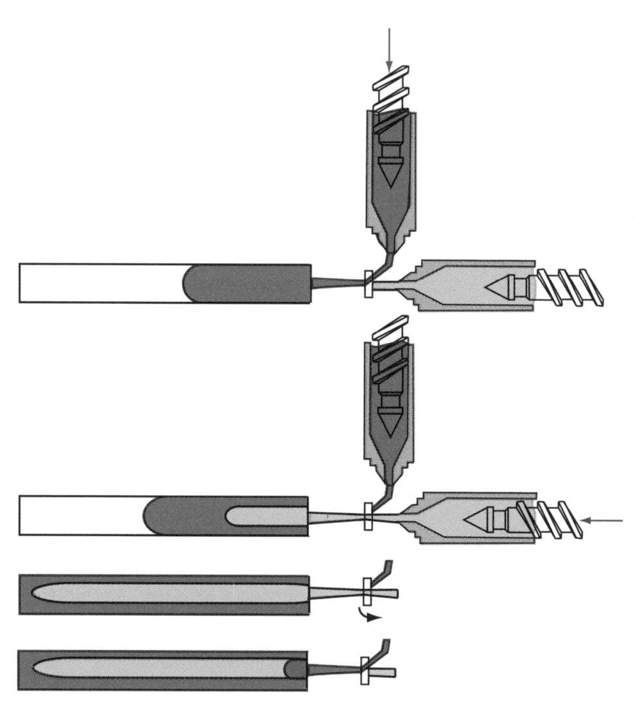

Figure 6.18 Sequential co-injection molding process

adjacent to the mold walls. This process continues until the cavity is nearly filled, with skin material appearing on the surface and the end of the part. Finally, a small additional amount of skin material is injected again to purge the core material away from the sprue so that it will not appear on the part surface in the next shot. When there is not enough skin material injected prior to the injection of core material, the skin material may sometimes eventually be depleted during the filling process and the core material will show up on portions of the surface and the end of the part that is last filled. This is referred to as *core surfacing* or *core breakthrough.* There are other variations to the sequential (namely, skin-core-skin, or A–B–A) co-injection molding process. In particular, one can start to inject the core material while the skin material is being injected (i. e., A–AB–B–A). That is, a majority of skin material is injected into a cavity, followed by a combination of both skin and core materials flowing into the same cavity, and then followed by the balance of the core material to fill the cavity. Again, an additional small amount of skin injection will cap the end of the sequence, as described previously. In addition to the one-channel technique configuration, two- and three-channel techniques have been developed that use nozzles with concentric flow channels to allow simultaneous injection of skin and core materials.

6.3.3 Gas-Assisted Injection Molding (GAIM)

The gas-assisted injection molding (GAIM) process consists of a partial or nearly full injection of polymer melt into the mold cavity, followed by injection of an inert gas (typically nitrogen) into the core of the polymer melt through the nozzle, sprue, runner, or directly into the cavity. The compressed gas takes the path of the least resistance, flowing toward the melt front where the pressure is lowest. As a result, the gas penetrates and hollows out a network of predesigned, thick-sectioned gas channels, displacing molten polymer at the hot core to fill and pack out the entire cavity. As depicted in Fig. 6.19, gas assisted injection, as well as other fluid assisted injection molding, work based on several variations of two principles. The first principle is based on partially filling a mold cavity, and completing the mold filling by displacing the melt with a pressurized fluid. Figure 6.20 presents the gas-assisted injection molding process cycle based on this principle. With the second principle, the cavity is nearly or completely filled and the molten core is evacuated into a secondary cavity. This secondary cavity can be either a side cavity that will be scrapped after demolding, a side cavity that will result in an actual part, or the melt shot cavity in front of the screw in the plasticating unit of the injection molding machine. In the latter, the melt is reused in the next molding cycle. In the so-called gas-pressure control process, the compressed gas is injected with a regulated gas pressure profile, either constant, ramped, or stepped. In the *gas-volume control process*, gas is initially metered into a compression cylinder at preset volume and pressure; then, it is injected under pressure generated from reducing

The main fluid assisted injection molding processes are gas-assisted injection molding and water-assisted injection molding

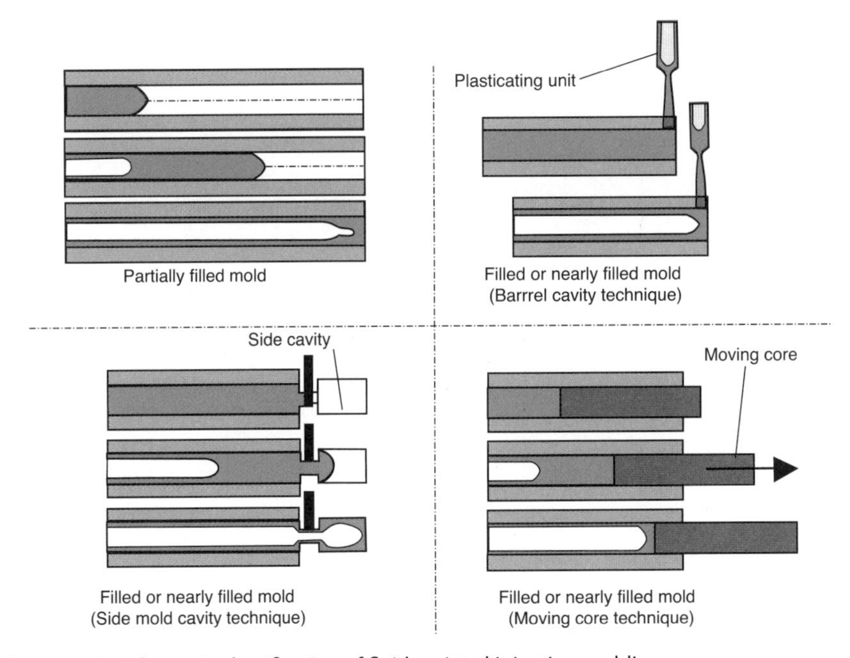

Figure 6.19 Schematic classification of fluid-assisted injection molding processes

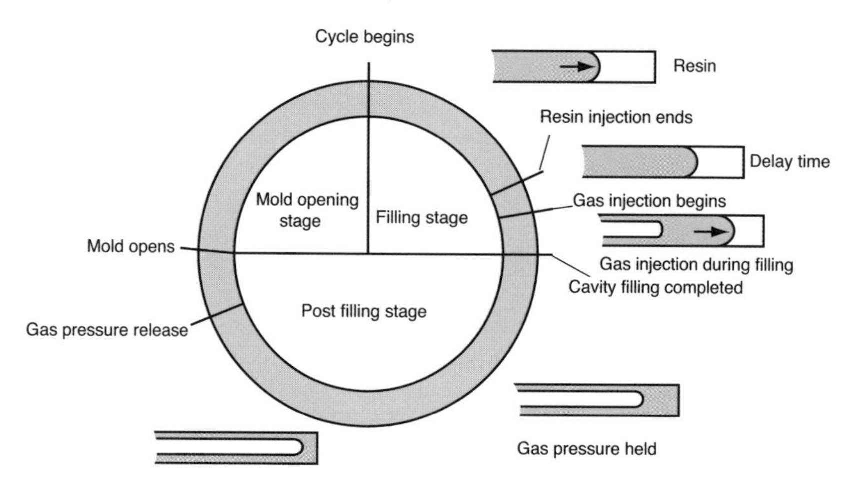

Figure 6.20 Gas-assisted injection molding cycle

the gas volume by movement of the plunger. Conventional injection molding machines with precise shot volume control can be adapted for gas-assisted injection molding with add-on conversion equipment, a gas source, and a control device for gas injection, as schematically depicted in Fig. 6.21. Gas-assisted injection molding, however, requires a different approach to product, tool, and process design due to the need for control of additional gas injection and the layout and sizing of gas channels to guide the gas penetration in a desirable fashion.

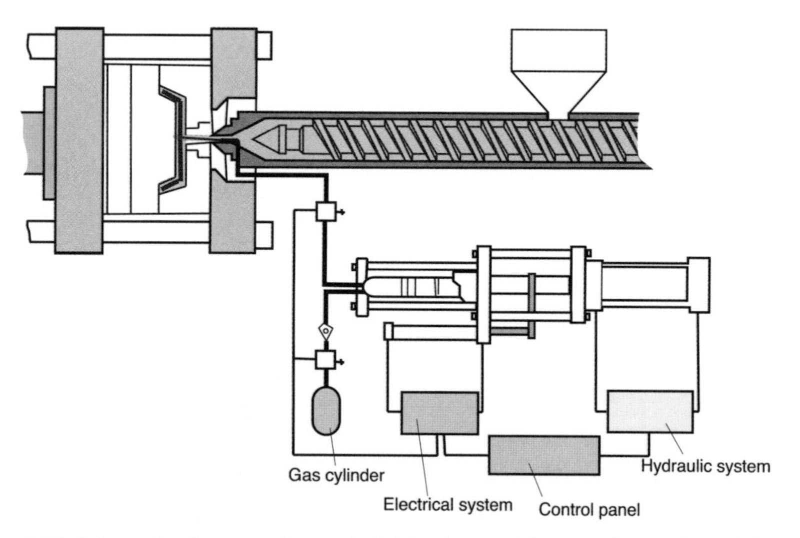

Figure 6.21 Schematic diagram of a typical injection molding machine adapted for gas-assisted injection molding with an add-on gas-compression cylinder and accessory equipment

The gas-assisted injection molding process is a special form of a more general category of *fluid-assisted injection molding*. Another process that falls under this category is *water-assisted injection molding*. The main difference of this latter process is that water is incompressible and has a much higher thermal conductivity and heat capacity than air. Consequently, this leads to significant reductions in cycle time.

6.3.4 Injection-Compression Molding

The injection-compression molding (ICM) is an extension of conventional injection molding by incorporating a mold compression action to compact the polymer material for producing parts with dimensional stability and surface accuracy. In this process, the mold cavity has an enlarged cross-section initially, which allows polymer melt to proceed readily to the extremities of the cavity under relatively low pressure. At some time during or after filling, the mold cavity thickness is reduced by a mold-closing movement, which forces the melt to fill and pack out the entire cavity. This mold compression action results in a more uniform pressure distribution across the cavity, leading to more homogenous physical properties and less shrinkage, warpage, and molded-in stresses than are possible with conventional injection molding. The injection-compression molding process is schematically depicted in Fig. 6.22. A potential drawback associated with the two-stage sequential ICM is the *hesitation* or *witness* mark resulting from flow stagnation during injection-compression transition. To avoid this surface defect and to facilitate continuous flow of the polymer melt, *simultaneous ICM* activates mold compression while resin is being injected. The primary advantage of ICM is the ability to produce dimensionally stable, relatively stress-free parts, at a low pressure, clamp tonnage (typically 20 to 50 % lower than with injection molding), and reduced cycle time. For thin-wall applications, difficult-to-flow materials, such as polycarbonate, have been molded as thin as 0.5 mm. Additionally, the compression of a relatively circular charge significantly lowers molecular orientation, consequently leading to reduced birefringence, improving the optical properties of a finished part. ICM

CD's are injection compression molded to reduce molecular orientation and cycle time

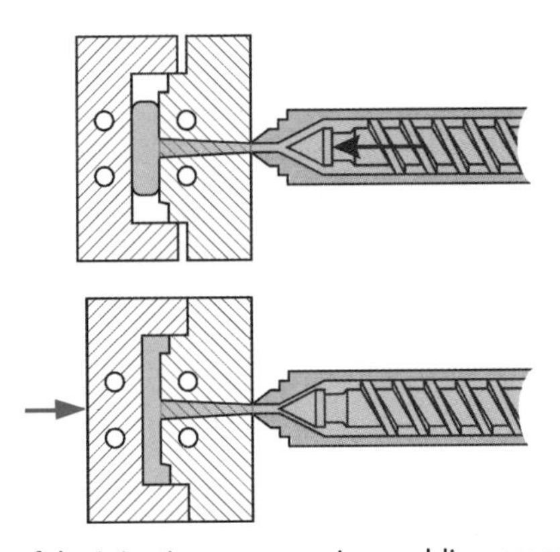

Figure 6.22 Schematic of the injection-compression molding process

is the most suitable technology for the production of high-quality and cost-effective CD-audio/ROMs as well as many types of optical lenses.

6.3.5 Reaction Injection Molding (RIM)

Reaction injection molding (RIM) involves mixing of two reacting liquids in a mixing head before injecting the low-viscosity mixture into mold cavities at relatively high injection speeds. The liquids react in the mold to form a cross-linked solid part. Figure 6.23 presents a schematic of a high pressure polyurethane injection system. The mixing of the two components occurs at high speeds in *impingement mixing heads.* Low pressure polyurethane systems, such as the one schematically presented in Fig. 6.24, require mixing heads with a mechanical stirring device. The short cycle times, low injection pressures, and clamping forces, coupled with superior part strength and heat and chemical resistance of the molded part make RIM well suited for the rapid production of large, complex parts, such as automotive bumper covers and body panels. Reaction injection molding is a process for rapid production of complex parts directly from monomers or oligomers. Unlike thermoplastic injection molding, the shaping of solid RIM parts occurs through polymerization (cross-linking or phase separation) in the mold rather than solidification of the polymer melts. RIM is also different from thermoset injection molding in that the polymerization in RIM is activated via chemical mixing rather than thermally activated by the warm mold. During the RIM process, the two liquid reactants (e.g., polyol and an isocyanate, which were the precursors for

High pressure RIM

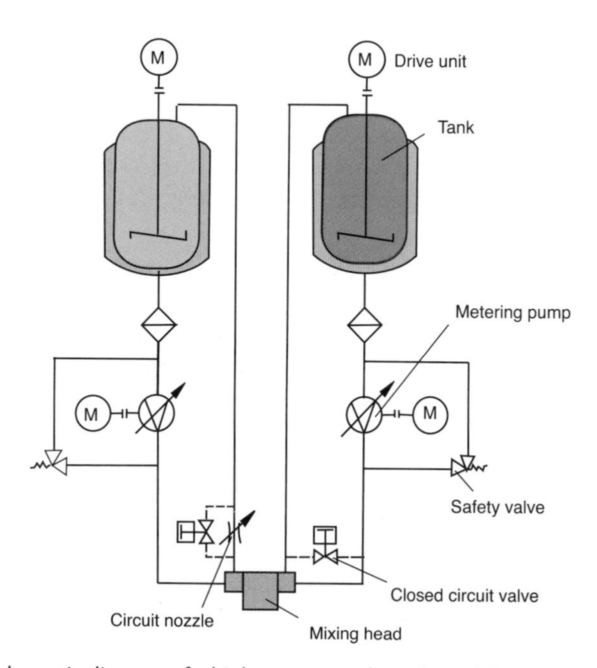

Figure 6.23 Schematic diagram of a high pressure polyurethane injection system

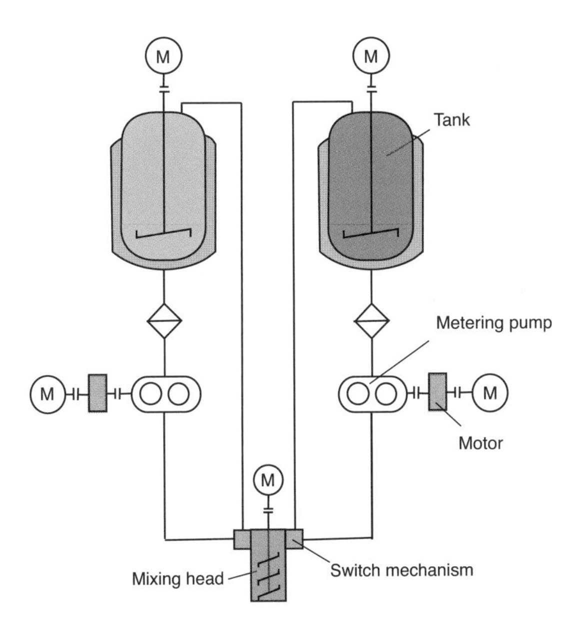

Low pressure RIM

Figure 6.24 Schematic diagram of a low pressure polyurethane injection system

polyurethanes) are metered in the correct proportion into a mixing chamber where the streams impinge at a high velocity and start to polymerize prior to being injected into the mold. Due to the low-viscosity of the reactants, the injection pressures are typically very low, even though the injection speed is fairly high. Because of the fast reaction rate, the final parts can be de-molded in typically less than one minute. There are a number of RIM variants. For example, in the so-called reinforced reaction injection molding (RRIM) process, fillers, such as short glass fibers or glass flakes, have been used to enhance the stiffness, maintain dimensional stability, and reduce material cost of the part. As another modification of RIM, structural reaction injection molding (SRIM) is used to produce composite parts by impregnating a reinforcing glass fiber-mat (preform) preplaced inside the mold with the curing resin. Resin transfer molding (RTM) is very similar to SRIM in that it also employs reinforcing glass fiber-mats to produce composite parts; however, the resins used in RTM are formulated to react more slowly, and the reaction is thermally activated as it is in thermoset injection molding. The capital investment for molding equipment for RIM is lower compared with that for injection molding machines. Finally, RIM parts generally exhibit greater mechanical and heat-resistant properties due to the resulting cross-linking structure. The mold and process designs for RIM become generally more complex because of the chemical reaction during processing. For example, slow filling may cause premature gelling, which results in short shots, whereas fast filling may induce turbulent flow, creating internal porosity. Moreover, the low viscosity of the material tends to cause flash that requires trimming. Another disadvantage of RIM is that the reaction with isocyanate requires special environmental precaution due to health issues. Finally, like many other thermosetting materials, the recycling of RIM parts is not as easy as that of thermoplastics. Polyurethane materials (rigid, foamed, or elastomeric) have traditionally been synonymous with RIM as they and urea urethanes account for more than 95 % of RIM production.

6.3.6 Liquid Silicone Rubber Injection Molding

Injection molding of liquid silicone rubber (LSR) has evolved over the past 35 years. Due to the thermosetting nature of the material, liquid silicone injection molding requires special treatment, such as intensive distributive mixing, while maintaining the material cool before it is pushed into the heated cavity and vulcanized. Figure 6.25 schematically depicts an LSR injection molding process. Liquid silicone rubbers are supplied in barrels or hobbocks. Because of their low viscosity, these rubbers can be pumped through pipelines and tubes to the vulcanization equipment. The two components (labeled component A and B in the figure) are pumped through a static mixer by a metering pump. One of the components contains the catalyst, typically platinum based. A coloring paste as well as other additives can also be added before the material enters the static mixer section. In the static mixer the components are well mixed and then transferred to the cooled metering section of the injection molding machine. The static mixer renders a very homogeneous material allowing for products that are not only very consistent throughout the part, but also from part to part. This is in

In liquid silicone rubber injection molding the screw-barrel assembly as well as the runner system are cooled to avoid curing before injection

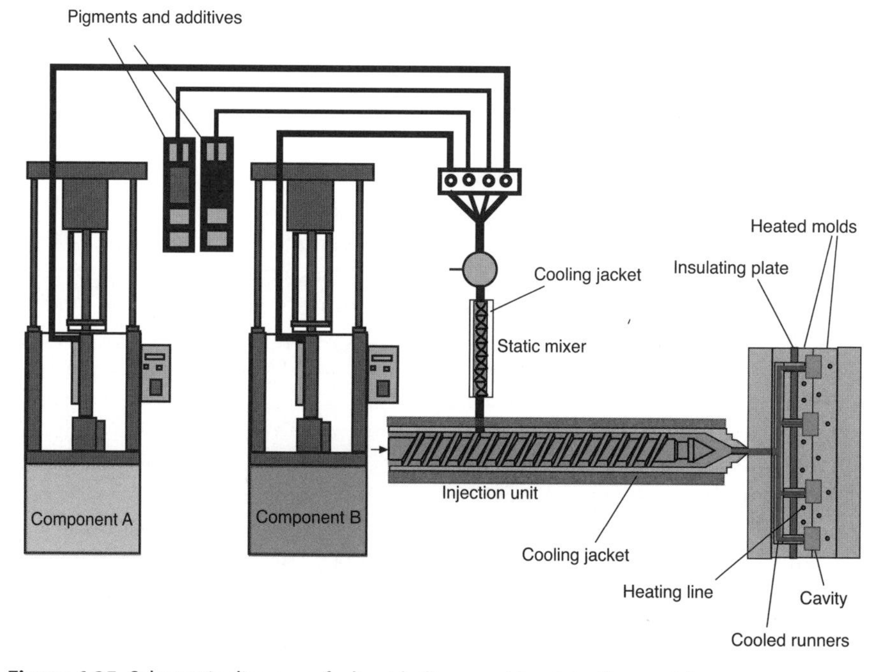

Figure 6.25 Schematic diagram of a liquid silicone rubber injection molding system

contrast to solid silicone rubber materials purchased pre-mixed and partially vulcanized. In contrast, hard silicone rubbers are processed by transfer molding and result in less material consistency and control, leading to higher part variability. Additionally, solid silicone rubber materials are processed at higher temperatures and require longer vulcanization times. From the metering section of the injection molding machine, the compound is pushed through cooled sprue and runner systems into a heated cavity where the vulcanization takes place. The cold runner and general cooling result in minimal loss of material in the feed lines. The cooling allows production of LSR parts with nearly zero material waste, eliminating trimming operations and yielding significant savings in material cost. Silicone rubber is a family of thermoset elastomers that have a backbone of alternating silicone and oxygen atoms and methyl or vinyl side groups. Silicone rubbers constitute about 30 % of the silicone family, making them the largest group of that family. Silicone rubbers maintain their mechanical properties over a wide range of temperatures and the presence of methyl-groups in silicone rubbers makes these materials extremely hydrophobic. Typical applications for liquid silicone rubber are products that require high precision such as seals, sealing membranes, electric connectors, multi-pin connectors, infant products where smooth surfaces are desired, such as bottle nipples, medical applications as well as kitchen goods such as baking pans, spatulas, etc.

6.4 Computer Simulation in Injection Molding

Computer simulation of polymer processes offer the tremendous advantage of enabling designers and engineers to consider virtually any geometric and processing option without incurring the expense associated with prototype mold or die making or the material waste of time-consuming trial-and-error procedures. The ability to try new designs or concepts on the computer gives the engineer the opportunity to detect and fix problems before beginning production. Additionally, the process engineer can determine the sensitivity of processing parameters on the quality and properties of the final part. For example, computer aided engineering (CAE) offers the designer the flexibility to determine the effect of different gating scenarios, runner designs, or cooling line locations when designing an injection mold.

However, process simulation is not a panacea. As with any modeling technique, there are limitations caused by assumptions in the constitutive material models, or geometric simplifications of the model cavity. For example, there is a tendency in the industry to continuously decrease the part thickness of injection molded parts. Thickness reductions increase the pressure requirements during mold filling, with typical pressures reaching 2 000 bar. Such pressures have a profound effect on the viscosity and thermal properties of the melt; effects that in great part are not accounted for in commercially available software.

The first step of CAE in process design and optimization is to transform a solid model, such as the PA6 housing presented in Fig. 6.26, into a finite element mesh that can be used by the simulation software package. Typically, a fairly three-dimensional

Modern injection molding simulation packages are used to design the thermal layout of a mold cavity

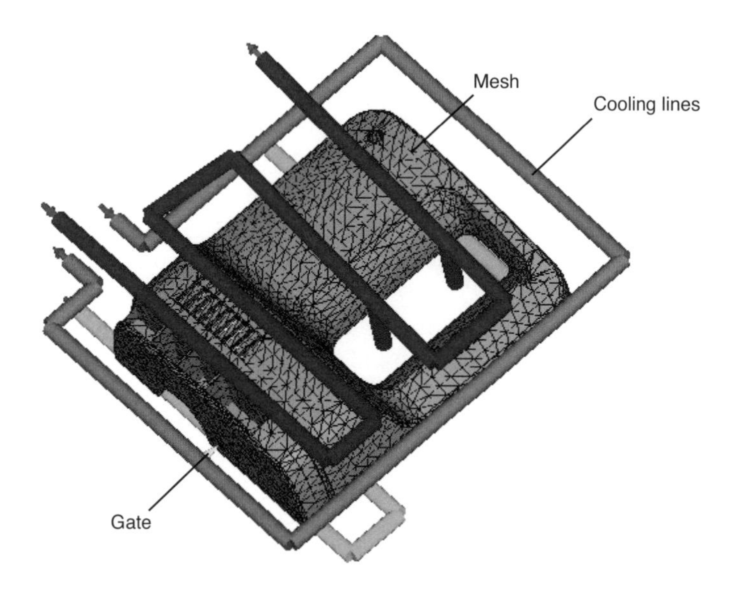

Figure 6.26 Finite element mesh of the mid-plane surface of a part and mold cooling line locations (Courtesy SIMCON Kunststofftechnische Software GmbH)

geometric model is transformed into a mid-plane model that essentially represents a two-dimensional geometry oriented in three-dimensional space. A finite element model is then generated on the mid-plane surface. Basically, the most common injection molding models use this approach to represent the geometry of the part. While most injection molded parts are thin and planar and would be well represented with such a model (Hele-Shaw model [7]), some injection molded parts are of smaller aspect ratios, or have three-dimensional features making these models invalid.

6.4.1 Mold Filling Simulation

Using a finite element mesh, such as the one presented in Fig. 6.26, in conjunction with the control volume approach, a simulation package solves a coupled energy and momentum balance, bringing as a result a mold filling pattern that not only includes the non-Newtonian effects present in the flow of polymer melts, but also the effect that the cooling has on the melt flow inside the mold cavity.

The mold filling analysis and the resulting filling pattern can be used to predict the formation of weld lines (knit lines when dealing with fiber reinforced composite parts) and gas entrapment. These can cause weak spots and surface finish problems that can lead to cracks and failure of the final part, as well as esthetic problems in the finished product. Figure 6.27 presents the predicted 60 % filled mold of the part and

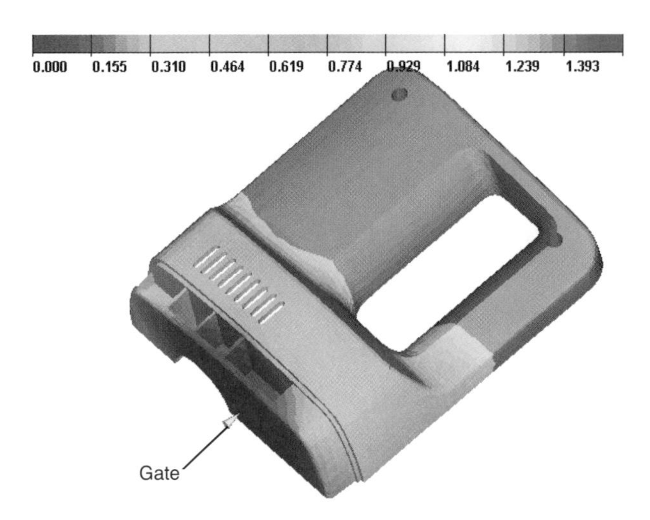

Figure 6.27 Short at 60 % fill of the part presented in Fig. 6.26 (Courtesy SIMCON Kunststofftechnische Software GmbH)

Mold filling simulations allow prediction of weldline and air entrapment locations. They are also used to optimize the gate location(s)

gate presented in Fig. 6.26. The pressure and clamping force requirements are also needed information during part and process design. Both are computed by commercial injection molding software. In simulations where the rheology of the material is well defined, along with proper geometry and processing conditions, the accuracy of a simulation can be quite high. Figure 6.28 presents a comparison of experimental and predicted short shots; as can be seen, prediction and simulation agree.

Today's injection molding CAE packages are very accurate when predicting mold filling patterns

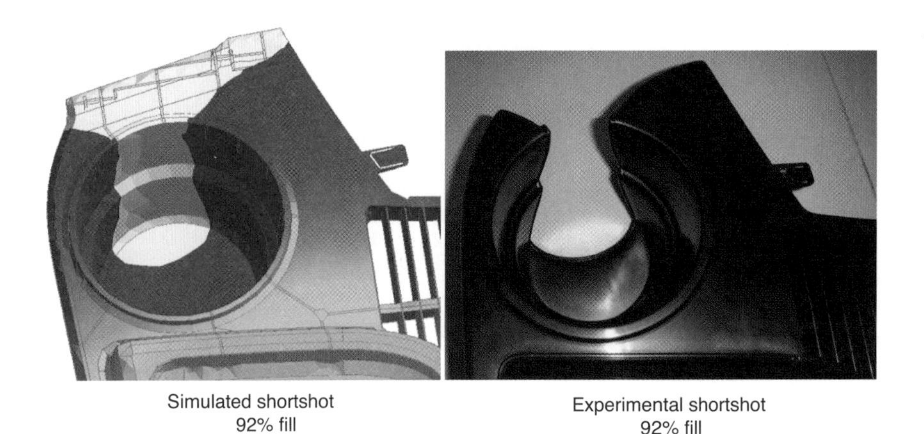

Figure 6.28 Experimental and simulated 92 % short shot of an automotive cup holder (Courtesy CoreTech System Co., Ltd.)

6.4.2 Orientation Predictions

Molecular and filler orientation have a profound effect on the properties of the finished part. Molecular orientation will not only influence the mechanical properties of the polymer but also its optical quality. For example, birefringence is controlled by molecular orientation, which must be kept low for products that require certain optical properties, such as lenses. The Folgar-Tucker model has been implemented into various, commercially available injection and compression mold filling simulation programs. Figure 6.29 presents a comparison between predicted and experimental birefringence patterns. For the polycarbonate lens shown in the figure, the birefringence pattern is directly related to molecular orientation.

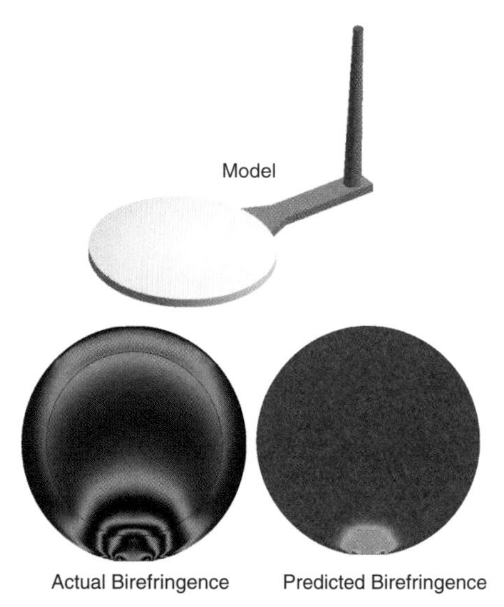

Figure 6.29 Experimental and simulated birefringence pattern in a polycarbonate lens (Courtesy CoreTech System Co., Ltd.)

6.4.3 Shrinkage and Warpage Predictions

Shrinkage and warpage are directly related to residual stresses that result from locally varying strain fields that occur during the curing or solidification stage of a manufacturing process. Such strain gradients are caused by nonuniform thermomechanical properties and temperature variations inside the mold cavity. Shrinkage due to cure can also play a dominant role in the residual stress development in thermosetting polymers and becomes important for fiber reinforced thermosets; shrinkage is also a concern when sink marks appear in thick sections or ribbed parts. When processing thermoplastic materials, shrinkage and warpage in a final product depend on the molecular orientation and residual stresses that form during processing. The molecular or fiber orientation and the residual stresses inside the part in turn depend on the flow and heat

Warpage is a very common problem when manufacturing plastic parts. Modern software packages can accurately predict warpage within the final part

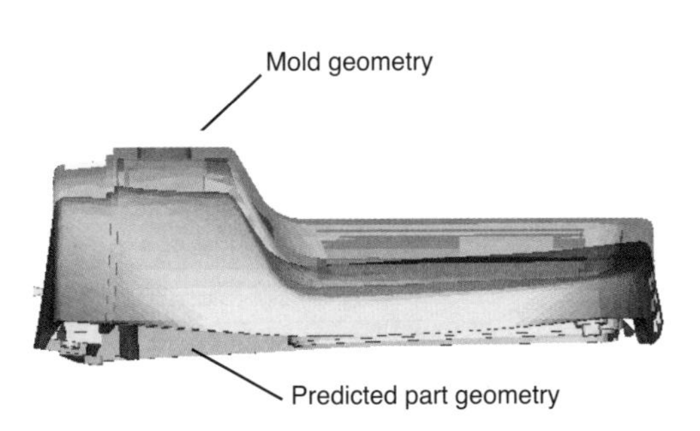

Figure 6.30 Predicted warped geometry after mold removal and cooling of the part presented in Fig. 6.26 (Courtesy SIMCON Kunststofftechnische Software GmbH)

transfer during the mold filling, packing, and cooling stage of the injection molding process. To predict the residual stress in the finished part, modern software packages characterize the thermomechanical response of the polymer from melt to room temperature using the PvT behavior of the material, in conjunction with the temperature dependent stress-strain behavior. Figure 6.30 presents the warped geometry of the part depicted in Fig. 6.26 after mold removal and cooling. The warpage is usually depicted graphically as total amount of deflection as well as superposing deflected part geometry and mold geometry. Minimizing warpage is one of the biggest concerns for the design engineer. This is sometimes achieved by changing the formulation of the resin. Further reduction in warpage can also be achieved by changing the number and location of gates. Although trial-and-error solutions, which are still the most feasible with today's technology, are commonly done, computer optimization often reduces cost.

Example 6.1 Balancing a runner system

Consider the multi-cavity injection molding process shown in Fig. 6.17. To achieve equal part quality, the filling time for all cavities must be balanced. For the case in question, balance your cavities by solving for the runner radius, R_2. For a balanced runner system, the flow rates into all cavities must match. For a given flow rate, Q, length L, and radius R_1, you must also solve for the pressures at the runner system junctures. Assume an isothermal flow of a Newtonian polymer with viscosity η. Compute the radius R_2 for a part molded of polycarbonate with a viscosity of 350 Pa·s. Use values of $L = 10$ cm, $R_1 = 4$ mm, and $Q = 20$ cm/s.

The flow through each runner section is governed by the Hagen-Poiseuille equation:

$$Q = \frac{\pi R^4}{8\eta}\left(-\frac{dp}{dz}\right)$$

The various sections can be represented using

$$\text{Section 1: } 4Q = \frac{\pi (2R_1)^4}{8\eta}\,(P_1 - P_2)$$

Section 2: $2Q = \frac{\pi (2R_1)^4}{8\eta} (P_2 - P_3)$

Section 3: $Q = \frac{\pi (R_2)^4}{8\eta} (P_2 - 0)$

Section 4: $Q = \frac{\pi (R_1)^4}{8\eta} (P_3 - 0)$

The unknown parameters, P_1, P_2, P_3, and R_2, can be obtained using the above equations. For the given values, a radius, R_2, of 3.78 mm would result in a balanced runner system, with pressures $P_1 = 104.4$ bar, $P_2 = 87.0$ bar and $P_3 = 69.6$ bar. More detail on this topic is given in Chapter 9, including the same example for a shear-thinning material.

■

References

1. Pötsch, G., and W. Michaeli, Injection Molding: An Introduction, Hanser Publishers (2008), Munich
2. Greener, J., *Polym. Eng. Sci.* (1986), *26*, 886
3. Michaeli, W., and M. Lauterbach, *Kunststoffe* (1989), *79*, 852
4. Osswald, T.A., and G. Menges, *Materials Science of Polymers for Engineers,* 2nd Edition, Hanser Publishers (2003), Munich
5. Wimberger-Friedl, R., *Polym. Eng. Sci.* (1990), *30*, 813
6. Ehrenstein, G.W., *Polymer Werkstoffe,* Hanser Publishers (1999), Munich
7. Hele-Shaw, H.S., *Proc. Roy. Inst.* (1899), *16*, 49

7 Other Plastics Processes

In addition to extrusion and injection molding there many other processes, such as secondary shaping operations, calendering, coating, compression molding, and many more. Secondary shaping operations such as extrusion blow molding, film blowing, and fiber spinning occur immediately after the extrusion profile emerges from the die. The thermoforming process is performed on sheets or plates previously extruded and solidified. In general, secondary shaping operations consist of mechanical stretching or forming of a preformed cylinder, sheet, or membrane.

7.1 Fiber Spinning

Fiber spinning is used to manufacture synthetic fibers. During fiber spinning, a filament is continuously extruded through an orifice and stretched to diameters of 100 μm and smaller. The process is schematically depicted in Fig. 7.1. The molten polymer is first extruded through a filter, or *screen pack*, to eliminate small contaminants. The melt is then extruded through a spinneret, a die composed of multiple orifices. A spinneret can have between one and 10 000 holes. The fibers are then drawn to their final diameter, solidified, and wound onto a spool. The solidification takes place either in a water bath or by forced convection. When the fiber solidifies in a water bath, the extrudate

Fiber spinning is dominated by elongational deformation (stretching). Further stretching is done after the point of solidification

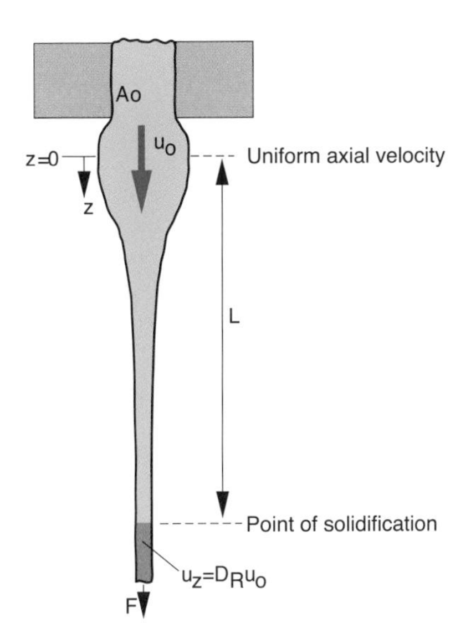

Figure 7.1 The fiber spinning process

undergoes an adiabatic stretch before cooling begins in the bath. The forced convection cooling, which is more commonly used, leads to a non-isothermal spinning process.

The drawing and cooling processes determine the morphology and mechanical properties of the final fiber. For example, ultra high molecular weight HDPE fibers with high degrees of orientation in the axial direction can have the stiffness of steel with today's fiber spinning technology.

Of major concern during fiber spinning are the instabilities that arise during drawing, such as brittle fracture, Rayleigh disturbances, and draw resonance. Brittle fracture occurs when the elongational stress exceeds the melt strength of the drawn polymer melt. The instabilities caused by Rayleigh disturbances are like those causing filament break-up during dispersive mixing, as discussed in Chapter 5. Draw resonance appears under certain conditions and manifests itself as periodic fluctuations that result in diameter oscillation.

The physics and theory behind the fiber spinning process [1] are covered in detail in Chapter 8 of this book.

7.2 Film Production

7.2.1 Cast Film Extrusion

In a cast film extrusion process, a thin film is extruded through a slit onto a chilled, highly polished turning roll, where it is quenched from one side. The speed of the roller controls the draw ratio and final film thickness. The film is then sent to a second roller for cooling of the other side. Finally, the film passes through a system of rollers and is wound onto a roll.A typical film casting process is depicted in Figs. 7.2 and 7.3. The cast film extrusion process exhibits stability problems similar to those encountered in fiber spinning [2].

Thicker polymer sheets are manufactured similarly. A sheet is distinguished from a film by its thickness; by definition a sheet has a thickness exceeding 250 μm; otherwise, it is called a film.

Multi-layer films used for meat packaging are manufactured using film casting operations

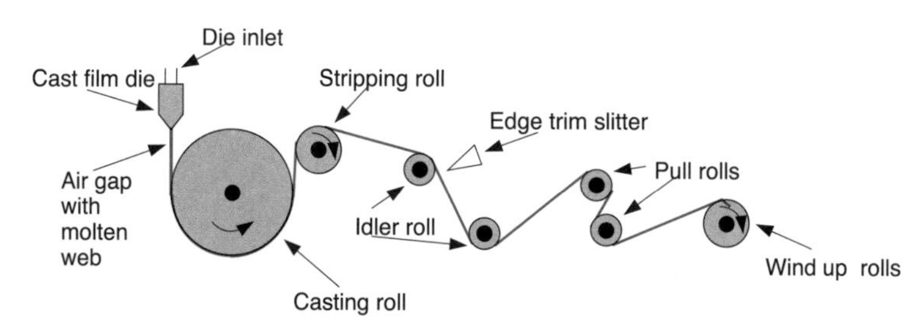

Figure 7.2 Schematic of a film casting operation

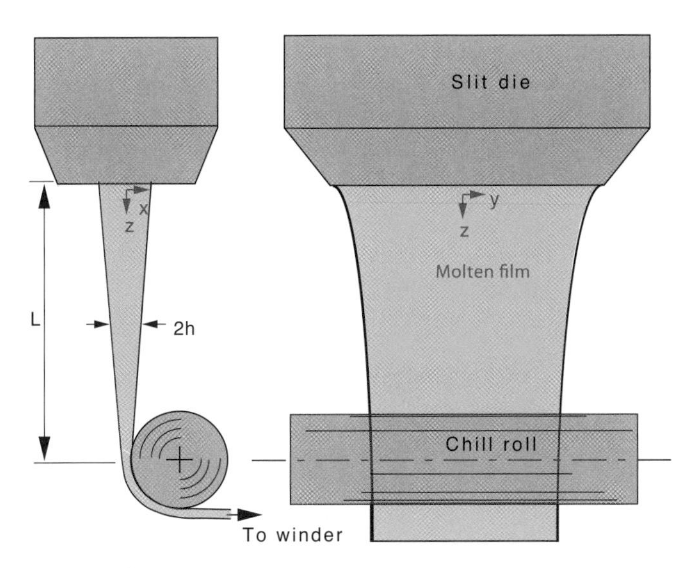

Figure 7.3 Detail in a film casting process

7.2.2 Film Blowing

In film blowing, a tubular cross-section is extruded through an annular die, normally a spiral die, and is drawn and inflated until the *freezing line* is reached. Beyond this point, the stretching is practically negligible. The process is schematically depicted in Fig. 7.4 [3]. The advantage of film blowing over casting is that the induced biaxial stretching renders a stronger and less permeable film.

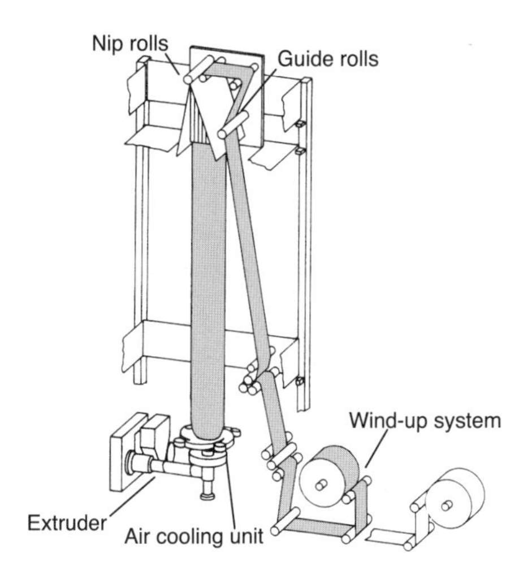

Figure 7.4 Film blowing process

The extruded tubular profile passes through one or two air rings to cool the material. The tube interior is maintained at a certain pressure by blowing air into the tube through a small orifice on the die mandrel. The air is retained in the tubular film, or bubble, by collapsing the film well above its freeze-off point and tightly pinching it between rollers. The size of the tubular film is calibrated between the air ring and the collapsing rolls.

The important outcome of a film blowing process (Fig. 7.5) is the size or diameter of the tubular film defined by the dimensionless blow-up ratio [4]

$$B_R = \frac{R_f}{R_0} \tag{7.1}$$

and the dimensionless draw ratio

$$D_R = \frac{v_f}{v_0} \tag{7.2}$$

The draw-down and blow-up results in biaxial molecular orientation, which is necessary to achieve best mechanical properties within the film

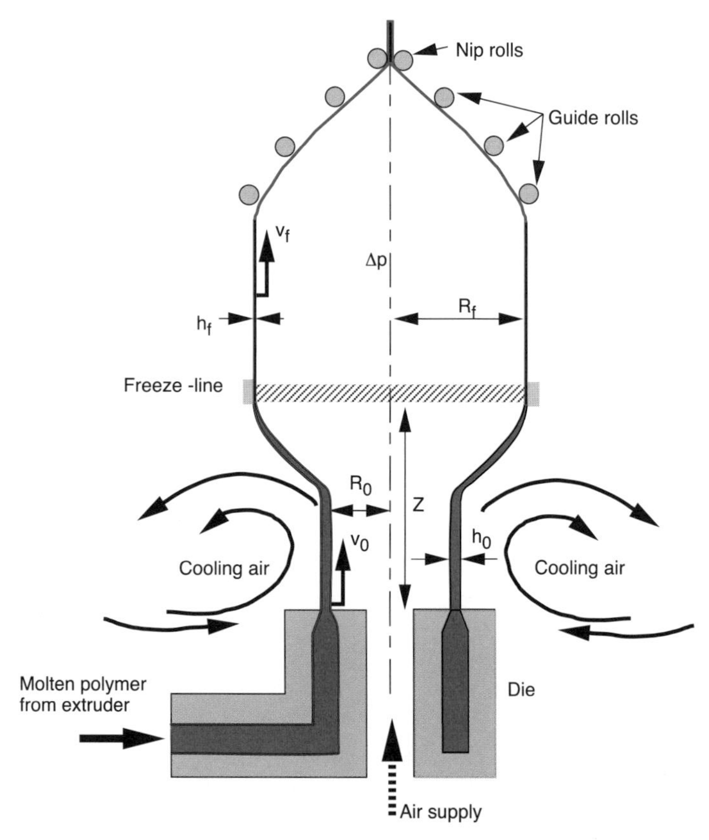

Figure 7.5 Film blowing variables

Conservation of mass leads to a dimensionless thickness defined by

$$\frac{h_0}{h_f} = B_R D_R \tag{7.3}$$

Pearson and Petrie [4] developed and solved a Newtonian model of film blowing and plotted their results in dimensionless graphs. In addition to dimensionless blow-up ratio, draw ratio, and thickness, they introduced a dimensionless pressure, P,

$$P = \frac{\pi R_0^3 \Delta p}{\mu Q} \tag{7.4}$$

a dimensionless take-up force, F,

$$F = \frac{R_0 f_z}{\mu Q} \tag{7.5}$$

and a dimensionless freeze-off point, X,

$$X = \frac{Z}{R_0} \tag{7.6}$$

where μ is the Newtonian viscosity, R_0 is the radius of the extruded tube, R_f is the radius of the final tubular film, v_0 is the velocity of the material at the exit of the die, v_f is the take-up speed, Q is the volumetric material throughput, Δp is the pressure difference across the film, and f_z is the film take-up force. The film blowing process is covered in more detail in Chapter 8 of this book, where a simplified Newtonian model is used to predict the outcome of a film blowing process [4].

7.3 Blow Molding

The predecessor of the blow molding process was the blowing press developed by Hyatt and Burroughs in the 1860s to manufacture hollow celluloid articles. Polystyrene was the first synthetic polymer used for blow molding during World War II and polyethylene was the first material to be implemented in commercial applications. Until the late 1950s, the main application for blow molding was the manufacture of LDPE articles such as squeeze bottles.

Blow molding [5] produces nearly hollow articles that do not require a homogeneous thickness distribution. Today, HDPE, LDPE, PP, PET, and PVC are the most common materials used for blow molding.

7.3.1 Extrusion Blow Molding

In extrusion blow molding, a *parison* or tubular profile is extruded and inflated into a cavity with the specified geometry. The blown article is held inside the cavity until it is sufficiently cool. Figure 7.6 presents a schematic of the steps in blow molding.

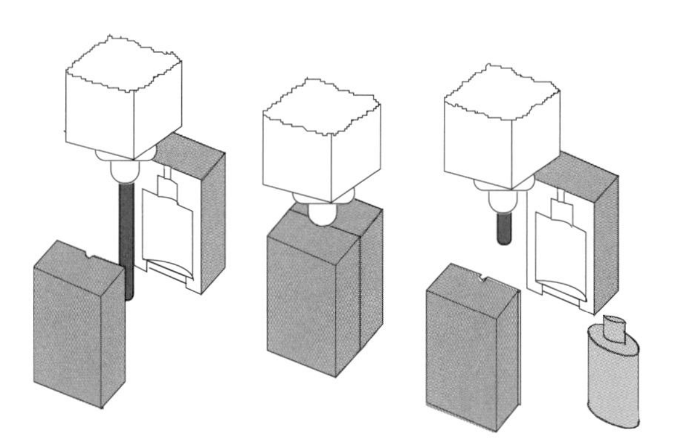

Figure 7.6 Schematic of the extrusion blow molding process

During blow molding, one must generate the appropriate parison length such that the trim material is minimized. Another means of saving material is by generating a parison of variable thickness, usually referred to as *parison programming*, such that an article with an evenly distributed wall thickness is achieved after stretching the material. An example of a programmed parison and finished bottle thickness distribution is presented in Fig. 7.7 [6].

It is possible to reduce thickness variation in the blown product by having a parison with an initial thickness distribution

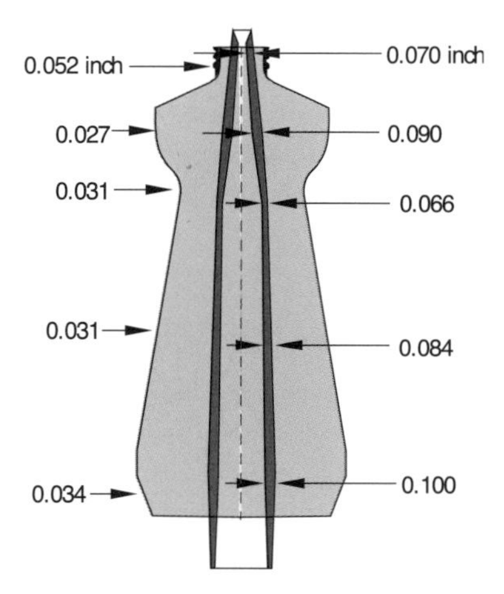

Figure 7.7 Wall thickness distribution in the parison and the part

A parison of variable thickness can be generated by moving the mandrel vertically during extrusion, as shown in Fig. 7.8. A thinner wall not only results in material savings but also reduces the cycle time due to the shorter required cooling times.

As expected, the largest portion of the cycle time is the cooling of the blow molded container in the mold cavity. For this reason, for high volume applications, rotary molds are often used in conjunction with vertical or horizontal rotating tables (Fig. 7.9 [5]).

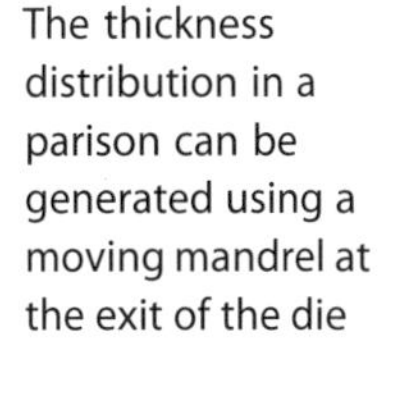
The thickness distribution in a parison can be generated using a moving mandrel at the exit of the die

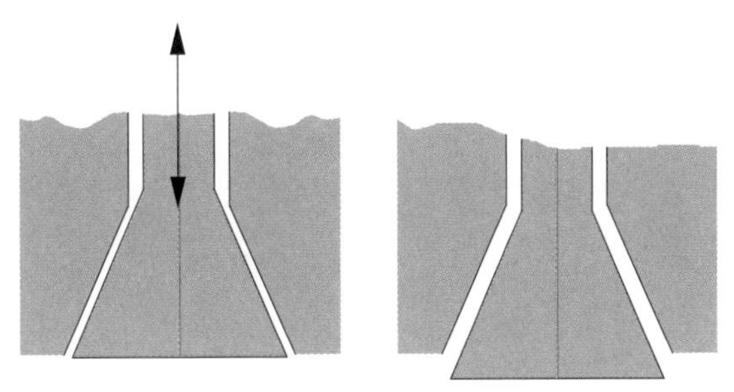
Figure 7.8 Moving mandrel used to generate a programmed parison

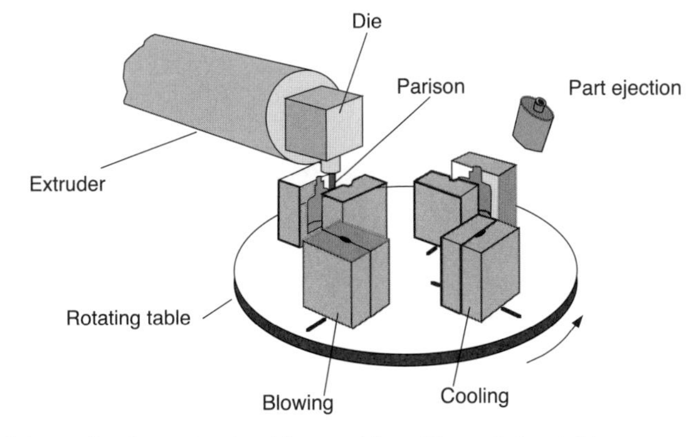

Figure 7.9 Schematic of an extrusion blow molder with a rotating table

The same procedure used for injection molding can be used to estimate the cooling time during blow molding. Because the article is cooled from one side only, the inner surface can be considered insulated. The modified Eq. 6.3 now becomes

$$t_{\text{cooling}} = \frac{4h^2}{\pi\alpha} \ln\left(\frac{8}{\pi^2}\frac{T_\text{m} - T_\text{w}}{T_\text{D} - T_\text{w}}\right) \tag{7.7}$$

7.3.2 Injection Blow Molding

Injection blow molding, depicted in Fig. 7.10 [5], begins by injection molding the parison onto a core and into a mold with finished bottle threads. The formed parison has a thickness distribution that leads to reduced thickness variations throughout the container. Before blowing the parison into the cavity, it can be mechanically stretched to orient molecules axially, Fig. 7.11 [5]. The subsequent blowing operation introduces tangential orientation. A container with biaxial molecular orientation exhibits higher optical (clarity) and mechanical properties and lower permeability. With the injection blow molding process one can either go directly from injection to blowing or a re-heating stage can be implemented between injection and blowing.

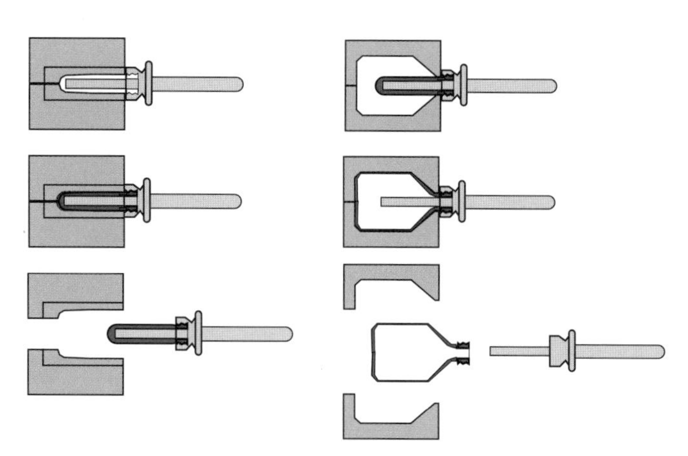

Figure 7.10 Injection blow molding

The sequence of stretching and blowing results in bi-axial molecular orientation, needed to reduce permeability

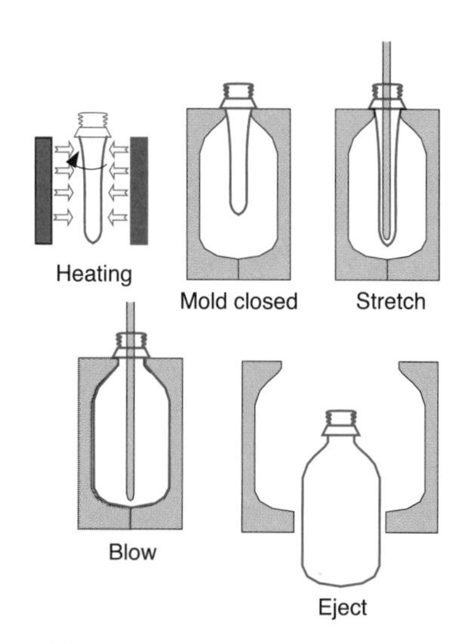

Figure 7.11 Stretch blow molding

The advantages of injection blow molding over extrusion blow molding are:

- Pinch-off and therefore post-mold trimming are eliminated
- Controlled container wall thickness
- Dimensional control of the neck and screw-top of bottles and containers

Disadvantages include higher initial mold cost and the need for both injection and blow molding units.

7.4 Thermoforming

Thermoforming is an important secondary shaping operation of plastic film and sheet. Thermoforming consists of warming the plastic sheet and forming it into a cavity or over a tool using vacuum, air pressure, and mechanical means. During the 18th century, tortoiseshells and hooves were thermoformed into combs and other shapes. The process was refined during the mid-19th century to thermoform various cellulose nitrate articles. During World War II, thermoforming was used to manufacture acrylic aircraft cockpit enclosures, canopies, and windshields, as well as translucent covers for outdoor neon signs. During the 1950s, the process made an impact in the mass production of cups, blister packs, and other packaging commodities. Today, in addition to packaging, thermoforming is used to manufacture refrigerator liners, pick-up truck cargo box liners, shower stalls, bathtubs, as well as automotive trunk liners, glove compartments, and door panels.

A typical thermoforming process is presented in Fig. 7.12 [7]. The process begins by heating the plastic sheet slightly above the glass transition temperature for amorphous polymers, or slightly below the melting point for semi-crystalline materials. Although, both amorphous and semi-crystalline polymers are used for thermoforming, the process is easiest with amorphous polymers because they have a wide rubbery temperature range above the glass their transition temperature. At these temperatures the polymer is easily shaped, but still has enough rigidity to hold the heated sheet without much sagging. Most semi-crystalline polymers lose their strength rapidly once the crystalline structure breaks up above the melting temperature.

The heating is achieved using radiative heaters and the temperature reached during heating must be high enough for sheet shaping, but low enough so the sheets do not droop into the heaters. One key requirement for successful thermoforming is to bring the sheet to a uniform forming temperature. The sheet is then shaped into the cavity over the tool. This can be accomplished in several ways. Most commonly, a vacuum sucks the sheet onto the tool, stretching the sheet until it contacts the tool surface.

A plug assist is used to reduce thickness variations in the thermoformed part

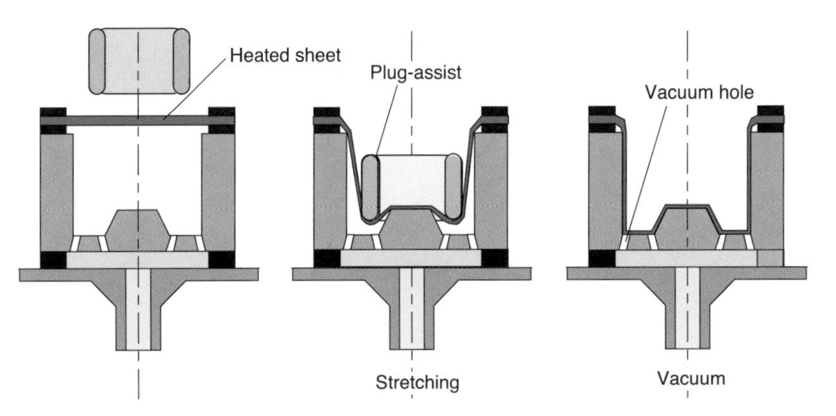

Figure 7.12 Plug-assist thermoforming using vacuum

The main problem here is the irregular thickness distribution that arises throughout the part. Hence, the main concern of the process engineer is to optimize the system such that the differences in thickness throughout the part are minimized. This can be accomplished in many ways, but most commonly is achieved by plug-assist. Here, as the plug pushes the sheet into the cavity, only the parts of the sheet not touching the plug-assist stretch. Since the unstretched portions of the sheet must remain hot for subsequent stretching, the plug-assist is made of a low thermal conductivity material, such as wood or hard rubber. The initial stretch is followed by a vacuum for final shaping. Once cooled, the product is removed.

To reduce thickness variations in the product, the sheet can be pre-stretched by forming a bubble at the beginning of the process. This is schematically depicted in Fig. 7.13 [3]. The mold is raised into the bubble, or a plug-assist pushes the bubble into the cavity, and a vacuum finishes the process.

One of the main reasons for the rapid growth and high volume of thermoformed products is that the tooling costs for a thermoforming mold are much lower than for injection molding. Some thermoforming processes are highly automated and form part of a larger assembly and packaging process, such as the one shown in Fig. 7.14. Here, the thermoforming of the container is integrated with the filling station, the labeling, the sealing, and the cutting of the filled container.

Pre-stretching the sheet in form of a bubble before thermoforming significantly reduces thickness variations within the thermoformed part

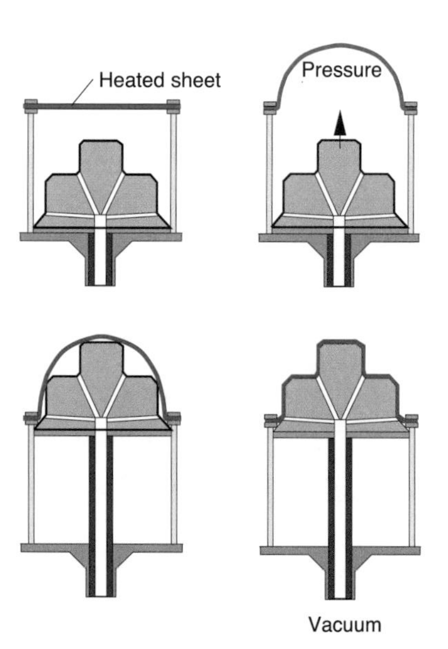

Figure 7.13 Reverse draw thermoforming with plug-assist and vacuum

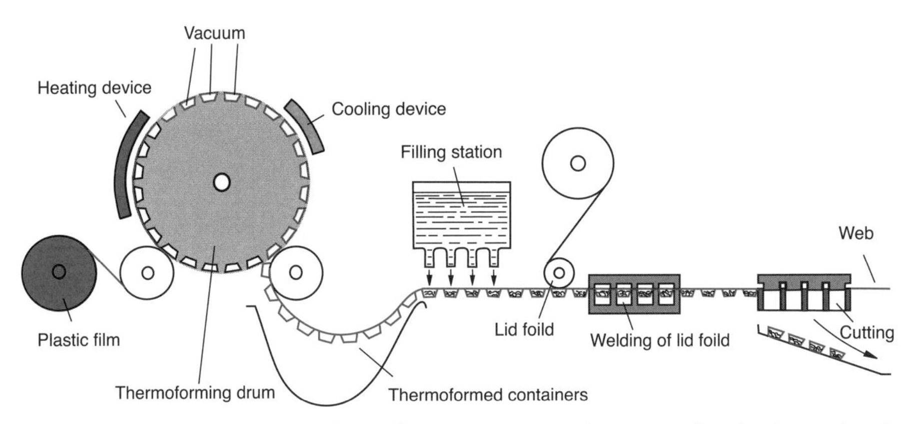

Figure 7.14 Rotational vacuum thermoforming process with integrated packaging and sealing stations

7.5 Calendering

In a calender line, the polymer melt is transformed into films and sheets by squeezing it between pairs of co-rotating high precision rollers. Calenders are also used to produce certain surface textures which may be required for different applications. Today, calendering lines are used to manufacture PVC sheet, floor coverings, rubber sheet, and rubber tires. They are also used to texture or emboss surfaces. When producing PVC sheet and film, calender lines have a great advantage over extrusion processes because of the shorter residence times, resulting in a lower requirement for stabilizer. This can be cost effective because stabilizers are a major part of the overall expense of processing these polymers.

Figure 7.15 [8–10] presents a typical calender line for manufacturing PVC sheet. A typical system is composed of:

- plasticating unit
- calender
- cooling unit
- accumulator
- wind-up station

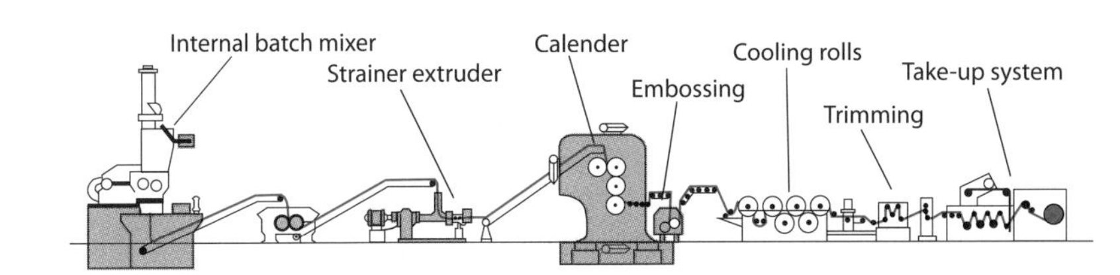

Figure 7.15 Schematic of a typical calendering process (Berstorff GmbH)

In the plasticating unit, the material is melted and mixed by an internal batch mixer or a roll-mill before it is fed between the nip of the first two rolls. Here, the first pair of rolls controls the feeding rate, while subsequent rolls in the calender calibrate the sheet thickness. Most calender systems have four rolls, as does the one in Fig. 7.15, which is an inverted L- or F-type system. Other typical roll arrangements are shown in Fig. 7.16. After the main calender, the sheet can be passed through a secondary calendering operation for embossing. The sheet is then passed through a series of chilling rolls, where it is cooled from both sides in an alternating fashion. After cooling, the film or sheet is wound.

One of the major concerns in a calendering system is generating a film or sheet with a uniform thickness distribution with tolerances as low as ± 0.005 mm. To achieve this, the dimensions of the rolls must be precise. It is also necessary to compensate for roll bowing resulting from high pressures in the nip region. Roll bowing is a structural problem that can be mitigated by placing the rolls in a slightly crossed pattern, rather than completely parallel, or by applying moments to the roll ends to counteract the separating forces in the nip region.

7.6 Coating

In coating a liquid film is continuously deposited on a moving, flexible or rigid substrate. Coating is done on metal, paper, photographic films, audio and video tapes, and adhesive tapes. Typical coating processes include *wire coating, dip coating, knife coating, roll coating, slide coating,* and *curtain coating.*

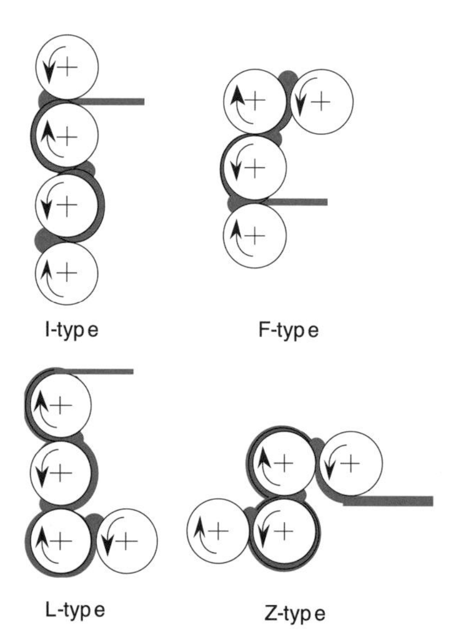

Figure 7.16 Calender arrangements

In wire coating, a wire is continuously coated with a polymer melt by pulling the wire through an extrusion die. The polymer resin is deposited onto the wire using the drag flow generated by the moving wire and sometimes by a pressure flow generated by the back pressure of the extruder. The process is schematically depicted in Fig. 7.17.[1)] The second normal stress differences, generated by the high shear deformation in the die, help keep the wire centered in the annulus [11].

Dip coating is the simplest and oldest coating operation. Here, a substrate is continuously dipped into a fluid and withdrawn with one or both sides coated with the fluid. Dip coating can also be used to coat individual objects that are dipped and withdrawn from the fluid. The fluid viscosity and density and the speed and angle of the surface determine the coating thickness.

Knife coating, depicted in Fig. 7.18, consists of metering the coating material onto the substrate from a pool of material, using a fixed rigid or flexible knife. The knife can be normal to the substrate or angled and the bottom edge can be flat or tapered. The thickness of the coating is nearly half the gap between the knife edge and the moving substrate or web. A major advantage of a knife edge coating system is its simplicity and relatively low maintenance.

Roll coating consists of passing a substrate and the coating simultaneously through the nip region between two rollers. The physics governing this process are similar to

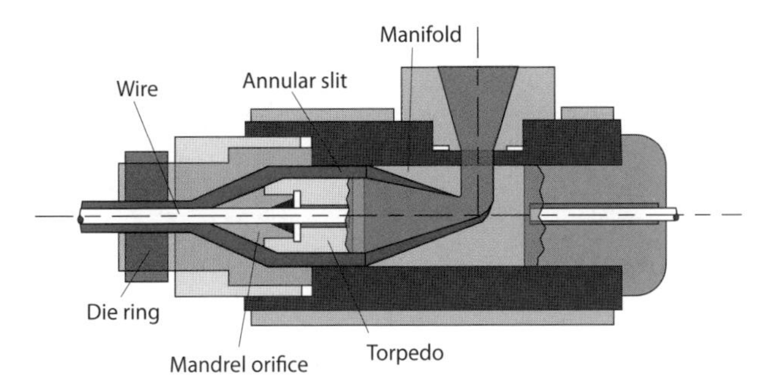

Figure 7.17 Wire coating process

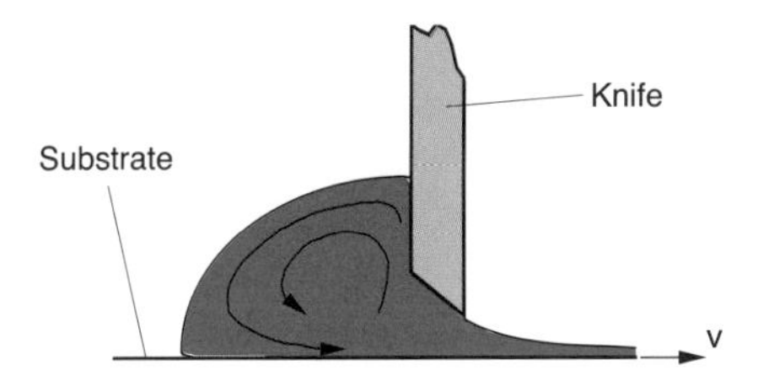

Figure 7.18 Schematic of a knife coating process

1) Other wire coating processes extrude a tubular sleeve which adheres to the wire via stretching and vacuum. This is called tube coating.

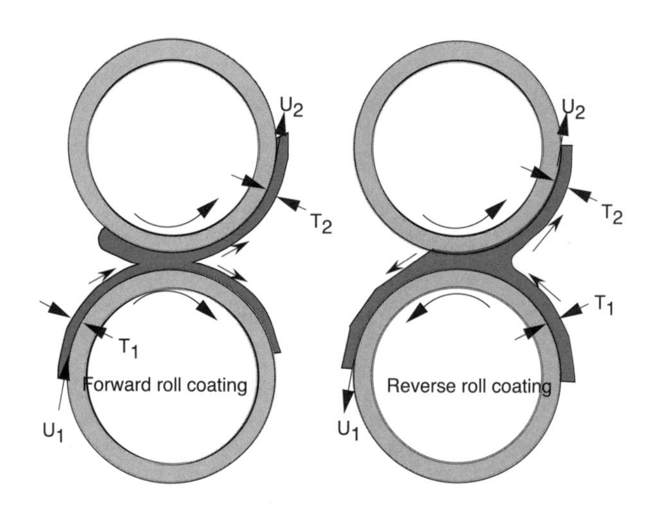

Figure 7.19 Schematic of forward and reverse roll coating processes

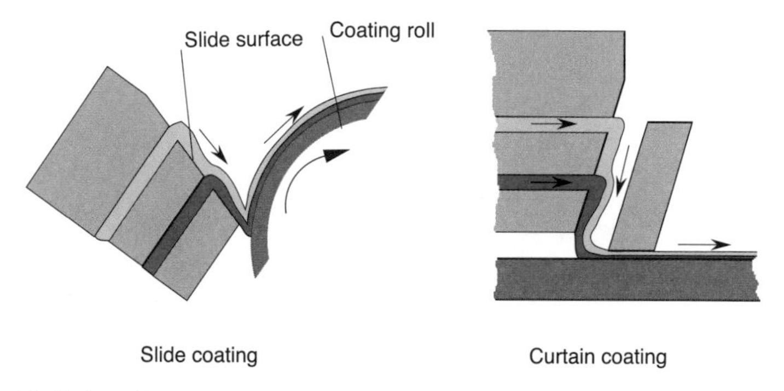

Figure 7.20 Slide and curtain coating

calendering, except that the fluid adheres to both the substrate and the opposing roll. The coating material is a low viscosity fluid, such as a polymer solution or paint, and is picked up from a bath by the lower roll and applied to one side of the substrate. The thickness of the coating can be as low as a few μm and is controlled by the viscosity of the coating liquid and the nip dimension. This process can be configured as either forward roll coating for co-rotating rolls or reverse roll coating for counter-rotating rolls (Fig. 7.19). The reverse roll coating process delivers the most accurate coating thicknesses.

Slide coating and curtain coating, schematically depicted in Fig. 7.20, are commonly used to apply multi-layered coatings. However, curtain coating has also been widely used to apply single layers of coatings to cardboard sheet. In both methods, the coating fluid is pre-metered.

7.7 Processing Reactive Polymers

Reactive polymers, such as thermosets, undergo a chemical reaction during solidification. In processing, thermosets are often grouped into three distinct categories, namely those that undergo a *heat activated cure,* those dominated by a *mixing activated cure,* and those activated by the absorption of humidity or ultraviolet radiation. Examples of heat activated thermosets are phenolics; examples of mixing activated cure are epoxy resins and polyurethane. Similarly to thermoplastics, thermosets either cure via *condensation polymerization* or *addition polymerization.*

Condensation polymerization is the growth process that results from combining two or more monomers with reactive end-groups; it leads to by-products such as alcohol, water and acid. The by-product of the reaction when making phenolics is water. Another well known example of a thermoset that cross-links via condensation polymerization is the co-polymerization of unsaturated polyester with styrene molecules, also called free radical reaction, as shown in Fig. 7.21 [12]. The molecules contain several carbon-carbon double bonds which act as cross-linking sites during curing. An example of the resulting network after the chemical reaction is shown in Fig. 7.22.

No matter which category of thermoset, its curing reaction can be described by the reaction between two chemical groups denoted by A and B that link two segments of a polymer chain. The reaction can be followed by tracing the concentration of unreacted A's or B's, C_A or C_B, respectively. If the initial concentrations of A's and B's are defined

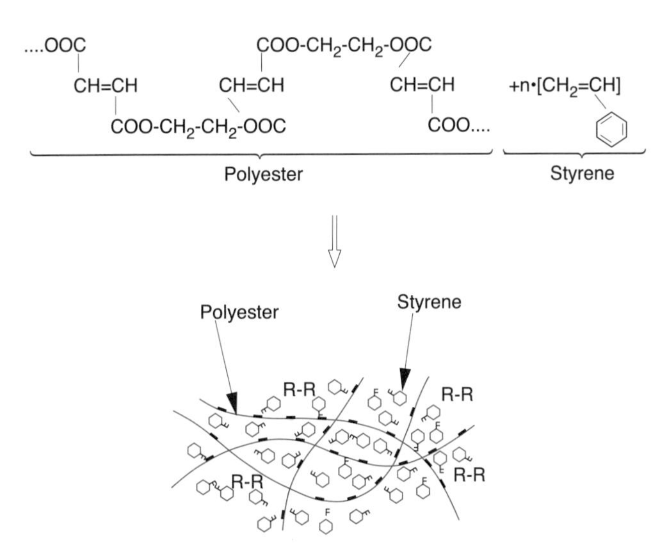

Double bonds within the unsaturated polyester molecule serve as cross-linking sites for the polystyrene bridges. During cure, the polystyrene polymerizes

Figure 7.21 Symbolic and schematic representations of uncured unsaturated polyester

The cross-linking process is exothermic

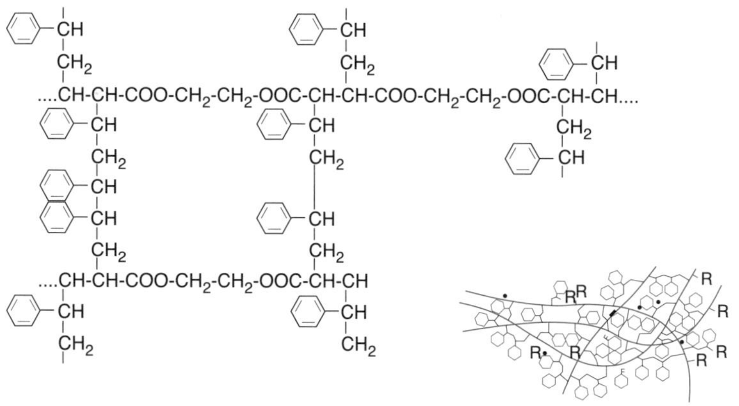

Figure 7.22 Symbolic and schematic representations of cured unsaturated polyester

as C_{A_0} and C_{B_0}, the degree of cure can be described as

$$c = \frac{C_{A_0} - C_A}{C_{A_0}} \tag{7.8}$$

The degree of cure or conversion, c, equals zero when there has been no reaction and equals one when all A's have reacted and the reaction is complete. It is difficult to monitor reacted and unreacted A's and B's during the curing reaction of a thermoset polymer. However, the heat released during curing can be used to monitor the conversion, c. When small samples of an unreacted thermoset polymer are placed in a differential scanning calorimeter (DSC), each at a different temperature, every sample releases nearly the same amount of heat, Q_T. This occurs because every cross-link that occurs during a reaction releases a little heat. For example, Fig. 7.23 [13] shows the heating rate during isothermal cure of a vinyl ester at various temperatures.

The degree of cure of a thermoset resin can be defined by the following relation

$$c = \frac{Q}{Q_T} \tag{7.9}$$

where Q is the heat released up to an arbitrary time t. DSC data is commonly fitted to empirical models that accurately describe the curing reaction. Hence, the rate of cure can be described by the exotherm, $\dot{Q}$, and the total heat released during the curing reaction, Q_T. The curing kinetics for many heat activated cure materials, such as vinyl esters and unsaturated polyesters, can be described fairly well using the Kamal-Sourour Model [14] given by

$$\frac{dc}{dt} = \frac{\dot{Q}}{Q_T} = \left(a_1 e^{\frac{-b_1}{RT}} + a_2 e^{\frac{-b_2}{RT}} c^m \right) (1 - c)^n \tag{7.10}$$

where the six constants can be fit to DSC data.

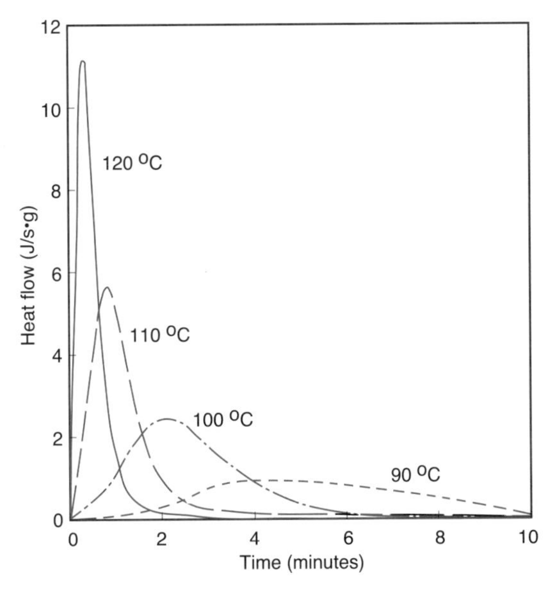

Figure 7.23 DSC scan of the isothermal curing reaction of vinyl ester at various temperatures

In an exothermic reaction, heat is released. This heat can be measured using a differential scanning calorimeter (DSC)

On the other hand, mixing activated cure materials, such as polyurethanes, instantly start releasing exothermic heat after the mixture of its two components. A model that accurately fits this behavior is the *Castro-Macosko Curing Model* [15]:

$$\frac{dc}{dt} = k_0 e^{-\frac{E}{RT}} (1 - c)^2 \qquad (7.11)$$

Heat Transfer During Cure

A well-known problem in thermoset components with thick sections is that the exothermic reaction encountered during curing leads to significant temperature rises within the material that may cause thermal degradation. A quick way of estimating the maximum temperature rise that can occur during curing is by using the adiabatic temperature rise given by

$$\Delta T = \frac{Q_T}{C_p} \qquad (7.12)$$

where Q_T is the total heat released during reaction and C_p is the specific heat of the resin. Figure 7.24 is a plot of the time to reach 80 % cure versus part thickness for various mold temperatures. The shaded area represents the conditions at which the internal temperature within the part exceeds 200 °C as a result of the exothermic reaction. A total heat of reaction of 84 000 J/kg was used in these calculations.

The exothermic reaction may result in overheating

Thick parts must be molded at lower temperatures to avoid overheating

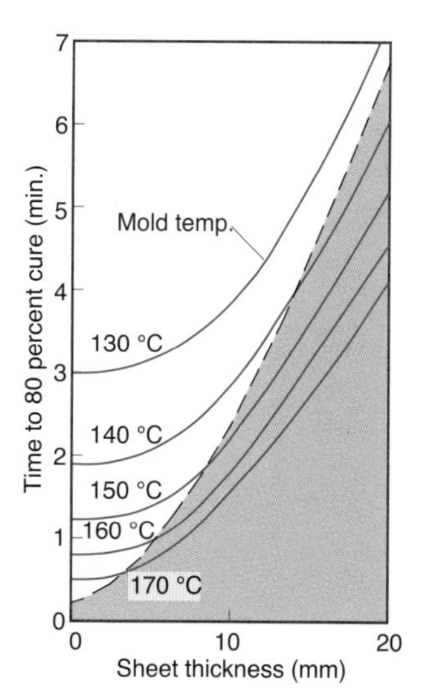

Figure 7.24 Cure times versus plate thickness for various mold temperatures (shaded region represents the conditions at which thermal degradation can occur [14])

7.8 Compression Molding

Compression molding is widely used in the automotive industry to produce parts that are large, thin, lightweight, strong, and stiff. It is also used in the household goods and electrical industries. Compression molded parts are formed by squeezing a glass fiber reinforced charge inside a mold cavity, as depicted in Fig. 7.25. The matrix can be either a thermoset or thermoplastic. The most common matrix used to manufacture compression molded articles is unsaturated polyester sheet reinforced with glass fibers, known as sheet molding compound (SMC).

The 25 mm long reinforcing fibers are randomly oriented in the plane of the sheet and make up for 20–30 % of the molding compound's volume fraction. A schematic diagram of an SMC production line is depicted in Fig. 7.26 [16]. When producing SMC, the chopped glass fibers are sandwiched between two carrier films previously coated with unsaturated polyester-filler matrix. A fiber reinforced thermoplastic charge is often called a glass mat reinforced thermoplastic (GMT) charge. The most common GMT matrix is polypropylene.

During processing of thermoset charges, the SMC blank is cut from a preformed roll and is placed between heated cavity surfaces. Generally, the mold is charged with 1 to 4 layers of SMC, each layer about 3 mm thick, which initially cover about half the mold cavity's surface. During molding, the initially randomly oriented glass fibers orient, leading to anisotropic properties in the finished product. When processing GMT charges, the preforms are cut and heated between radiative heaters. Once heated, they are placed inside a cooled mold that rapidly closes and squeezes the charges before they cool and solidify.

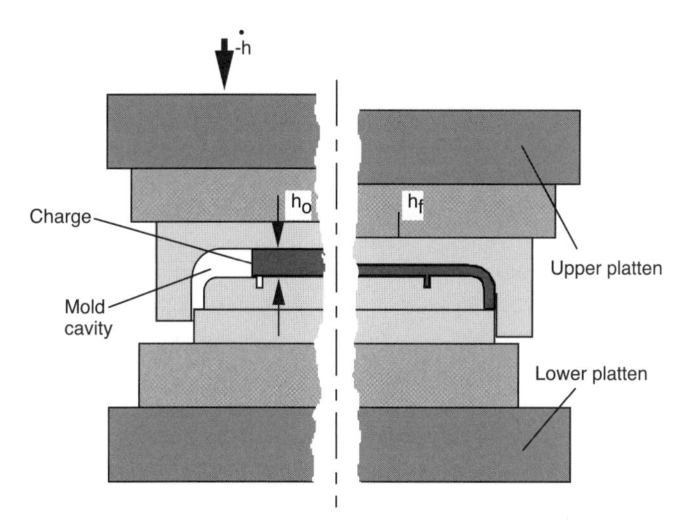

Figure 7.25 Compression molding process (h_0 = charge thickness, h_f = part thickness, and $\dot{h}$ = closing speed)

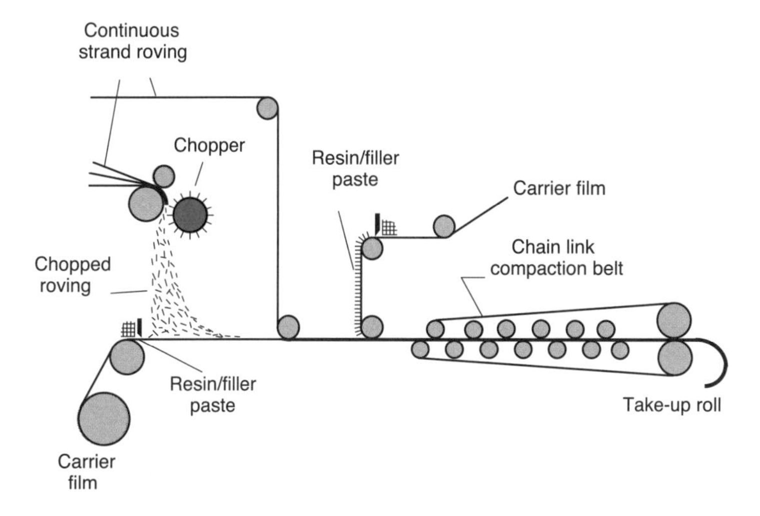

Figure 7.26 SMC production line

One of the main advantages of the compression molding process is the low fiber attrition during processing. Here, relatively long fibers can flow in the melt without the fiber damage commonly seen during plastication and cavity filling in injection molding.

An alternate process is injection-compression molding. Here, a charge is injected through a large gate followed by a compression cycle. The material used in the injection compression molding process is called bulk molding compound (BMC), which is reinforced with shorter fibers, generally 10 mm long, with an unsaturated polyester matrix. The main benefit of injection compression molding over compression molding is automation. The combination of injection and compression molding leads to a lower degrees of fiber orientation and fiber attrition compared to injection molding.

As in any polymer process, in compression molding there is a direct relationship between deformation and final orientation in the part. Figure 7.27 depicts the fiber orientation distribution within a plate where the initial charge coverage was 33 % [17]. Such distribution functions are very common in compression or transfer molding and lead to high degrees of anisotropy throughout a product. To illustrate the effect of fiber orientation on material properties of the final part, Fig. 7.28 [18] shows how the deformation and resulting orientation from 33, 50, 66 and 100 % mold coverage affects the stiffness of the plate.

Similar to injection molding, there are commercially available codes that can be used to predict mold filling, fiber orientation, and warpage of compression molded parts. To predict fiber orientation in realistic parts, the Folgar–Tucker [19] model has been implemented into commercially available compression mold filling simulation programs. The predicted fiber orientation distribution field for a compression molded automotive fender is shown in Fig. 7.29 [20].

To calculate the residual stress development during the manufacturing process and shrinkage and warpage of the finished product, commercially available programs use models where the heat transfer equation is coupled to the stress–strain analysis through constitutive equations. Figure 7.30 compares the mold geometry with the part geometry for the truck fender shown in Fig. 7.29 after mold removal and cooling, computed using numerical models [17].

High degrees of fiber orientation are common in compression molded parts

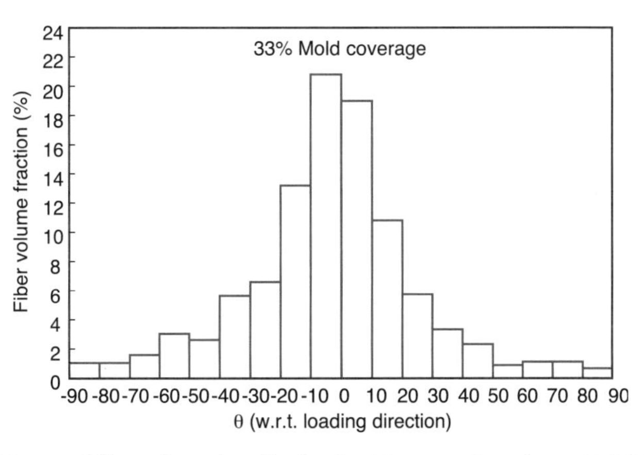

Figure 7.27 Measured fiber orientation distribution histogram in a plate with 33 % initial mold coverage and extensional flow during mold filling

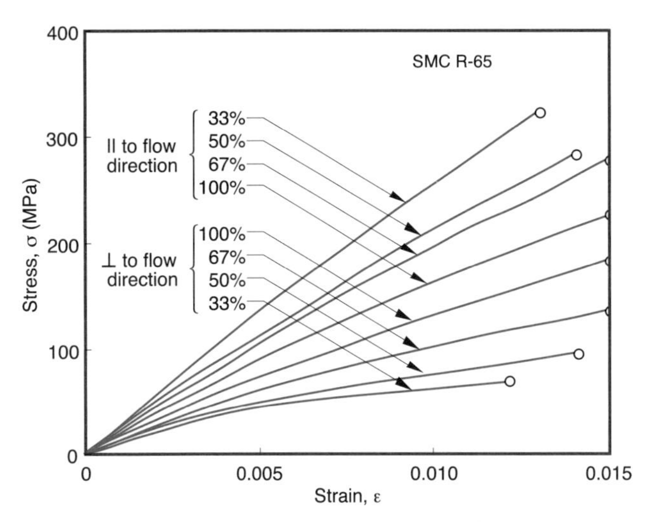

Figure 7.28 Stress–strain curves of 65 % glass by volume SMC for various degrees of deformation

Fiber orientation results in parts with large variations in thermomechanical properties

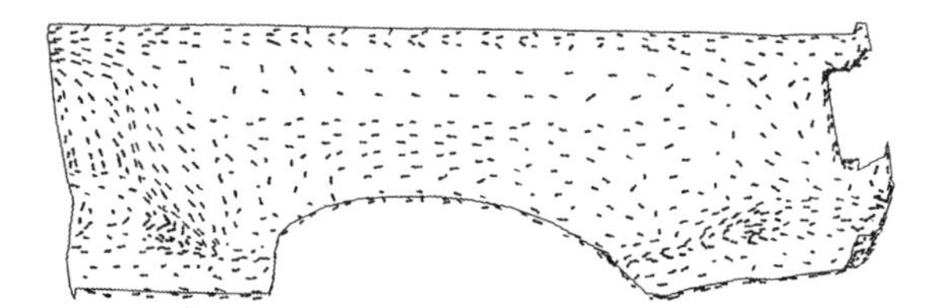

Figure 7.29 Fiber orientation distribution in a compression molded automotive fender

7.9 Foaming

Foamed polymers exhibit a cellular or porous structure that has been generated through the addition and reaction of *physical* or *chemical blowing agents*. The basic steps of foaming are cell nucleation, expansion or cell growth, and cell stabilization. Nucleation occurs when, at a given temperature and pressure, the solubility of a gas is reduced, leading to saturation, expelling the excess gas to form bubbles. Nucleating agents, such as powdered metal oxides, are used for initial bubble formation. The bubbles reach an equilibrium shape when their inside pressure balances their surface tension and surrounding pressures. The cells formed can be completely enclosed (closed cell) or can be interconnected (open cell).

A physical foaming process is one where a gas, such as nitrogen or carbon dioxide, is introduced into the polymer melt. Physical foaming also occurs after heating a melt that contains a low boiling point fluid, causing it to vaporize. For example, the heat-induced volatilization of low-boiling point liquids, such as pentane and heptane, is used to produce polystyrene foams. Also, foaming occurs during volatilization from the

Most of the warpage in a compression molded part comes from the anisotropy that results from fiber orientation. Poor thermal mold layout can also be a major source of warpage

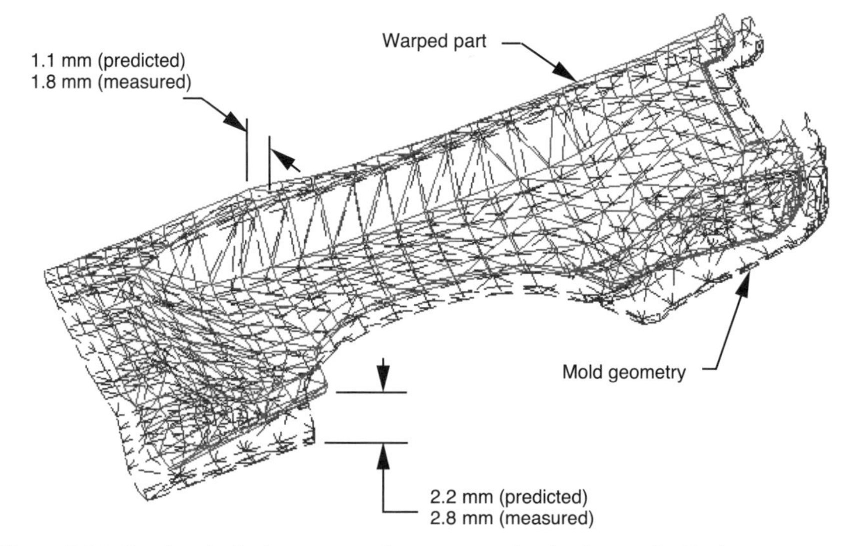

Figure 7.30 Simulated displacements of an automotive body panel. Displacements were magnified by a factor of 20

exothermic reaction of gases produced during polymerization, such as the production of carbon dioxide during the reaction of isocyanate with water. Physical blowing agents are added to the plasticating zone of the extruder or molding machine. The most widely used physical blowing agent is nitrogen. Liquid blowing agents are often added to the polymer in the hopper of the plasticating unit.

Chemical blowing agents are usually powders introduced in the hopper of the molding machine or extruder. Chemical foaming occurs when the blowing agent thermally decomposes, releasing large amounts of gas. The most widely used chemical polyolefin blowing agent is azodicarbonamide.

In mechanical foaming, a gas dissolved in a polymer expands upon reduction of the processing pressure.

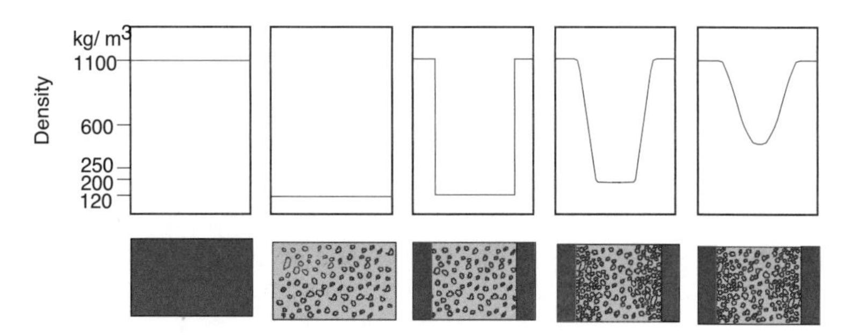

Figure 7.31 Schematic of various foam structures

The foamed structures commonly generated are either homogeneous foams or integral foams. Figure 7.31 [21] presents the various types of foams and their corresponding characteristic density distributions. In integral foam, the unfoamed skin surrounds the foamed inner core. This type of foam can be achieved during injection molding and extrusion and it replaces the sandwiched structure also shown in Fig. 7.31.

7.10 Rotational Molding

Rotational molding is used to make hollow objects. In rotational molding, a carefully measured amount of powdered polymer, typically polyethylene, is placed in a mold. The mold is then closed and placed in an oven, where the mold turns about two axes as the polymer melts, as depicted in Fig. 7.32. During heating and melting, which occur at oven temperatures between 250 and 450 °C, the polymer is deposited evenly on the mold's surface. To ensure uniform thickness, the axes of rotation should not coincide with the centroid of the molded product. The mold is then cooled and the part is removed from the mold cavity. The parts can be as thick as 1 cm, and still be manufactured with relatively low residual stresses. The reduced residual stress and the controlled dimensional stability of the rotational molded product depend in great part on the cooling rate after the mold is removed from the oven. A mold that is cooled too fast yields warped parts. Usually, a mold is first cooled with air to start the cooling slowly, followed by a water spray for faster cooling.

The main advantages of rotational molding over blow molding are the uniform part thickness and the low cost involved in manufacturing the mold. In addition, large parts such as play structures or kayaks, can be manufactured more economically than with

Large parts, such as children slides and kayaks, can be rotational molded at much lower cost than with injection molding or blow molding

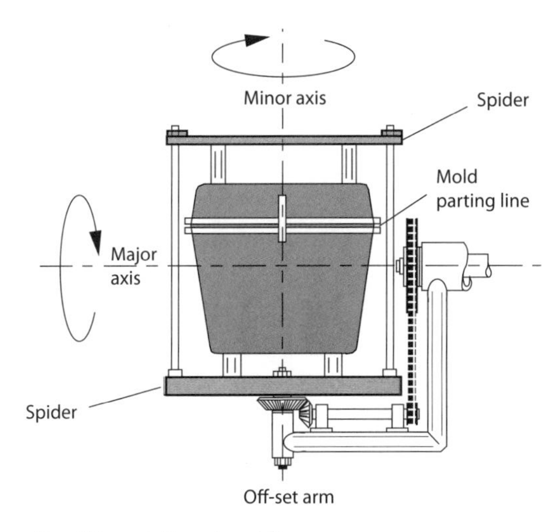

Figure 7.32 Schematic of the rotational molding process

injection molding or blow molding. The main disadvantage of the process is the long cycle time for heating and cooling of the mold and polymer.

Figure 7.33 presents the air temperature inside the mold in a typical rotational molding cycle for polyethylene powders [22]. The process can be divided into six distinct phases:

(1) Induction or initial air temperature rise

(2) Melting and sintering

(3) Bubble removal and densification

(4) Pre-cooling

(5) Crystallization of the polymer melt

(6) Final cooling

The induction time can be significantly reduced by pre-heating the powder, and the bubble removal and cooling stage can be shortened by pressurizing the material inside the mold. The melting and sintering of the powder during rotational molding depends on the rheology and geometry of the particles. This phenomenon was studied in depth by Bellehumeur and Vlachopoulos [23].

The cycle time of the rotational molding process is governed by the time it takes to heat and cool the mold and the polymer. Because the mass of the polymer is negligible

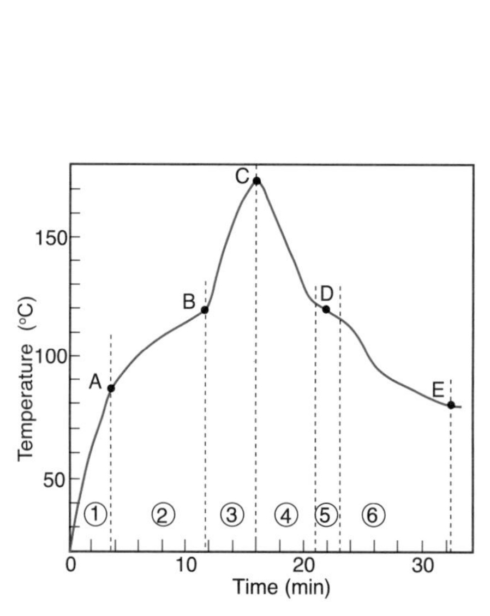

Figure 7.33 Temperature history inside the mold in a rotational molding process

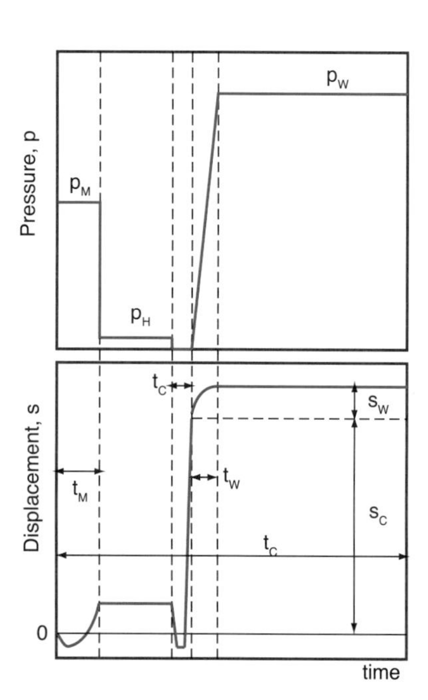

Figure 7.34 Process steps of heated tool butt welding. Subscripts: M = mating, C = change-over, W = welding and C = cooling

compared to the mass of the mold, the temperature can be estimated using

$$\left(\frac{T_f - T}{T_f - T_0}\right) = e^{-\frac{hA}{mC_p}t} \quad (7.13)$$

where T is temperature, T_f is the oven temperature (or cooling medium during cooling), T_0 is the initial temperature, h is the convective heat transfer coefficient, A is the outside surface area of the mold, C_p is the specific heat of the mold material, m is the mass of the mold and t is time.

7.11 Welding

Joining or assembly of plastics and polymeric composites is an important step in manufacturing of parts from these materials. Joining methods for plastics can be divided into three major groups:

- Mechanical joining,
- Adhesive bonding, and
- Welding (also called fusion bonding)

In this section we will concentrate on welding. Welding of plastics is used to join large (e.g., vessels, tanks, and pipelines) and small (e.g. lighters, electronic dip switches, and cameras) parts. It is used to join simple and complex structures and in batch and mass production. It is used in numerous industries, including automotive, aerospace, toy, medical, electronic, and infrastructure. Examining most welding processes, one can identify five distinct steps that make up these processes. For some welding processes, these steps are sequential, while for other processes some of the steps may occur simultaneously. The basic welding steps are:

- surface preparation
- heating
- pressing
- intermolecular diffusion
- cooling

Figure 7.34 present the pressure and displacement during a heated tool butt welding process (Fig. 7.35). While butt welding is perhaps the better known plastics welding process, it is only one of many. Welding processes are often categorized and identified by the heating method that is used. All processes can be divided into two general categories, depending on whether internal heating or external heating is employed. Internal heating methods are further divided into two categories, internal mechanical heating and internal electromagnetic heating. Internal mechanical heating methods rely on the conversion of mechanical energy into heat through surface friction and intermolecular friction. These processes include: ultrasonic, vibration, and spin welding. Internal

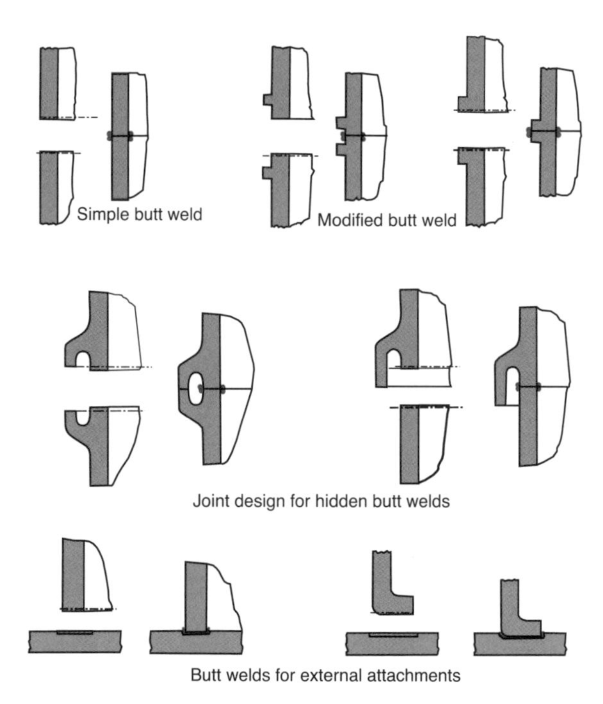

Figure 7.35 Variations of the hot tool butt welding process

electromagnetic heating methods rely on the absorption and conversion of electromagnetic radiation into heat. These processes include: infrared/laser, radio frequency, and microwave welding. External heating methods rely on convection and/or conduction to heat the weld surface. These processes include: hot tool, hot gas, extrusion, implant induction, and implant resistance welding.

References

1. Osswald, T. A., and J. P. Hernández, *Polymer Processing — Modeling and Simulation*, Hanser Publishers (2006), Munich
2. Anturkar, N. R., and A. Co, *J. Non-Newtonian Fluid Mech.* (1988), *28*, 287
3. Menges, G., *Einführung in die Kunststoffverarbeitung*, Hanser Publishers (1986), Munich
4. Pearson, J. R. A., and C. J. S. Petrie, *Plastics & Polymers* (1970), *38*, 85
5. Lee, N. C., *Blow Molding Design Guide*, Hanser Publishers (2008), Munich
6. Modern Plastics Encyclopedia, 53, McGraw-Hill (1976), New York
7. Throne, J. L., *Understanding Thermoforming*, Hanser Publishers (2008), Munich
8. Osswald, T. A., E. Baur, S. Brinkmann and E. Schmachtenberg, *International Plastics Handbook*, Hanser Pubishers (2006), Munich

9. Middleman, S., *Fundamentals of Polymer Processing*, McGraw-Hill Book Company (1977), New York
10. Brazinsky, I., H. F. Cosway, C. F. Valle, Jr., R. C. Jones, and V. Story, *J. Applied Polym. Sci.* (1970), *14*, 2771
11. Tadmor, Z., and R. B. Bird, *Polym.Eng.Sci.* (1973), *14*, 124
12. Osswald, T. A., and G. Menges, *Materials Science of Polymers for Engineers*, Hanser Publishers (2003), Munich
13. Palmese, G. R., O. Andersen, and V. M. Karbhari, *Advanced Composites X: Proceedings of the 10th Annual ASM/ESD Advance Composites Conference*, Dearborn, MI, ASM International (1994), Material Park
14. Kamal, M. R., S. Sourour, *Polym. Eng. Sci.* (1973), *13*, 59
15. Macosko, C. W., *RIM Fundamentals of Reaction Injection Molding*, Hanser Publishers (1989), Munich
16. Davis, B. A., P. J. Gramann, T. A. Osswald, and A. C. Rios, *Compression Molding*, Hanser Publishers (2003), Munich
17. Lee, C.-C., F. Folgar, and C. L. Tucker III, *J. Eng. Ind.* (1984), 186
18. Jackson, W. C., S. G. Advani, and C. L. Tucker III, *J. Comp. Mat.* (1986), *20*, 539
19. Folgar, F. and C. L. Tucker, *J. Reinf. Plast. Comp.* (1984), *3*,98
20. Gramann, P. J., E. M. Sun, and T. A. Osswald, *SPE 52nd Antec*, (1994)
21. Shutov, F. A., *Integral/Structural Polymer Foams*, Springer-Verlag (1986), Berlin
22. Crawford, R. J., and J. L. Throne, *Rotational Molding Technology*, Plastics Design Library (2001), Norwich
23. Bellehumeur, C. T., and J. Vlachopoulos, *SPE 56th Antec*, (1998)

Part III
Modeling

8 Transport Phenomena in Polymer Processing

The field of transport phenomena is the basis of modeling in polymer processing. Modeling often begins with a dimensional analysis of the system, which provides insight into the meaningful parameters that govern that system or process. The resulting dimensionless groups or numbers, in conjunction with experiments and models, can help the engineer scale a pilot or model process to industrial dimensions. This chapter presents a clear matrix technique to perform the dimensional analysis, as well as the derivation of the balance equations and combines them with constitutive models to allow modeling of polymer processes. The chapter also presents ways to simplify the complex equations in order to model basic systems, such as flow in a tube or Hagen-Poiseuille flow, pressure flow between parallel plates, flow between two rotating concentric cylinders or Couette flow, and many more. These simple systems, or their combinations, can be used to model actual systems in order to gain insight into the processes and predict pressures, flow rates, rates of deformation, and much more.

8.1 Dimensional Analysis and Scaling

Dimensional analysis is used by engineers to gain insight into a problem by allowing presentation of theoretical and experimental results in a compact manner. This is done by reducing the number of variables in a system by lumping them into meaningful dimensionless numbers. For example, if a flow system is dominated by the fluid's inertia as well as the viscous effects, it may best to present the results, i. e., pressure requirements, in terms of the Reynolds number, which is the ratio of both effects. As one checks the order of these dimensionless numbers and compares them to one another, one can gain insight into what parameters, such as process conditions and material properties, are most important. Many researches also use dimensional analysis in theoretical studies. Often, dimensional analysis, in combination with experiments, results in fundamental relations that govern a process. In polymer processing, as well as other manufacturing techniques or operations, one often works on a laboratory scale when developing new processes or materials, and when testing and optimizing a certain system. This laboratory operation, often referred to as a *pilot* plant, is a physical model of the actual or final system. The transition from this laboratory model, which probably produces only a few cubic centimeters of material per hour, to the actual production process, which can generate hundreds of kilograms per hour, is what is called *scale-up*. On some occasions, such as when trying to push the envelope in injection molding, where the thickness of the part is always being reduced, the term *scale-down* is also used. Because the methods mentioned in this chapter work for both, here we will simply call them scaling.

8.1.1 Dimensional Analysis

The classic technique to determine dimensionless numbers using the so-called Buckingham Pi-Theorem is cumbersome to use in cases where the list of related physical quantities becomes large. Pawlowski [1] developed a matrix transformation technique that offers a systematic approach to the generation of P-sets. To demonstrate Pawlowski's matrix transformation technique, an example will be used. Here, we will consider a forced convection problem: the cooling of an extrudate or of the blown film, where a fluid with a viscosity η, a density ϱ, a specific heat C_p and a thermal conductivity k, is forced past a surface with a characteristic size D at an average speed u. The temperature difference between the fluid and the surface is described by $\Delta T = T_f - T_s$ and the resulting heat transfer coefficient is defined by h. Again, the first step is to generate the relevance list. Here, the relevant list of physical quantities is:

- Geometric variable: D
- Process variables: u and ΔT
- Physical or material properties: η, ϱ, C_p, and k
- Target quantity: h

The first step in generating the dimensionless variables is to set-up a dimensional matrix with the physical quantities and their respective units,

$$\begin{array}{c|cccccccc} & D & u & \Delta T & \eta & \varrho & C_p & k & h \\ M & 0 & 0 & 0 & 1 & 1 & 0 & 1 & 1 \\ L & 1 & 1 & 0 & -1 & -3 & 2 & 1 & 0 \\ T & 0 & -1 & 0 & -1 & 0 & -2 & -3 & -3 \\ \Theta & 0 & 0 & 1 & 0 & 0 & -1 & -1 & -1 \end{array} \quad (8.1)$$

where, M, L, T and Q represent units of mass, length, time and temperature, respectively. The above dimensional matrix must be rearranged and divided into two parts, a square core matrix, which contains the dimensions pertaining to the repeating variables, and a residual matrix. Using the rules given in the previous section, the repeating variables are D, η, ϱ and C_p and the dimensional matrix can be written as

The dimensional matrix is composed of a core matrix and a residual matrix

$$\begin{array}{c|cccc|cccc} & \eta & D & \varrho & C_p & k & u & \Delta T & h \\ M & 1 & 0 & 1 & 0 & 1 & 0 & 0 & 1 \\ L & -1 & 1 & -3 & 2 & 1 & 1 & 0 & 0 \\ T & -1 & 0 & 0 & -2 & -3 & -1 & 0 & -3 \\ \Theta & 0 & 0 & 0 & -1 & -1 & 0 & 1 & -1 \\ & \multicolumn{4}{c|}{\text{Core Matrix}} & \multicolumn{4}{c}{\text{Residual Matrix}} \end{array} \quad (8.2)$$

The next step is to transform the core matrix into a unity matrix. Hence, the order of the physical variables in the core matrix should be such that a minimum amount of linear transformations is required. Adding the M row to the L and T rows eliminates the non-zero term below the diagonal in the core matrix, i. e.,

$$\begin{array}{c|cccc|cccc} & \eta & D & \varrho & C_p & k & u & \Delta T & h \\ M & 1 & 0 & 1 & 0 & 1 & 0 & 0 & 1 \\ L+M & 0 & 1 & -2 & 2 & 2 & 1 & 0 & 1 \\ T+M & 0 & 0 & 1 & -2 & -2 & -1 & 0 & -2 \\ \Theta & 0 & 0 & 0 & -1 & -1 & 0 & 1 & -1 \\ & \multicolumn{4}{c}{\text{Core Matrix}} & \multicolumn{4}{c}{\text{Residual Matrix}} \end{array} \tag{8.3}$$

Performing the same operation with the upper portion of the core and residual matrices, for example, multiplying the $T + M$ row by 2 and add it to the $L + M$ row above, leads to a unity core matrix,

Reducing the dimensional matrix until the core matrix is a unit matrix. The coefficients of the dimensionless groups are stored within the residual matrix

$$\begin{array}{c|cccc|cccc} & \eta & D & \varrho & C_p & k & u & \Delta T & h \\ -T+2\Theta & 1 & 0 & 0 & 0 & 1 & 1 & 2 & 1 \\ L+3M+2T-2\Theta & 0 & 1 & 0 & 0 & 0 & -1 & -2 & -1 \\ T+M-2\Theta & 0 & 0 & 1 & 0 & 0 & -1 & -2 & 0 \\ -\Theta & 0 & 0 & 0 & 1 & 1 & 0 & -1 & 1 \\ & \multicolumn{4}{c}{\text{Core Matrix}} & \multicolumn{4}{c}{\text{Residual Matrix}} \end{array} \tag{8.4}$$

With the above matrix set, the dimensionless numbers can be generated, in this case, 4 dimensionless groups, by placing the physical quantities in the residual matrix in the numerator and the quantities in the core matrix in the denominator with the coefficients in the residual matrix as their exponent. Hence,

$$\begin{aligned} \Pi_1 &= \frac{k}{\eta^1 D^0 \varrho^0 C_p^1} = \frac{k}{\eta C_p} \\ \Pi_2 &= \frac{u}{\eta^1 D^{-1} \varrho^{-1} C_p^0} = \frac{uD\varrho}{\eta} \\ \Pi_3 &= \frac{\Delta T}{\eta^2 D^{-2} \varrho^{-2} C_p^{-1}} = \frac{\Delta T D^2 \varrho^2 C_p}{\eta^2} \\ \Pi_4 &= \frac{h}{\eta^1 D^{-1} \varrho^0 C_p^1} = \frac{hD}{\eta C_p} \end{aligned} \tag{8.5}$$

Taking the set of dimensionless numbers we can generate numbers that are more meaningful. For example, one such group of dimensionless numbers that can be generated is

The Brinckman number reveals the significance of viscous heating. Viscous heating is very important in polymer processing

$$
\begin{aligned}
Nu &= \frac{k}{\eta C_\mathrm{p}} = \frac{\Pi_4}{\Pi_1} && \text{Nusselt number} \\
Re &= \frac{\varrho D u}{\eta} = \Pi_2 && \text{Reynolds number} \\
Br &= \frac{\eta u^2}{k \Delta T} = \frac{\Pi_2^2}{\Pi_1 \Pi_3} && \text{Brinckman number} \\
Pr &= \frac{\eta C_\mathrm{p}}{k} = \frac{1}{\Pi_1} && \text{Prandtl number}
\end{aligned}
\tag{8.6}
$$

Example 8.1 Flow in a tube

Consider the classical problem of pressure drop during flow in a smooth straight pipe, ignoring the inlet effects. The first step is to list all possible variables or quantities that are related to the problem under consideration. In this case, we have:

- Target quantity: Pressure drop Δp
- Geometric variables: Pipe diameter D, and pipe length L
- Physical or material properties: the viscosity η, and the density ϱ of the fluid
- Process variable: average fluid velocity u

If we choose D, u and ϱ as the repeating variables, the dimensional matrix is written as,

$$
\begin{array}{ccccccc}
 & D & u & \varrho & \Delta p & L & \eta \\
M & 0 & 0 & 1 & 1 & 0 & 1 \\
L & 1 & 1 & -3 & -1 & 1 & -1 \\
T & 0 & -1 & 0 & -2 & 0 & -1
\end{array}
\tag{8.7}
$$

After reducing the core matrix to an identity matrix, the dimensional matrix becomes

$$
\begin{array}{ccccccc}
 & D & u & \varrho & \Delta p & L & \eta \\
M & 1 & 0 & 0 & 0 & 1 & 1 \\
L & 0 & 1 & 0 & 2 & 0 & 1 \\
T & 0 & 0 & 1 & 1 & 0 & 1
\end{array}
\tag{8.8}
$$

which results in the 3 dimensionless groups,

$$
\begin{aligned}
\Pi_1 &= \frac{\Delta p}{u^2 \varrho} = Eu \text{ (Euler number)} \\
\Pi_2 &= \frac{L}{D} \\
\Pi_3 &= \frac{\eta}{D u \varrho} = Re^{-1} \text{ (Reynolds number)}
\end{aligned}
\tag{8.9}
$$

The following relationship can be written

$$
f\left(Eu, Re, \frac{L}{D}\right) = 0
\tag{8.10}
$$

Tube flow – laminar and turbulent ranges

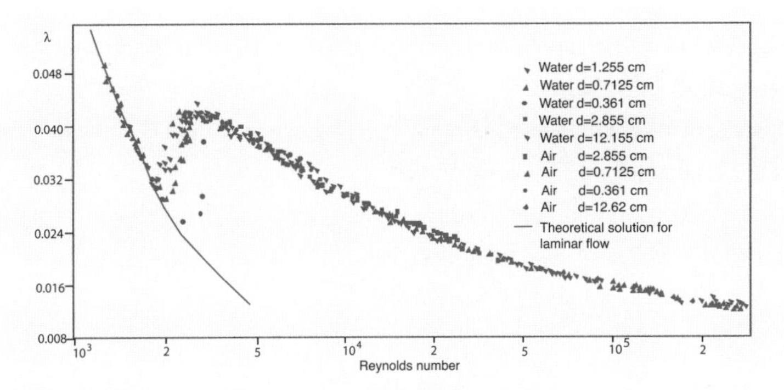

Figure 8.1 Pressure drop characteristic of a straight smooth tube

which, of course, by itself cannot produce the nature of the relation; however, the form of the function f can be generated experimentally. Figure 8.1 presents results from such experiments performed by Stanton and Pannell [2, 3], where they plot $\lambda = 2Eu\,D/L$ as a function of Re. This figure clearly demonstrates the usefulness of dimensional analysis. In the figure note the line that denotes laminar flow, which is represented by pressure flow in a tube, derived in this chapter.

■

Example 8.2 **Mixing time of two compatible fluids with the same density, viscosity, and diffusivity**

During mixing operations, it is often important to know when the blend can be considered homogeneous. In this example, consider t the time it takes for two compatible fluids of similar density and viscosity to be molecularly homogeneous [4]. Figure 8.2 depicts the

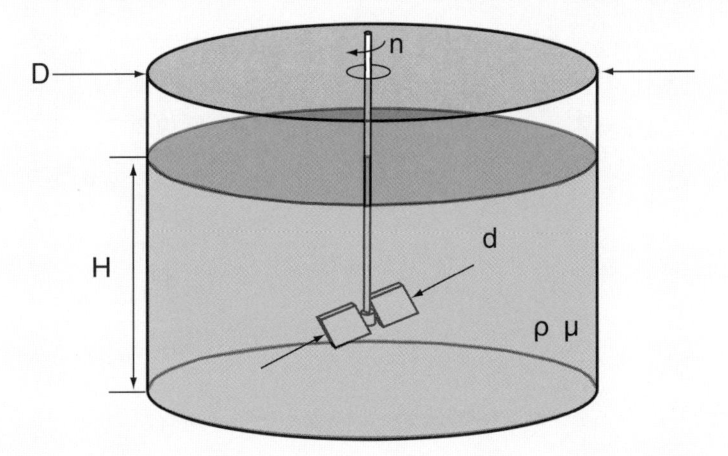

Figure 8.2 Schematic diagram of a stirring tank

set-up for this mixing operation. Here, the relevant parameters can be defined as:

- Target quantity: mixing time t
- Process variables: rotational speed of the stirrer n, a tank with or without baffles
- Geometric variables: stirrer diameter d
- Physical or material properties: density ϱ, diffusivity D and kinematic viscosity ν

The corresponding dimensional matrix can be written as

$$\begin{array}{c|cccccc} & \varrho & d & D & t & n & \nu \\ M & 1 & 0 & 0 & 0 & 0 & 0 \\ L & -3 & 1 & 2 & 0 & 0 & 2 \\ t & 0 & 0 & -1 & 1 & -1 & -1 \end{array} \tag{8.11}$$

From the dimensional matrix, it is clear that the mass unit appears only in the density term. Hence, density must be eliminated from the list along with the row corresponding to the mass unit, leaving a system with only 2 repeating parameters. In fact, the density is fully accounted for in the dynamic viscosity. Choosing ν and n as the repeating parameters results in

$$\begin{array}{c|ccccc} & \nu & n & d & D & t \\ L & 2 & 0 & 1 & 2 & 0 \\ T & -1 & -1 & 0 & -1 & 1 \end{array} \tag{8.12}$$

The matrix transformation results in

$$\begin{array}{c|ccccc} & \nu & n & d & D & t \\ L & 1 & 0 & 1/2 & 1 & 0 \\ T & 0 & 1 & -1/2 & 0 & -1 \end{array} \tag{8.13}$$

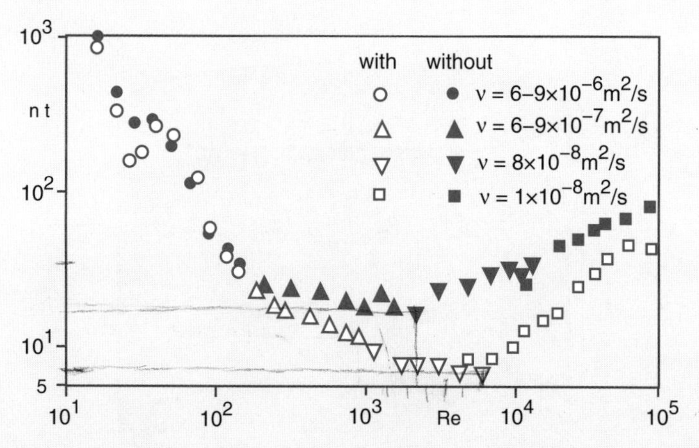

Figure 8.3 Dimensionless mixing time inside a stirring tank with and without baffles as a function of Reynolds number

Here, $\Pi_1 = dn^{1/2}/\nu^{1/2}$, $\Pi_2 = D/\nu$ and $\Pi_3 = nt$ are the resulting dimensionless groups. These three dimensionless numbers reduce to

$$Re = \frac{d^2 n}{\nu} = \Pi_1^2 = \text{Reynolds number}$$

$$Sc = \frac{\nu}{D} = \Pi_2^{-1} = \text{Schmidt number} \tag{8.14}$$

$$\tau = nt = \Pi_3 = \text{Dimensionless mixing time}$$

The plot presented in Fig. 8.3 shows how the Reynolds number plays an important role on mixing time. The graph shows two sets of points, one for a mixing tank with baffles and the other for a mixing tank without baffles.

Schmidt number is the ratio of kinetic mixing to diffusion mixing – it is typically very large, in other words, diffusion is very slow when compared to mechanical mixing

Example 8.3 Single screw extruder operating curves

The conveying characteristics of a single screw extruder can also be analyzed by use of dimensional analysis. Pawlowski [5, 6] used dimensional analysis and extensive experimental work to fully characterize the conveying and heat transfer characteristics of single screw extruders, schematically depicted in Fig. 8.4. The relevant physical quantities that

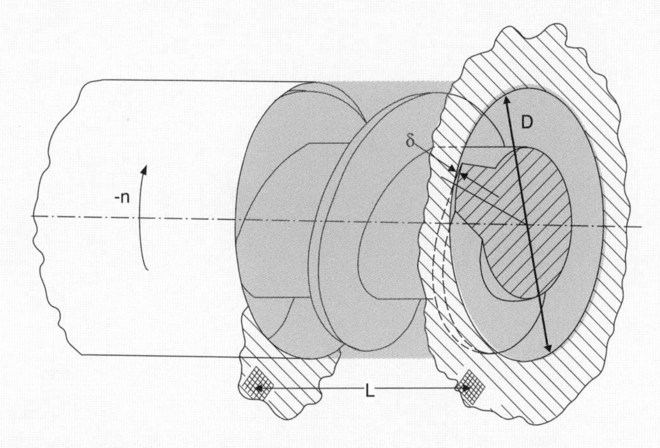

Figure 8.4 Schematic diagram of a screw geometry

may be considered when characterizing a single screw extruder are:

- Target quantities: power consumption P, axial screw force F, pumping pressure Δp, temperature of the extrudate expressed in temperature difference $\Delta T = T - T_0$, and volumetric throughput Q
- Process variables: processing or heater temperature expressed in temperature difference $\Delta T_p = T_h - T_0$, and screw speed n
- Geometric variables: screw or inner barrel diameter D, axial screw length L, channel depth h, and clearance between the screw flight and the barrel δ

- Physical or material quantities: thermal conductivity k, density ϱ, specific heat C_p, viscosity $\eta_0 = \eta(T_0)$ and viscosity temperature dependence a, from $\eta = \eta_0 e^{-a\Delta T}$

With this list of relevant parameters a dimensional matrix can be set up. Choosing η_0, D, n, and ΔT as the repeating parameters, the following dimensional matrix is set

$$\begin{array}{c|cccc|cccccccccccc} & \eta_0 & D & n & \Delta T_p & P & F & \Delta p & Q & C_p & \varrho & k & a & h & L & \delta & \Delta T \\ M & 1 & 0 & 0 & 0 & 1 & 1 & 1 & 0 & 0 & 1 & 1 & 0 & 0 & 0 & 0 & 0 \\ L & -1 & 1 & 0 & 0 & 2 & 1 & -1 & 3 & 2 & -3 & 1 & 0 & 1 & 1 & 1 & 0 \\ T & -1 & 0 & -1 & 0 & -3 & -2 & -2 & -1 & -2 & 0 & -3 & 0 & 0 & 0 & 0 & 0 \\ \Theta & 0 & 0 & 0 & 1 & 0 & 0 & 0 & 0 & -1 & 0 & -1 & -1 & 0 & 0 & 0 & 1 \end{array} \tag{8.15}$$

which can be reduced to the following dimensionless groups

$$\begin{aligned} \Pi_1 &= \frac{P}{\eta_0 D^3 n^2} \quad \Pi_2 = \frac{F}{\eta_0 D^2 n} \quad \Pi_3 = \frac{\Delta p}{\eta_0 n} \\ \Pi_4 &= \frac{Q}{D^3 n} \quad \Pi_5 = \frac{C_p \Delta T}{D^2 n^2} \quad \Pi_6 = \frac{\varrho D^2 n}{\eta_0} \\ \Pi_7 &= \frac{k \Delta T}{\eta_0 D^2 n^2} \quad \Pi_8 = a\Delta T \quad \Pi_9 = \frac{h}{D} \\ \Pi_{10} &= \frac{L}{D} \quad \Pi_{11} = \frac{\delta}{D} \quad \Pi_{12} = \frac{\Delta T}{\Delta T_p} \end{aligned} \tag{8.16}$$

The first three and the last dimensionless groups are extruder operation characteristic values for power consumption, axial screw force, pumping pressure, and the extrudate temperature, respectively, and depend on the process, material, and geometry dimensionless groups. Π_4

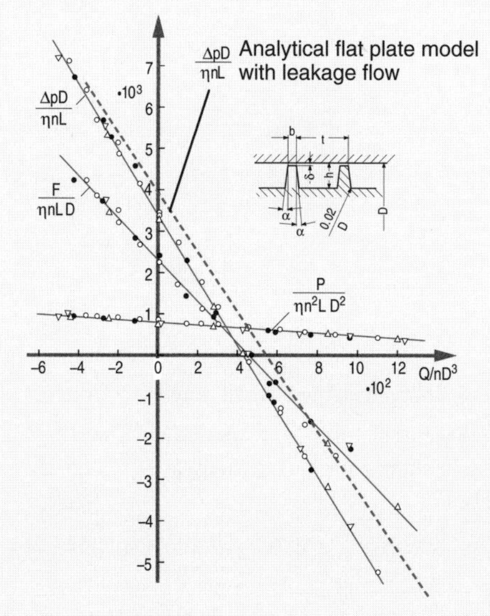

Figure 8.5 Throughput, power, and axial force characteristic curves for a single screw extruder

is a dimensionless volumetric throughput and Π_6 the Reynolds number related to the rotational speed of the screw. Π_9 through Π_{11} are geometry dependent dimensionless groups. The remaining are

$$\begin{aligned} Br &= \frac{\eta_0 D^2 n^2}{k \Delta T} = \frac{1}{\Pi_7} = \text{Brinkman number} \\ Pr &= \frac{C_p \eta_0}{k} = \frac{\Pi_5}{\Pi_7} = \text{Prandtl number} \\ Na &= a \Delta T Br = \frac{\Pi_8}{\Pi_7} = \text{Nahme-Griffith number} \end{aligned} \tag{8.17}$$

The Nahme-Griffith number is a measure of the effect viscous heating has on the viscosity of the polymer

The following relation between the dimensionless extruder operation curves and the other dimensionless groups can be expressed

$$\frac{P}{\eta_0 D^3 n^2}, \frac{F}{\eta_0 D^2 n}, \frac{\Delta p}{\eta_0 n}, \frac{\Delta T}{\Delta T_p} = f\left(\frac{Q}{D^3 n}, Re, Br, Pr, Na, \frac{h}{D}, \frac{L}{D}, \frac{\delta}{D}\right) \tag{8.18}$$

The above equation can be simplified assuming a Newtonian isothermal problem. For such a case, Pawlowski reduced the above equations to a set of characteristic functions that describe the conveying properties of a single screw extruder under isothermal and creeping flow ($Re \approx 0$) assumptions. These relationships are illustrated in the experimental measurements performed by Pawlowski [5, 6] and presented in Figs. 8.5 and 8.6. Figure 8.5 also presents an analytical solution for the screw characteristic curve of a single screw extruder with leakage flow effects. The discrepancies between analytical solution and experimental results arise due to the fact that the screw curvature, the flight angle, and the fillet radii are not included in the analytical model. The analytical solution given by Tadmor and Klein [7] was used.

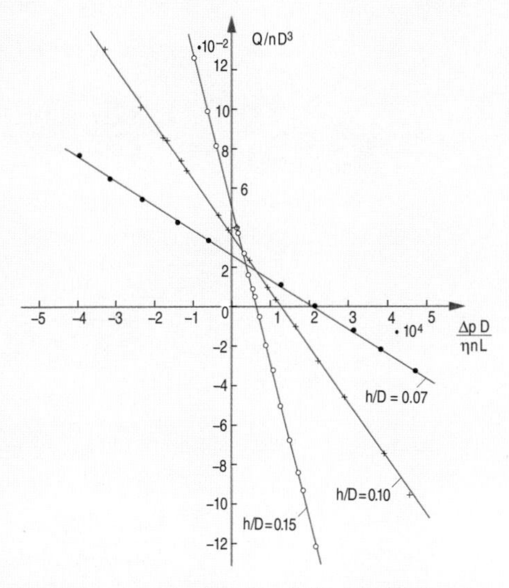

Figure 8.6 Screw characteristic curves as a function of channel depth

Table 8.1 lists several dimensionless numbers useful in polymer processing.

Table 8.1 Dimensionless numbers in polymer processing

Name	Symbol	Definition	Meaning
Biot	Bi	$\frac{h\,L}{k}$	$\frac{\text{Convection from surface}}{\text{Conduction through body}}$
Brinkman	Br	$\frac{\eta\,u^2}{k\Delta T}$	$\frac{\text{Viscous heating}}{\text{Conduction}}$
Capillary	Ca	$\frac{\tau\,R}{\sigma_s}$	$\frac{\text{Deviatoric stresses}}{\text{Surface tension stresses}}$
Damköhler	Da	$\frac{c\,\Delta H_r}{\varrho\,C_p\,T_0}$	$\frac{\text{Reaction energy}}{\text{Internal energy}}$
Deborah	De	$\frac{\lambda}{t}$	$\frac{\text{Relaxation time}}{\text{Process time}}$
Fourier	Fo	$\frac{\alpha t}{L^2}$	$\frac{\text{Process time}}{\text{Thermal diffusion time}}$
Graetz	Gz	$\frac{u\,L}{\alpha}\left(\frac{d}{L}\right)$	$\frac{\text{Lengthwise convection}}{\text{Transverse conduction}}$
Manas-Zloczower	Mz	$\frac{\dot{\gamma}}{\dot{\gamma}+\omega}$	$Mz = 0.0$ — No deformation $Mz = 0.5$ — Shear flow $Mz = 1.0$ — Elongational flow
Nahme-Griffith	Na	$a\,\Delta T\,Br$	Effect of viscous heating on flow field
Nusselt	Nu	$\frac{h\,L}{k_{\text{fluid}}}$	$\frac{\text{Convective heat transfer}}{\text{Conductive heat transfer}}$
Péclet	Pe	$\frac{U\,L}{\alpha}$	$\frac{\text{Rate of advection heat transfer}}{\text{Rate of diffusion heat transfer}}$
Prandtl	Pr	$\frac{\nu}{\alpha}$	$\frac{\text{Momentum diffussivity}}{\text{Thermal diffussivity}}$
Reynolds	Re	$\frac{\varrho\,u\,L}{\eta}$	$\frac{\text{Inertia forces}}{\text{Viscous forces}}$
Schmidt	Sc	$\frac{\nu}{D}$	$\frac{\text{Mechanical mixing}}{\text{Diffusion mixing}}$
Weissenberg	We	$\lambda\dot{\gamma}$ or $\frac{N_1}{\tau}$	$\frac{\text{Elastic stresses}}{\text{Viscous stresses}}$

8.1.2 Scaling and Similarity

When designing a new polymer processing operation to produce a product or to blend or compound a new material, it is often desirable or necessary to work on a smaller scale, such as a laboratory extruder, internal batch mixer, stirring tank, etc. The evolving model must then be scaled up or down to the actual operation. When scaling a process, similarity between the various sizes and processes is sought. As a rule, a perfectly scalable prototype is one that is perfectly similar to its scaled system. A perfectly similar system is one where all the dimensionless numbers or Π-groups produced, have the same numerical value when scaling the system up or down. A rather simple system, for which this can be easily demonstrated, is the smooth pipe pressure drop experiments presented in the previous section. Here, the same dimensionless pressure drop, λ, versus Reynolds number curve was developed using pipes whose diameter varied from 3.61 mm to 126.2 mm and with viscosities of water and air that differ from each other by a factor of 100. Hence, if the system is a smooth pipe and the diameter were increased, the velocity determined would render the same Reynolds number and would then adjust the L/D to render the same λ. For example, doubling the diameter of the original system that had a Reynolds number of 5 and therefore a $\lambda = 2.1$, leads to a reduction of the speed by half, and an increase in the length of the pipe by a factor of 8. This results in a perfectly similar system. In a similar way, scaling between two types of fluids with different viscosities is also possible. In such a case, the speed or the diameter as well as the length of the pipe must be adjusted to obtain constant dimensionless numbers and therefore perfectly similar systems. Of course, not all systems are as straight forward as the smooth pipe flow system.

A perfectly scaled or similar system is one where the dimensionless groups remain the same

In most cases, scale-up by similarity is not always fully achieved. A process may be geometrically similar, but not thermally similar. Depending on the type of process involved, one or several kinds of similarities may be required. These may be geometric, kinematic, dynamic, thermal, kinetic, or chemical similarities. However, in one or more of these a compromise must be made.

Example 8.4 Single screw extruder

Consider the case of a single screw extruder section that works well when dispersing a liquid additive within a polymer matrix. Let use the single screw extruder discussed in the previous section. However, the effect of surface tension, which is important in dispersive mixing, was not included in that analysis. Hence, if we also add surface tension as a relevant physical quantity, it would add one more column to the dimensional matrix. To find the additional dimensionless group associated with surface tension, σ_s, and size of the dispersed phase, *R*, two new columns to the matrix in Eq. 8.19 must be added, resulting in

$$\begin{array}{lrrrrrr} & \eta_0 & D & n & \Delta T_p & \sigma_s & R \\ M & 1 & 0 & 0 & 0 & 1 & 0 \\ L & -1 & 1 & 0 & 0 & 0 & 1 \\ T & -1 & 0 & -1 & 0 & -2 & 0 \\ \Theta & 0 & 0 & 0 & 1 & 0 & 0 \end{array} \tag{8.19}$$

which after transformation results in $\Pi_{13} = \sigma_s/(\eta_0 D n)$ and $\Pi_{14} = R/D$ as the additional dimensionless groups. Combining these dimensionless along with Π_9 forms the well-known capillary number

$$Ca = \frac{\Pi_{14}}{\Pi_{13}\Pi_9} = \frac{\eta_0 D\, n\, R}{\sigma_s h} = \frac{\eta_0\, \dot{\gamma}\, R}{\sigma_s} \tag{8.20}$$

In addition to the geometric parameters of this problem, of interest to us in this dispersive mixing process are the capillary number, *Ca*, which must be maintained constant in order to achieve the same amount of dispersion and the Brinkman number, *Br*, which must also be maintained constant so that the material is not overheated during mixing. Since our scaling factor, $\Re$, is determined by an increase in diameter

$$D_{\text{scaled}} = D\Re \tag{8.21}$$

all our other dimensionless groups must be adjusted accordingly. Hence, if the Brinkman number is to remain constant, the rotational speed of the screw must be scaled by

$$n_{\text{scaled}} = \frac{n}{\Re} \tag{8.22}$$

Because the capillary number dominates the dispersion of the fluids, that dimensionless group must also be maintained constant. Since the rotational speed and the diameter have already been dealt with, the only remaining parameter in the capillary number is the channel depth, which must be maintained constant. Hence

$$h_{\text{scaled}} = h \tag{8.23}$$

which leads to a system that is economically unfeasible, because the material throughput increases proportionally to the increase in diameter, instead of the expected cubic relation. Hence, there is no geometric similarity between the model and the scaled process. The scaling of extruders remains a very complex and controversial art. One form of scaling proposed by Rauwendaal [8] leads to a system of constant Brinkmann number in order to control the viscous heating within the system. The scaled system, which compromises the dispersive stresses, is given by

When up- or down-scaling a system we must often make compromises, and allow for variations in some of the dimensionless groups

$$\begin{aligned} D_{\text{scaled}} &= D\Re \\ n_{\text{scaled}} &= \frac{n}{\Re} \\ h_{\text{scaled}} &= h\sqrt{\Re} \end{aligned} \tag{8.24}$$

■

8.2 Balance Equations

When solving flow and heat transfer problems in polymer processing, we must satisfy conservation of mass, forces or momentum, and energy. Momentum and energy balances, in combination with material properties through constitutive relations, sometimes result in governing equations that are highly non-linear. This chapter presents the balance equations, making use of constitutive relations presented in Chapter 3 of this book.

8.2.1 The Mass Balance or Continuity Equation

The most basic aspect of modeling polymer processing is to satisfy the conservation of mass. When modeling the flow of polymers we can assume incompressibility,[1)] making a volume balance equivalent to a mass balance. The resulting equation is what is referred to as the continuity equation. In order to derive the continuity equation, we place an imaginary wire frame of dimensions $\Delta x \times \Delta y \times \Delta z$ inside a flowing system, as schematically depicted in Fig. 8.7.

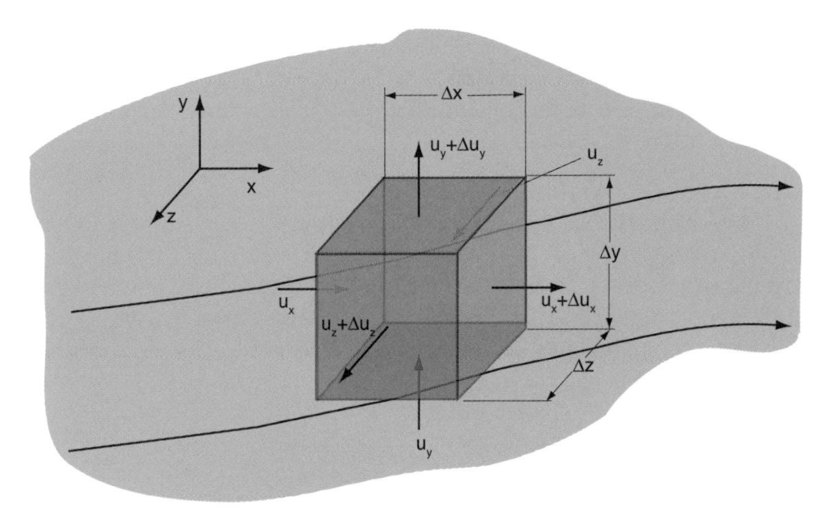

Figure 8.7 Differential frame immersed in a flow and fixed in space

Using the notation introduced in Fig. 8.7, we can perform a volumetric balance, in and out of the differential element, in a volume specific form, by dividing the balance by the element's volume $\Delta x \times \Delta y \times \Delta z$,

$$\frac{\Delta u_z}{\Delta z} + \frac{\Delta u_y}{\Delta y} + \frac{\Delta u_x}{\Delta x} = 0 \tag{8.25}$$

1) From the PvT behavior of a polymer melt we know that, in principle, a polymer is not an incompressible fluid. However, the changes of volume with respect to pressure variations within a process are not significant enough to affect the flow field.

Table 8.2 Continuity equation

Cartesian Coordinates (x, y, z):
$\frac{\partial}{\partial x}(\varrho u_x) + \frac{\partial}{\partial y}(\varrho u_y) + \frac{\partial}{\partial z}(\varrho u_z) = 0$
Cylindrical Coordinates (r, θ, z):
$\frac{1}{r}\frac{\partial}{\partial r}(\varrho r u_r) + \frac{1}{r}\frac{\partial}{\partial \theta}(\varrho u_\theta) + \frac{\partial}{\partial z}(\varrho u_z) = 0$

Letting the size of the differential element go to zero results in

The continuity equation represents conservation of mass

$$\frac{\partial u_z}{\partial z} + \frac{\partial u_y}{\partial y} + \frac{\partial u_x}{\partial x} = \frac{\partial u_i}{\partial x_i} = 0 \tag{8.26}$$

which states that the divergence of the velocity vector must equal zero when the mass or the volume is conserved. We can also write this equation as

$$\nabla \cdot \boldsymbol{u} = 0 \tag{8.27}$$

For cases where the flow is compressible, the continuity equation must be written as

$$\nabla \cdot (\varrho\, \boldsymbol{u}) = 0 \tag{8.28}$$

Table 8.2 presents the continuity equation in the Cartesian and the cylindrical coordinate systems. Note that for most cases where density is constant, the density, ϱ, can be dropped from the equation.

8.2.2 The Material or Substantial Derivative

It is possible to describe a flowing system from a fixed or moving observer point of view. A fixed observer, such as described in Fig. 8.8, feels the transient effects; changes in the variables during the time before the system reaches steady state.

In a non-isothermal flow, a fixed observer feels

$$\frac{\partial u_z}{\partial t}, \frac{\partial u_y}{\partial t}, \frac{\partial u_x}{\partial t}, \frac{\partial T}{\partial t}, \quad \text{etc}$$

Once the system reaches steady state, the fixed observer feels a constant velocity, temperature, and other field variables.

On the other hand, a moving observer, such as the one shown in Fig. 8.9, not only feels the transient effects, but also the changes that the variables undergo as the material element travels through a gradient of velocity, temperature, concentration, etc.

A substantial or material derivative is a time derivative that contains variations due to flow (convective terms)

The moving observer, described by a fluid particle, feels

$$\frac{\partial u_x}{\partial t} + u_x \frac{\partial u_x}{\partial x} + u_y \frac{\partial u_x}{\partial y} + u_z \frac{\partial u_x}{\partial z} = \frac{\partial u_i}{\partial t} + u_j \frac{\partial u_i}{\partial x_j} \tag{8.29}$$

as the change of u_x. Equation 8.29 is often written in short form as Du_x/Dt and is referred to as the *material derivative* or the *substantial derivative*.

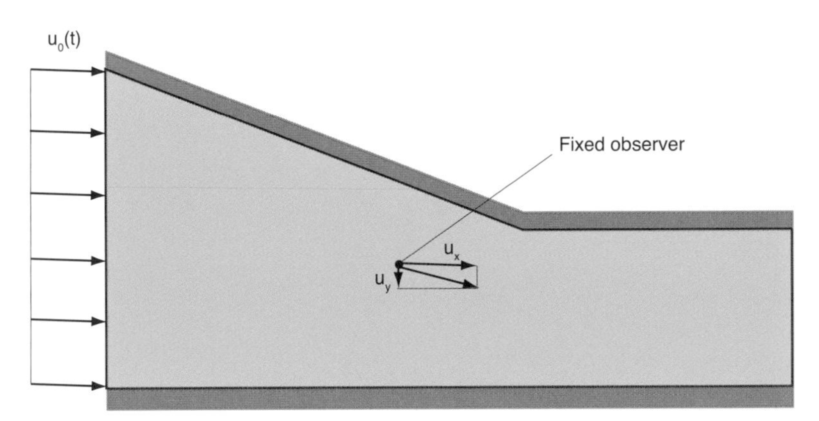

Figure 8.8 Flow system with a fixed observer

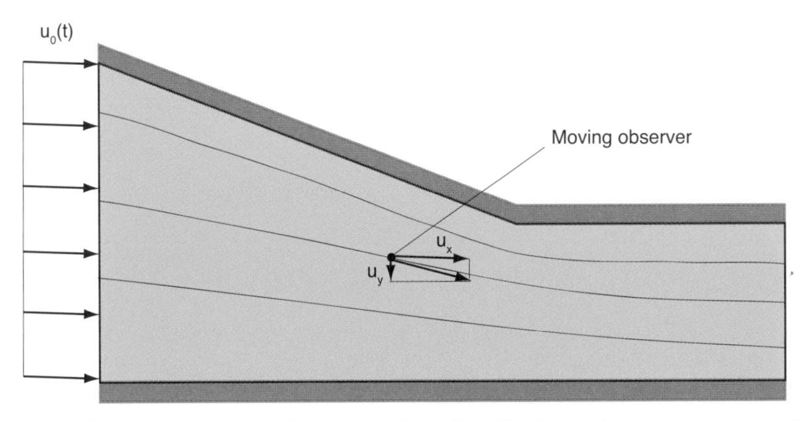

Figure 8.9 Flow system with an observer moving with a fluid particle on a given streamline

8.2.3 The Momentum Balance or Equation of Motion

For a momentum balance, we take the same flow system as presented in Fig. 8.10, but instead of submerging an imaginary frame into the melt, we take an actual fluid element of dimensions $\Delta x \times \Delta y \times \Delta z$ (Fig. 8.11) and perform a force balance with the forces acting on its surfaces.

The force balance can be written as

$$\sum \boldsymbol{f} = m\,\boldsymbol{a} \tag{8.30}$$

where the terms in the equation define force, $\boldsymbol{f}$, mass, m, and acceleration, $\boldsymbol{a}$, respectively. For simplicity, here we will only show the balance of forces in the x-direction. The balance in the y- and z-directions are left to the reader as a short exercise. The forces acting in the x-direction on a small fluid element are described in Fig. 8.10. Because the element in Fig. 8.10 is a fluid particle that moves with the flow, the change

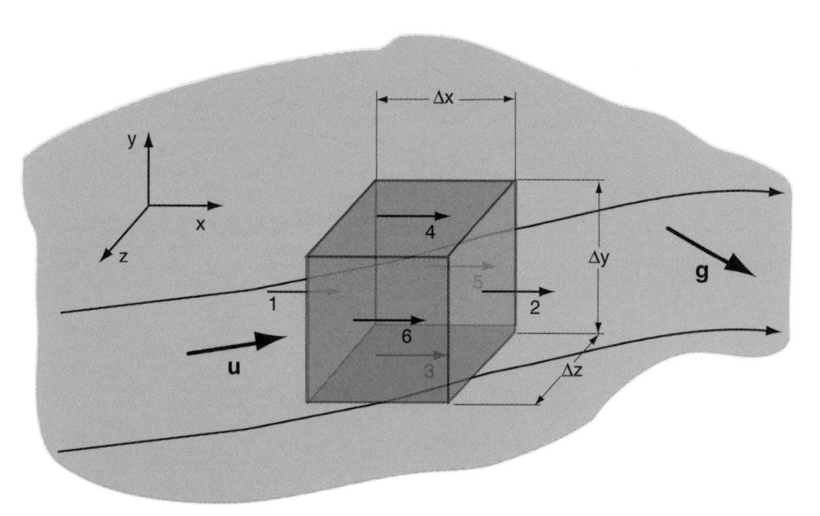

Figure 8.10 Differential fluid element traveling along its streamline x-direction forces that act on its surfaces

Deviatoric stresses lead to deformation

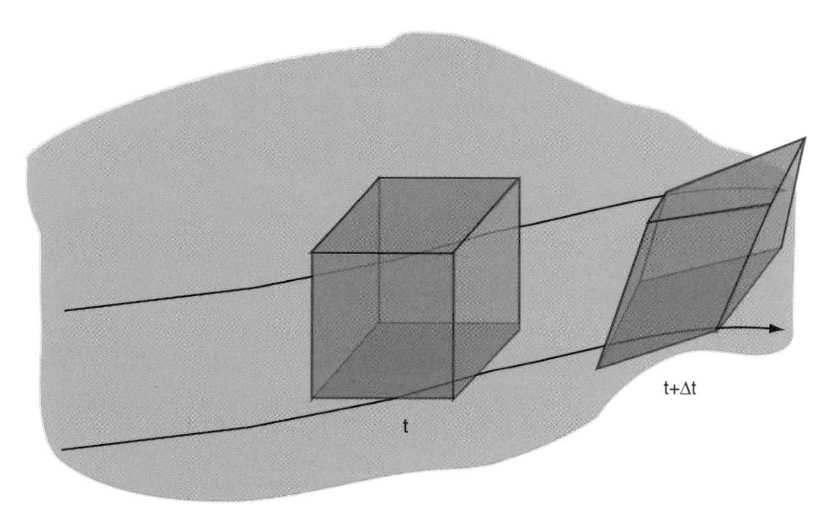

Figure 8.11 Effect of deviatoric stresses as the fluid element travels along its streamline

of its velocity components is described by the material derivative. Hence, the force balance in the x-direction is given by

$$\sum f = m\frac{Du_x}{Dt} \tag{8.31}$$

where $m = \varrho\Delta x \times \Delta y \times \Delta z$.

After adding the forces, dividing by the element's volume, and letting the volume go to zero, the force balance in the x-direction results in

$$\varrho\frac{Du_x}{Dt} = \frac{\partial\sigma_{xx}}{\partial x} + \frac{\partial\sigma_{yx}}{\partial y} + \frac{\partial\sigma_{zx}}{\partial z} + \varrho\, g_x \tag{8.32}$$

which for all three directions can be written as

$$\begin{aligned}\varrho\frac{Du_i}{Dt} &= \frac{\partial\sigma_{ji}}{\partial x_j} + \varrho\, g_i\\ \varrho\frac{Du_i}{Dt} &= \nabla\cdot\sigma + \varrho\, g\end{aligned} \tag{8.33}$$

In fluid flow, however, it is necessary to split the total stress, σ_{ij}, into a deviatoric stress, τ_{ij}, and a hydrostatic stress, σ_{H}. The deviatoric stress is the one that leads to deformation (Fig. 8.11) and the hydrostatic stress is the one that is described by pressure (Fig. 8.12).

Hydrostatic stresses lead to pressure but not deformation in an incompressible fluid

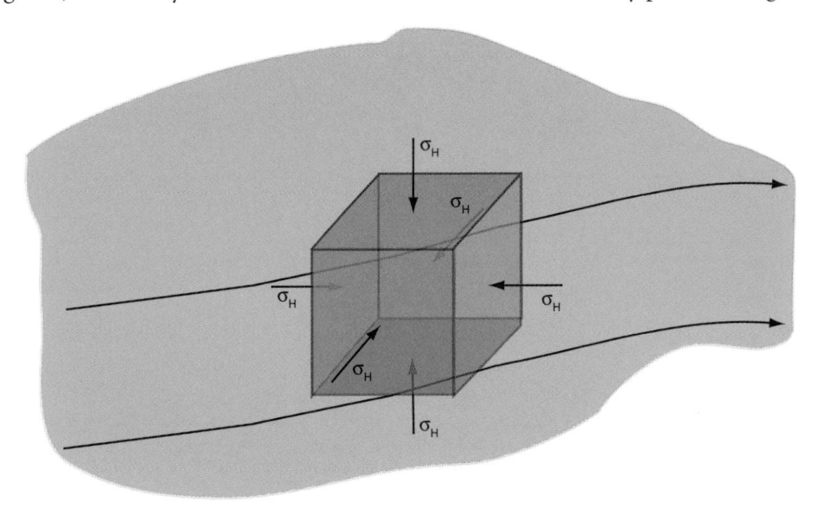

Figure 8.12 Hydrostatic stresses acting on a differential element

We can write

$$\sigma_{ji} = \sigma_{\mathrm{H}}\delta_{ij} + \tau_{ij} \tag{8.34}$$

where δ_{ij} is the Kronecker delta. As the above equation reveals, the hydrostatic stress can only act in the normal direction of a surface and it is equal in all three direction. Hence, we can write

$$\sigma_{\mathrm{H}} = -p \tag{8.35}$$

where p defines the pressure. The negative pressure is due to the fact that a positive pressure causes a compressive stress. The total stress can be written as,

$$\sigma_{ji} = -p\delta_{ij} + \tau_{ij} \tag{8.36}$$

Table 8.3 Momentum equation in terms of τ

Cartesian Coordinates (x, y, z):

$$\varrho\left(\frac{\partial v_x}{\partial t}+v_x\frac{\partial v_x}{\partial x}+v_y\frac{\partial v_x}{\partial y}+v_z\frac{\partial v_x}{\partial z}\right)=-\frac{\partial p}{\partial x}\left(\frac{\partial \tau_{xx}}{\partial x}+\frac{\partial \tau_{yx}}{\partial y}+\frac{\partial \tau_{zx}}{\partial z}\right)+\varrho g_x$$

$$\varrho\left(\frac{\partial v_y}{\partial t}+v_x\frac{\partial v_y}{\partial x}+v_y\frac{\partial v_y}{\partial y}+v_z\frac{\partial v_y}{\partial z}\right)=-\frac{\partial p}{\partial y}+\left(\frac{\partial \tau_{xy}}{\partial x}+\frac{\partial \tau_{yy}}{\partial y}+\frac{\partial \tau_{zy}}{\partial z}\right)+\varrho g_y$$

$$\varrho\left(\frac{\partial v_z}{\partial t}+v_x\frac{\partial v_z}{\partial x}+v_y\frac{\partial v_z}{\partial y}+v_z\frac{\partial v_z}{\partial z}\right)=-\frac{\partial p}{\partial z}+\left(\frac{\partial \tau_{xz}}{\partial x}+\frac{\partial \tau_{yz}}{\partial y}+\frac{\partial \tau_{zz}}{\partial z}\right)+\varrho g_z$$

Cylindrical Coordinates (r, θ, z):

$$\varrho\left(\frac{\partial v_r}{\partial t}+v_r\frac{\partial v_r}{\partial r}+\frac{v_\theta}{r}\frac{\partial v_r}{\partial \theta}-\frac{v_\theta^2}{r}+v_z\frac{\partial v_r}{\partial z}\right)=$$
$$-\frac{\partial p}{\partial r}+\left(\frac{1}{r}\frac{\partial}{\partial x}(r\tau_{rr})+\frac{1}{r}\frac{\partial \tau_{r\theta}}{\partial \theta}-\frac{\tau_{\theta\theta}}{r}+\frac{\partial \tau_{rz}}{\partial z}\right)+\varrho g_r$$

The most general form of the momentum equation is the equation of motion in terms of stress

$$\varrho\left(\frac{\partial v_\theta}{\partial t}+v_r\frac{\partial v_\theta}{\partial r}+\frac{v_\theta}{r}\frac{\partial v_\theta}{\partial \theta}+\frac{v_r v_\theta}{r}+v_z\frac{\partial v_\theta}{\partial z}\right)=$$
$$-\frac{1}{r}\frac{\partial p}{\partial \theta}+\left(\frac{1}{r^2}\frac{\partial}{\partial r}(r^2\tau_{r\theta})+\frac{1}{r}\frac{\partial \tau_{\theta\theta}}{\partial \theta}+\frac{\partial \tau_{\theta z}}{\partial z}\right)+\varrho g_\theta$$

$$\varrho\left(\frac{\partial v_z}{\partial t}+v_r\frac{\partial v_z}{\partial r}+\frac{v_\theta}{r}\frac{\partial v_z}{\partial \theta}+v_z\frac{\partial v_z}{\partial z}\right)=-\frac{\partial p}{\partial z}+\left(\frac{1}{r}\frac{\partial}{\partial r}(r\tau_{rz})+\frac{1}{r}\frac{\partial \tau_{\theta z}}{\partial \theta}+\frac{\partial \tau_{zz}}{\partial z}\right)+\varrho g_z$$

Using the definition of total stress given above, the momentum balance can now be written as,

$$\varrho\frac{Du_i}{Dt}=-\frac{\partial p}{\partial x_i}+\frac{\partial \tau_{ji}}{\partial x_j}+\varrho\, g_i \tag{8.37}$$

$$\varrho\frac{D\boldsymbol{u}}{Dt}=-\nabla p+\nabla\cdot\tau+\varrho\,\boldsymbol{g} \tag{8.38}$$

Table 8.3 presents the momentum balance in terms of deviatoric stress in the Cartesian, cylindrical, and spherical coordinate systems.

These forms of the equation of motion are commonly called *the Cauchy momentum equations*. For generalized Newtonian fluids we can define the terms of the deviatoric stress tensor as a function of a generalized Newtonian viscosity, η, and the components of the rate of deformation tensor, as described in Chapter 3.

In fluid mechanics, one common description of the deviatoric stress tensor is the Newtonian model given by

$$\tau_{ij}=\mu\dot{\gamma}_{ij} \tag{8.39}$$

which reduces the Cauchy momentum equations to,

$$\begin{aligned}\varrho\frac{Du_i}{Dt}&=-\frac{\partial p}{\partial x_i}+\mu\frac{\partial^2 u_i}{\partial x_j\partial x_j}+\varrho g_i\\ \varrho\frac{D\boldsymbol{u}}{Dt}&=-\nabla p+\mu\nabla^2\boldsymbol{u}+\varrho\boldsymbol{g}\end{aligned} \tag{8.40}$$

Table 8.4 Navier-Stokes equations

Cartesian Coordinates (x, y, z):

$$\varrho\left(\frac{\partial v_x}{\partial t}+v_x\frac{\partial v_x}{\partial x}+v_y\frac{\partial v_x}{\partial y}+v_z\frac{\partial v_x}{\partial z}\right)=-\frac{\partial p}{\partial x}+\mu\left(\frac{\partial^2 v_x}{\partial x^2}+\frac{\partial^2 v_x}{\partial y^2}+\frac{\partial^2 v_x}{\partial z^2}\right)+\varrho g_x$$

$$\varrho\left(\frac{\partial v_y}{\partial t}+v_y\frac{\partial v_y}{\partial x}+v_y\frac{\partial v_y}{\partial y}+v_z\frac{\partial v_y}{\partial z}\right)=-\frac{\partial p}{\partial y}+\mu\left(\frac{\partial^2 v_y}{\partial x^2}+\frac{\partial^2 v_y}{\partial y^2}+\frac{\partial^2 v_y}{\partial z^2}\right)+\varrho g_y$$

$$\varrho\left(\frac{\partial v_z}{\partial t}+v_x\frac{\partial v_z}{\partial x}+v_y\frac{\partial v_z}{\partial y}+v_z\frac{\partial v_z}{\partial z}\right)=-\frac{\partial p}{\partial z}+\mu\left(\frac{\partial^2 v_z}{\partial x^2}+\frac{\partial^2 v_z}{\partial y^2}+\frac{\partial^2 v_z}{\partial z^2}\right)+\varrho g_z$$

Cylindrical Coordinates (r, θ, z):

$$\varrho\left(\frac{\partial v_r}{\partial t}+v_r\frac{\partial v_r}{\partial r}+\frac{v_\theta}{r}\frac{\partial v_r}{\partial \theta}-\frac{v_\theta^2}{r}+v_z\frac{\partial v_r}{\partial z}\right)=$$
$$-\frac{\partial p}{\partial r}+\mu\left(\frac{\partial}{\partial r}\left(\frac{1}{r}\frac{\partial}{\partial r}(r v_r)\right)+\frac{1}{r^2}\frac{\partial^2 v_r}{\partial \theta^2}-\frac{2}{r^2}\frac{\partial v_\theta}{\partial \theta}+\frac{\partial^2 v_r}{\partial z^2}\right)+\varrho g_r$$

$$\varrho\left(\frac{\partial v_\theta}{\partial t}+v_r\frac{\partial v_\theta}{\partial r}+\frac{v_\theta}{r}\frac{\partial v_\theta}{\partial \theta}+\frac{v_r v_\theta}{r}+v_z\frac{\partial v_\theta}{\partial z}\right)=$$
$$-\frac{1}{r}\frac{\partial p}{\partial \theta}+\mu\left(\frac{\partial}{\partial r}\left(\frac{1}{r}\frac{\partial}{\partial r}(r v_\theta)\right)+\frac{1}{r^2}\frac{\partial^2 v_\theta}{\partial \theta^2}-\frac{2}{r^2}\frac{\partial v_r}{\partial \theta}+\frac{\partial^2 v_\theta}{\partial z^2}\right)+\varrho g_\theta$$

$$\varrho\left(\frac{\partial v_z}{\partial t}+v_r\frac{\partial v_z}{\partial r}+\frac{v_\theta}{r}\frac{\partial v_z}{\partial \theta}+v_z\frac{\partial v_z}{\partial z}\right)=$$
$$-\frac{\partial p}{\partial z}+\mu\left(\frac{1}{r}\frac{\partial}{\partial r}\left(r\frac{\partial v_z}{\partial r}\right)+\frac{1}{r^2}\frac{\partial^2 v_z}{\partial \theta^2}+\frac{\partial^2 v_z}{\partial z^2}\right)+\varrho g_z$$

The Navier-Stokes equation is the equation of motion for a simple Newtonian fluid

which is often referred to as the *Navier-Stokes equations.* Table 8.4 presents the full form of the Navier-Stokes equations.

With a few exceptions one can say that a flowing polymer melt does not follow the model presented in Eq. 8.40. To properly model the flow of a polymer we must take into account the effects of rate of deformation, temperature, and often time, making the partial differential equations that govern a system non-linear.

8.2.4 The Energy Balance or Equation of Energy

Using Fourier's law for heat conduction

$$q_i = -k_i\frac{\partial T}{\partial x_i} \tag{8.41}$$

and assuming an isotropic material, $k_x = k_y = k_z = k$, an energy balance around a moving fluid element, as shown in Fig. 8.11, can be written as,

$$\varrho C_p\frac{DT}{Dt} = k\left(\frac{\partial^2 T}{\partial x^2}+\frac{\partial^2 T}{\partial y^2}+\frac{\partial^2 T}{\partial z^2}\right)+\dot{Q}+\dot{Q}_{\text{viscous heating}} \tag{8.42}$$

where an arbitrary heat source $\dot{Q}$, and viscous dissipation $\dot{Q}_{\text{viscous heating}}$ terms were included.

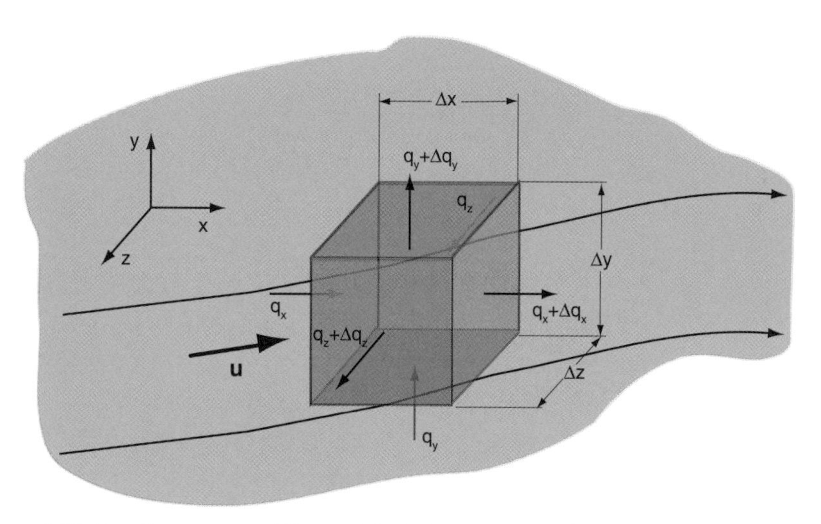

Figure 8.13 Heat flux across a differential fluid element during flow

As an illustration, we will derive the viscous dissipation terms in the energy balance using a simple shear flow system, such as the one shown in Fig. 8.13.

Here, the stresses within the system can be calculated using

$$\tau_{yx} = \mu \frac{\partial u_x}{\partial y} \tag{8.43}$$

which in terms of the parameters depicted in Fig. 8.14, such as force, F, area, A, gap height, h, and plate speed, u_0, can be written as

$$\frac{F}{A} = \mu \frac{u_0}{h} \tag{8.44}$$

Moving forces acting on a closed system lead to viscous heating within the flowing polymer

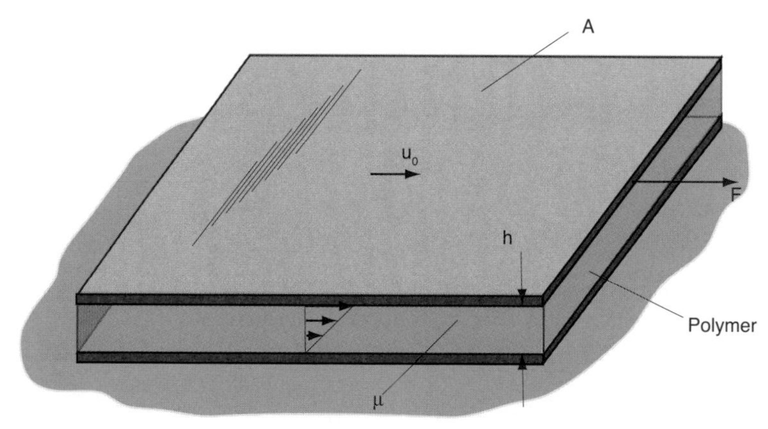

Figure 8.14 Schematic of a simple shear flow system used to illustrate viscous dissipation terms in the energy balance

In the system, the *rate of energy input* is given by

$$Fu_0 = \mu \frac{u_0}{h} A u_0 \tag{8.45}$$

and the rate of energy input per unit volume is represented by

$$\frac{Fu_0}{Ah} = \mu \left(\frac{u_0}{h}\right)\left(\frac{u_0}{h}\right) \tag{8.46}$$

or

$$\dot{Q}_{\text{viscous heating}} = \mu \left(\frac{\partial u_x}{\partial y}\right)\left(\frac{\partial u_x}{\partial y}\right) \tag{8.47}$$

From the above equation, we can deduce that for a Newtonian fluid the general term for viscous dissipation is given by $\mu\,(\dot{\gamma} : \dot{\gamma})$, where

$$\dot{\gamma} : \dot{\gamma} = \sum_{i=1}^{3}\sum_{j=1}^{3} \dot{\gamma}_{ij}\dot{\gamma}_{ji} \tag{8.48}$$

and for a non-Newtonian material, the viscous heating is written as $\tau : \dot{\gamma}$. Hence, the energy balance becomes

$$\begin{aligned} \varrho C_\mathrm{p} \frac{\partial T}{\partial t} + \varrho C_\mathrm{p} u_j \frac{\partial T}{\partial x_j} &= \frac{\partial}{\partial x_j} k \frac{\partial T}{\partial x_j} + \tau_{ij}\dot{\gamma}_{ji} + \dot{Q} \\ \varrho C_\mathrm{p} \frac{\partial T}{\partial t} + \varrho C_\mathrm{p} \boldsymbol{u} \cdot \nabla T &= \nabla \cdot k \nabla T + \tau : \dot{\gamma} + \dot{Q} \end{aligned} \tag{8.49}$$

Table 8.5 presents the energy equation in the Cartesian and cylindrical coordinate systems for a Newtonian fluid.

8.3 Model Simplification

In order to be able to obtain analytical solutions we must first simplify the balance equations. Although the balance equations are fundamental and rigorous, they are nonlinear, non-unique, complex, and difficult to solve. In other words, they do not have a general solution and so far, only particular solutions for special problems have been found.

Therefore, the balance equations must be simplified sufficiently in order to arrive at an analytical solution of the problem under consideration. The simplifications done on a system are typically based on the scale of the variables, an estimate of its maximum order of magnitude. As discussed in the previous section, scaling is the process of identifying the correct order of magnitude of the various unknowns. These magnitudes are often referred to as characteristic values, i.e., characteristic times, characteristic length, etc. When a variable is scaled with respect to its characteristic magnitude (scale),

Energy equation; from left to right:

- Transient term
- Convective terms
- Conductive terms
- Viscous heating terms
- Heat generation term, i.e., exothermic reaction

Table 8.5 Energy equation for a Newtonian fluid

Cartesian Coordinates (x, y, z):

$$\varrho C_v\left(\frac{\partial T}{\partial t}+v_x\frac{\partial T}{\partial x}+v_y\frac{\partial T}{\partial y}+v_z\frac{\partial T}{\partial z}\right)=K\left(\frac{\partial^2 T}{\partial x^2}+\frac{\partial^2 T}{\partial y^2}+\frac{\partial^2 T}{\partial z^2}\right)+$$
$$+2\mu\left(\left(\frac{\partial v_x}{\partial x}\right)^2+\left(\frac{\partial v_y}{\partial y}\right)^2+\left(\frac{\partial v_z}{\partial z}\right)^2\right)+$$
$$+\mu\left(\left(\frac{\partial v_x}{\partial y}+\frac{\partial v_y}{\partial x}\right)^2+\left(\frac{\partial v_x}{\partial z}+\frac{\partial v_z}{\partial x}\right)^2+\left(\frac{\partial v_y}{\partial z}+\frac{\partial v_z}{\partial y}\right)^2\right)+\dot{Q}$$

Cylindrical Coordinates (r, θ, z):

$$\varrho C_v\left(\frac{\partial T}{\partial t}+v_r\frac{\partial T}{\partial r}+\frac{v_\theta}{r}\frac{\partial T}{\partial \theta}+v_z\frac{\partial T}{\partial z}\right)=K\left(\frac{1}{r}\frac{\partial}{\partial r}\left(r\frac{\partial T}{\partial r}\right)+\frac{1}{r^2}\frac{\partial^2 T}{\partial \theta^2}+\frac{\partial^2 T}{\partial z^2}\right)+$$
$$+2m\left(\left(\frac{\partial v_r}{\partial r}\right)^2+\left(\frac{1}{r}\left(\frac{\partial v_\theta}{\partial \theta}+v_r\right)\right)^2+\left(\frac{\partial v_z}{\partial z}\right)^2\right)+$$
$$+\mu\left(\left(\frac{\partial v_\theta}{\partial z}+\frac{1}{r}\frac{\partial v_z}{\partial \theta}\right)^2+\left(\frac{\partial v_z}{\partial r}+\frac{\partial v_r}{\partial z}\right)^2+\left(\frac{1}{r}\frac{\partial v_r}{\partial \theta}+r\frac{\partial}{\partial r}\left(\frac{v_\theta}{r}\right)\right)^2\right)+\dot{Q}$$
$$\varrho\left(\frac{\partial v_z}{\partial t}+v_r\frac{\partial v_z}{\partial r}+\frac{v_\theta}{r}\frac{\partial v_z}{\partial \theta}+v_z\frac{\partial v_z}{\partial z}\right)=$$
$$-\frac{\partial p}{\partial z}+\mu\left(\frac{1}{r}\frac{\partial}{\partial r}\left(r\frac{\partial v_z}{\partial r}\right)+\frac{1}{r^2}\frac{\partial^2 v_z}{\partial \theta^2}+\frac{\partial^2 v_z}{\partial z^2}\right)+\varrho g_z$$

the new dimensionless variable will be of order 1, i.e. ($\sim O(1)$). For example, if we scale the x-velocity field, u_x, within a system, with respect to a characteristic velocity, U_0, we can generate a dimensionless velocity, or scaled velocity, given by

$$\hat{u}_x=\frac{u_x}{U_0} \tag{8.50}$$

Using the above relation, the original variable can be expressed in terms of the dimensionless variable and its characteristic value as,

$$u_x=U_0\hat{u}_x \tag{8.51}$$

By substituting the new variables into the original equations we will acquire information that allows the simplification of a specific model. Length and time scales, for example, can lead to geometrical simplifications, such as a reduction in dimensionality.

Example 8.5 Object submerged in a fluid

Consider an object with a characteristic length L and a thermal conductivity k that is submerged in a fluid of constant temperature T_∞ and convection coefficient h (see Fig. 8.15). If a heat balance is made on the surface of the object, it must be equivalent to the

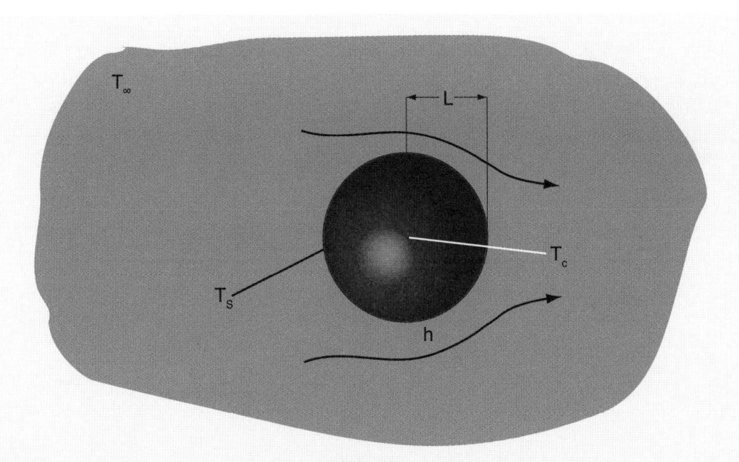

Figure 8.15 Schematic of a body submerged in a fluid

heat by conduction, i. e.,

$$-k \left. \frac{\partial T}{\partial n} \right|_S = h \left(T_S - T_\infty \right) \tag{8.52}$$

The maximum value possible for the temperature gradient must be the difference between the central temperature, T_c, and the surface temperature, T_S,

Order of magnitude analysis

$$\Delta T \sim T_c - T_S \tag{8.53}$$

giving us a characteristic temperature difference.[2] Here, the length variable is the normal distance Δn and has a characteristic length L. We can now approach the scaling of these problems in two ways. The first and quickest is to simply substitute the variables into the original equations, often referred to an *order of magnitude analysis*. The second is to express the original equations in terms of dimensionless variables. The order of magnitude analysis results in a scaled conduction given by,

$$k \left. \frac{\partial T}{\partial n} \right|_S \sim k \frac{T_c - T_S}{L} \tag{8.54}$$

reducing the problem to,

$$k \frac{T_c - T_S}{L} \sim h \left(T_S - T_\infty \right) \tag{8.55}$$

or in a more convenient way,

$$Bi = \frac{hL}{k} \sim \frac{T_c - T_S}{T_S - T_\infty} \tag{8.56}$$

where Bi is the *Biot number*.

2) Characteristic temperatures are always given in terms of temperature differences. For example, the characteristic temperature of the melt of an amorphous polymer in an extrusion operation is $\Delta T = T_h - T_g$, or the difference between the heater temperature and the glass transition temperature of the polymer.

When $Bi \ll 1$, the solid can be considered isothermal, which means that we *reduced the dimensionality* of the problem from (x, y, z), to a zero dimensional or *lumped model* [9]. On the other hand, if $Bi \gg 1$, the fluid can be considered isothermal and $T_S = T_\infty$, which changes the convection boundary condition to a thermal equilibrium condition.

The same can be deduced if we scale the problem by expressing the governing equations in dimensionless form. Again, we choose the same characteristic values for normal distance and temperature, allowing us to generate dimensionless variables as

$$\hat{T} = \frac{T}{T_c - T_S}, \quad \hat{n} = \frac{n}{L} \tag{8.57}$$

which can be solved to give

$$T = (T_c - T_S)\,\hat{T}, \quad n = L\hat{n} \tag{8.58}$$

Substituting these into the original equations results in

$$-\frac{k}{Lh}\frac{\partial \hat{T}}{\partial \hat{n}} = \frac{T_S - T_\infty}{T_c - T_S} \tag{8.59}$$

or

$$Bi \frac{\partial \hat{T}}{\partial \hat{n}} = \Theta \tag{8.60}$$

Again, because $\partial \hat{T}/\partial \hat{n}$ is of order one, the same analysis done above applies here.

8.3.1 Reduction in Dimensionality

The number of special coordinates, or dimensionality of a problem, can be reduced using three basic strategies: symmetry, aspect ratio, and series resistances.

Symmetry is the easiest to apply. It is based on the correct selection of the coordinate system for a given problem. For example, a temperature field with circular symmetry can be described using just the coordinates (r, z), instead of (x, y, z). In addition, symmetry can help to get rid of special variables that are not required by the conservation equations and interfacial conditions. For example, the velocity field in a tube, according to the Navier-Stokes and continuity equations, can have the functional form $u_z(r)$.

The ratio of two linear dimensions of an object is called the *aspect ratio*. There are a number of possible simplifications when the aspect ratio of an object or region is large (or small). For example, for the classical *fin approximation*, the thickness of the fin is small compared with the length, therefore the temperature will be assumed to change in the direction of the length only.

For example in problems where $Bi \ll 1$, the convection controls the cooling process and conduction is so fast that the solid is considered isothermal, reducing the dimensionality from (x, y, z) to a zero dimensional problem or lumped mass method.

Characteristic times are a key factor in formulating conduction or diffusion models, because they determine how fast a system can respond to changes imposed at a boundary. In other words, if the temperature or concentration is perturbed at some location, it is important to estimate the finite time required for the temperature or concentration changes to be noticed at a given distance from the original perturbation. The time involved in a stagnant medium is the characteristic time for conduction or diffusion, therefore this is the most widely used characteristic time in transport models [10, 13].

Example 8.6 Temperature development in an extruder channel during melting

In this example, we illustrate reduction in dimensionality of the energy equation to find an equation that would reveal the change of the melt temperature through the gap between the solid bed and extruder barrel during melting, as schematically depicted in Fig. 8.16. To simplify the problem, we assume constant properties and a Newtonian viscosity.

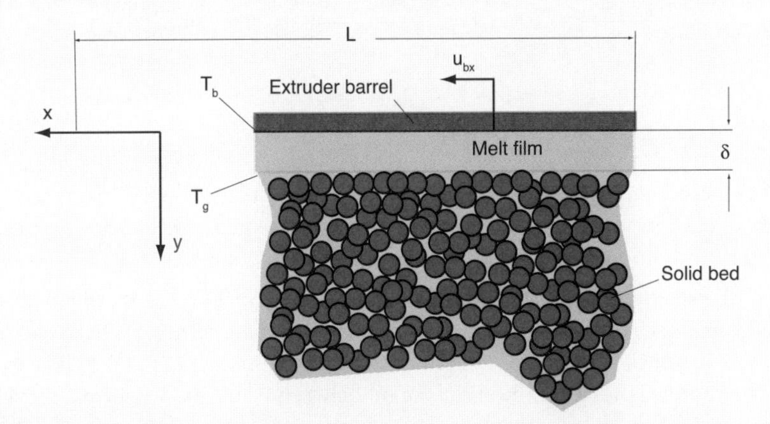

Figure 8.16 Schematic diagram of the melt film during melting in extruders

The thickness between the solid and the barrel is small compared to the screw channel, which indicates that a reduction in dimensionality can be performed. Initially, it can be assumed that the velocity field is unidirectional, i.e., $u_x(y)$. The energy equation is then reduced to

$$\varrho C_p u_x \frac{\partial T}{\partial t} = k\left(\frac{\partial^2 T}{\partial x^2} + \frac{\partial^2 T}{\partial y^2}\right) + \mu\left(\frac{\partial u_x}{\partial y}\right)^2 \tag{8.61}$$

By choosing characteristic variables for temperatures, velocities, and lengths we can reduce the dimensionality even further. The temperature is scaled based on the maximum gradient, the length with the gap thickness and screw channel depth and the velocity with the barrel x-velocity

Dimensionless variables

$$\Theta = \frac{T - T_g}{T_b - T_g} \qquad \eta = \frac{y}{\delta}$$

$$\xi = \frac{x}{L} \qquad \bar{u} = \frac{u_x}{u_{bx}}$$

and the energy equation will be,

$$\varrho C_p u_{bx} \frac{\delta^2}{L} \bar{u} \frac{\partial \theta}{\partial \xi} = \frac{\delta^2}{L^2} \frac{\partial^2 \theta}{\partial \xi^2} + \frac{\partial^2 \theta}{\partial \eta^2} + \frac{\mu u_{bx^2}}{k\left(T_b - T_g\right)} \left(\frac{\partial \bar{u}}{\partial \eta}\right)^2 \tag{8.62}$$

which indicates that for the small aspect ratio, δ/L, two extra terms can be neglected, the conduction and convection in the x-direction,

Dimensionless equation

$$\frac{\partial^2 \theta}{\partial \eta^2} + \frac{\mu u_{bx^2}}{k\left(T_b - T_g\right)} \left(\frac{\partial \bar{u}}{\partial \eta}\right)^2 = 0 \tag{8.63}$$

The last step is to compare the two remaining terms: conduction and viscous dissipation. The two derivatives, according to the scaling parameter, are of order 1. The remaining term, $Br = \mu u_{bx}^2 / k\left(T_b - T_g\right)$, is the Brinkman number, which indicates whether the viscous dissipation is important or not. For $Br \ll 1$, the conduction is dominant, while for $Br > 1$, the viscous dissipation has to be included, which is the case in most polymer processing operations.

■

8.3.2 Lubrication Approximation

Now, let's consider flows in which a second component and the inertial effects are nearly zero. Liquid flows in long, narrow channels or thin films often have these characteristics of being nearly unidirectional and dominated by viscous stresses.

Let's use the steady, two-dimensional flow in a thin channel or a narrow gap between solid objects, as schematically represented in Fig. 8.17. The channel height or gap width varies with the position, and there may be a relative motion between the solid surfaces. This type of flow is very common for the oil between bearings. The original solution came from the field of tribology and is therefore often referred to as the *lubrication approximation.*

Lubrication approximation

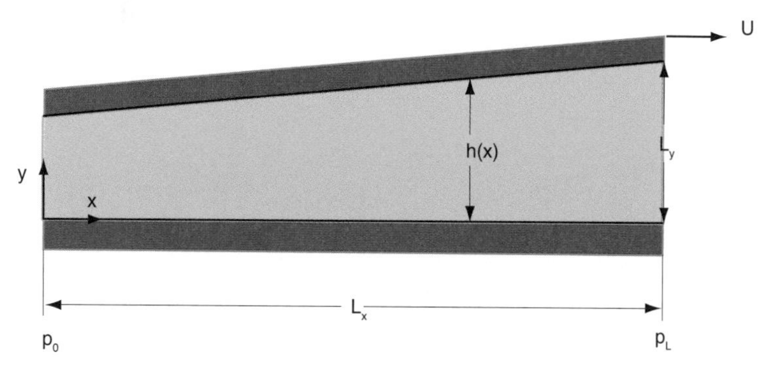

Figure 8.17 Schematic diagram of the lubrication problem

For this type of flow, the momentum equations (for a Newtonian fluid) are reduced to the steady Navier-Stokes equations, i. e.,

$$\frac{\partial u_x}{\partial x} + \frac{\partial u_y}{\partial y} = 0 \tag{8.64}$$

$$\begin{aligned} \varrho\left(u_x \frac{\partial u_x}{\partial x} + u_y \frac{\partial u_x}{\partial y}\right) &= -\frac{\partial p}{\partial x} + \mu\left(\frac{\partial^2 u_x}{\partial x^2} + \frac{\partial^2 u_x}{\partial y^2}\right) \\ \varrho\left(u_x \frac{\partial u_y}{\partial x} + u_y \frac{\partial u_y}{\partial y}\right) &= -\frac{\partial p}{\partial y} + \mu\left(\frac{\partial^2 u_y}{\partial x^2} + \frac{\partial^2 u_y}{\partial y^2}\right) \end{aligned} \tag{8.65}$$

The lubrication approximation depends on two basic conditions, one *geometric* and one *dynamic*. The geometric requirement is revealed by the continuity equation. If L_x and L_y represent the length scales for the velocity variations in the x-and y-directions, respectively, and let U and V be the respective scales for u_z and u_y. From the continuity equation we obtain

$$\frac{V}{U} \sim \frac{L_y}{L_x} \tag{8.66}$$

In order to neglect pressure variation in the y-direction all the terms in the y-momentum equation must be small, in other words $V/U \ll 1$. From the continuity scale analysis we get that the geometric requirement is

$$\frac{L_y}{L_x} \ll 1 \tag{8.67}$$

which holds for thin films and channels. The consequences of this geometric constrain in the Navier-Stokes equations are

$$\frac{\partial p}{\partial y} \ll \frac{\partial p}{\partial x} \quad \text{and} \quad \frac{\partial^2 u_x}{\partial x^2} \ll \frac{\partial^2 u_x}{\partial y^2} \tag{8.68}$$

In addition, the continuity equation also tells us that the two inertia terms in the x-momentum equation are of similar magnitude, i. e.,

$$u_y \frac{\partial u_x}{\partial y} \sim \frac{VU}{L_y} \sim \frac{U^2}{L_x} \sim u_x \frac{\partial u_x}{\partial x} \tag{8.69}$$

These inertia effects can be neglected, i. e.,

$$\varrho u_x \frac{\partial u_x}{\partial x} \ll \mu \frac{\partial^2 u_x}{\partial y^2} \quad \text{and} \quad \varrho u_y \frac{\partial u_x}{\partial x} \ll \mu \frac{\partial^2 u_x}{\partial y^2} \tag{8.70}$$

only if $\varrho U^2/L_x \ll \mu U/L_y^2$ or

$$\left(\frac{\varrho U L_y}{\mu}\right)\left(\frac{L_y}{L_x}\right) = Re\left(\frac{L_y}{L_x}\right) \ll 1 \tag{8.71}$$

which is the dynamic requirement for the lubrication approximation. The x-momentum (Navier-Stokes) equation is then reduced to,

$$\frac{\partial^2 u_x}{\partial y^2} = \frac{1}{\mu}\frac{dp}{dx} \tag{8.72}$$

for $p = p(x)$ only.

8.4 Simple Models in Polymer Processing

There are only a few exact or analytical solutions of the momentum balance equations, and most of those are for situations in which the flow is unidirectional; that is, the flow has only one non-zero velocity component. Some of these are illustrated in the following.

8.4.1 Pressure Driven Flow of a Newtonian Fluid through a Slit

One of the most common flows in polymer processing is the pressure driven flow between two parallel plates. When deriving the equations that govern *slit flow,* we use the notation presented in Fig. 8.18 and consider a steady, fully developed flow, i.e., a flow where the entrance effects are ignored.

Pressure flow through a slit:

- Extrusion die flow
- Flow inside an injection molding cavity

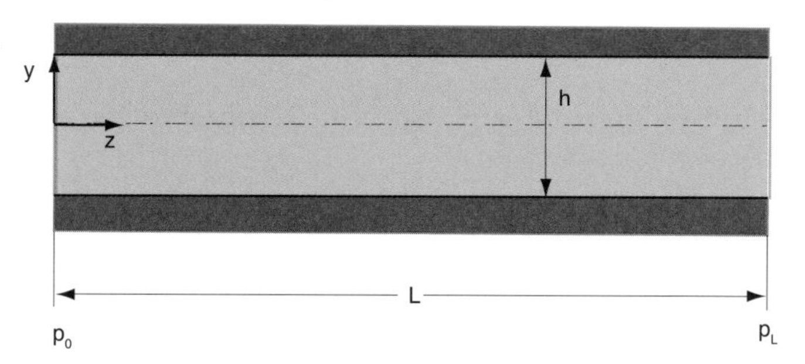

Figure 8.18 Schematic diagram of pressure flow through a slit

This flow is unidirectional, that is, there is only one non-zero velocity component. The continuity for an incompressible flow is reduced to

$$\frac{\partial u_z}{\partial z} = 0 \tag{8.73}$$

The z-momentum equation for a Newtonian, incompressible flow (Navier-Stokes equations) is

$$-\frac{\partial p}{\partial z} + \mu\frac{\partial^2 u_z}{\partial y^2} = 0 \tag{8.74}$$

and the x- and y-components of the equations of motion are reduced to

$$-\frac{\partial p}{\partial x} = -\frac{\partial p}{\partial y} = 0 \tag{8.75}$$

This relation indicates that for this fully developed flow, the total pressure is a function of z alone. Additionally, because u does not vary with z, the pressure gradient, $\partial p/\partial z$, must be a constant. Therefore,

$$\frac{\partial p}{\partial z} = \frac{\Delta p}{L} = 0 \tag{8.76}$$

The momentum equation can now be written as

$$\frac{1}{\mu}\frac{\Delta p}{L} = \frac{\partial^2 u_z}{\partial y^2} \tag{8.77}$$

As boundary conditions, two no-slip conditions given by $u_z(\pm h/2) = 0$ are used in this problem. Integrating twice and evaluating the two integration constants with the boundary conditions gives

$$u_z(y) = \frac{h^2}{8\mu}\frac{dp}{dz}\left[1 - \left(\frac{2y}{h}\right)^2\right] = \frac{h^2}{8\mu}\frac{\Delta p}{L}\left[1 - \left(\frac{2y}{h}\right)^2\right] \tag{8.78}$$

Also note that the same profile will result, if one of the non-slip boundary conditions is replaced by a symmetry condition at $y = 0$, namely $du_z/dy = 0$. The mean velocity in the channel is obtained integrating the above equation

$$\bar{u}_z = \frac{1}{h}\int_0^h u_z(y)dy = \frac{h^2}{12\mu}\frac{dp}{dz} \tag{8.79}$$

and the volumetric flow rate

$$Q = hW\bar{u}_z = \frac{Wh^3}{12\mu}\frac{\Delta p}{L} \tag{8.80}$$

Pressure flow through a slit:

- Extrusion die flow
- Flow inside an injection molding cavity

where W is the width of the channel.

8.4.2 Flow of a Power Law Fluid in a Straight Circular Tube (Hagen-Poiseuille Equation)

Tube flow is encountered in several polymer processes, such as in extrusion dies and sprue and runner systems inside injection molds. When deriving the equations for pressure driven flow in tubes, also known as Hagen-Poiseuille flow, we assume that the flow is steady, fully developed, with no entrance effects, and axis-symmetric (see Fig. 8.19).

Pressure flow through a tube:

- Flow through a runner system
- Extrusion die flow

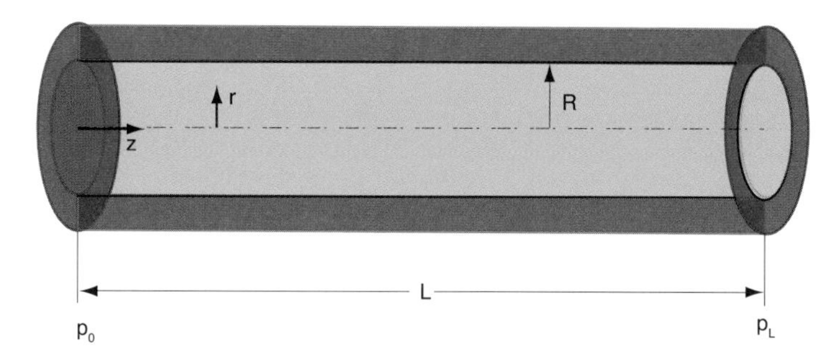

Figure 8.19 Schematic diagram of pressure flow through a tube

Thus, we have $u_z = u_z(r)$, $u_r = u_\theta = 0$ and $p = p(z)$. With this type of velocity field, the only non-vanishing component of the rate-of-deformation tensor is the zr-component. It follows that for the generalized Newtonian flow, τ_{zr} is the only non-zero component of the viscous stress and that $\tau_{zr} = \tau_{zr}(r)$. The z-momentum equation is then reduced to

$$\frac{1}{r}\frac{d}{dr}(r\tau_{zr}) = \frac{dp}{dz} \tag{8.81}$$

However, because $p = p(z)$ and $\tau_{zr} = \tau_{zr}(r)$, the above equation is satisfied only if both sides are constant and can be integrated to obtain

$$r\tau_{zr} = \frac{dp}{dz}\frac{r^2}{2} + c_1 \tag{8.82}$$

At this point, a symmetry argument at $r = 0$ leads to the conclusion that $\tau_{zr} = 0$ because the stress must be finite. Hence, we must satisfy $c_1 = 0$. For a power law fluid it is found that

$$\tau_{zr} = -m\left|\frac{du_z}{dr}\right|^n \tag{8.83}$$

The minus sign in this equation is required due to the fact that the pressure flow is in the direction of the flow ($dp/dz < 0$), indicating that $\tau_{zr} \leqslant 0$. Combining the above equations and solving for the velocity gradient gives

$$\frac{du_z}{dr} = -\left(-\frac{1}{2m}\frac{dp}{dz}\right)^{1/n} r^{1/n} \tag{8.84}$$

Integrating this equation and using the no-slip condition, at $r = R$, to evaluate the integration constant, the velocity as a function of r is obtained

$$u_z(r) = \left(\frac{3n+1}{n+1}\right)\left[1 - \left(\frac{r}{R}\right)^{(n+1)/n}\right]\bar{u}_z \tag{8.85}$$

where the mean velocity, $\bar{u}_z = 0$, is defined as

$$\bar{u}_z = \frac{2}{R^2}\int_0^R u_z r dr = \left(\frac{n}{3n+1}\right)\left[-\frac{R^{n+1}}{2m}\frac{dp}{dz}\right]^{1/n} \tag{8.86}$$

Finally, the volumetric flow rate is given by

Hagen-Poiseuille flow

$$Q = \pi R^2 \bar{u}_z = \left(\frac{n\pi}{3n+1}\right)\left[-\frac{R^{3n+1}}{2m}\frac{dp}{dz}\right]^{1/n} = \left(\frac{n\pi}{3n+1}\right)\left[-\frac{R^{3n+1}}{2m}\frac{\Delta p}{L}\right]^{1/n} \tag{8.87}$$

8.4.3 Volumetric Flow Rate of a Power Law Fluid in Axial Annular Flow

Annular flow is encountered in pipe extrusion dies, wire coating dies, and film blowing dies. In the problem under consideration, a Power law fluid is flowing through an annular gap between two coaxial cylinders of radius κR and R, with $\kappa < 1$, as schematically depicted in Fig. 8.20. The maximum in the velocity profile is located at $r = \beta R$, where β is a constant to be determined. Due to the geometrical characteristics and ignoring entrance effects, the flow is unidirectional, i. e., $\boldsymbol{u} = (u_r, u_\theta, u_z) = (0, 0, u_z(r))$.

Pressure flow through an annulus:

- Wire coating
- Extrusion die flow
- Injection molding cavity flow

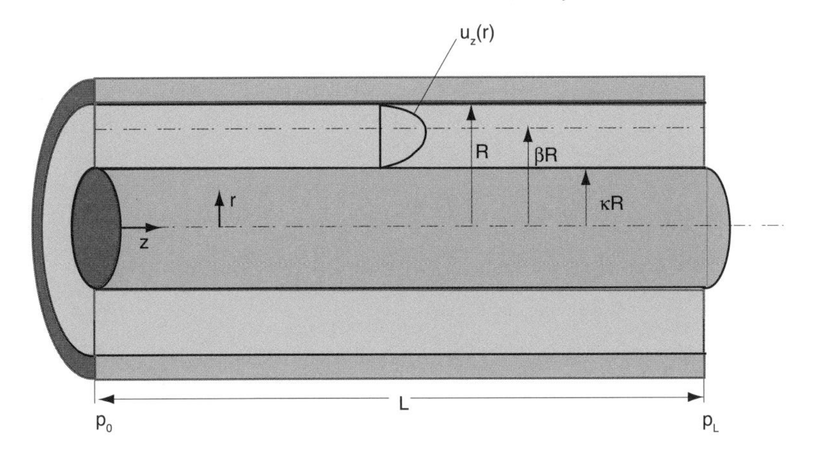

Figure 8.20 Schematic diagram of pressure flow through an annulus

The z-momentum equation is then reduced to

$$\frac{1}{r}\frac{d}{dr}(r\tau_{zr}) = \frac{dp}{dz} \tag{8.88}$$

Integrating this equation we obtain

$$r\tau_{zr} = \frac{dp}{dz}\frac{r^2}{2} + c_1 \tag{8.89}$$

The constant c_1 cannot be set equal to zero, because $\kappa R \leqslant r \leqslant R$. However, β can be used rather than c_1,

$$r\tau_{zr} = \frac{\Delta p\, R}{2L}\left(\frac{r}{R} - \beta^2\frac{R}{r}\right) \tag{8.90}$$

which makes β the new integration constant. The power-law expression for the shear stress is given by

$$\begin{aligned}\tau_{zr} &= -m\left(\frac{du_z}{dr}\right)^n \text{ if } \kappa R \leq r \leq \beta R \\ \tau_{zr} &= m\left(-\frac{du_z}{dr}\right)^n \text{ if } \beta R \leq r \leq R\end{aligned} \tag{8.91}$$

Substitution of these expressions into the momentum equation leads to differential equations for the velocity distribution in the two regions. Integrating these equations with boundary conditions, $uz = 0$ at $r = \kappa R$ and at $r = R$, leads to

$$\begin{aligned}u_z &= R\left[\frac{\Delta p\, R}{2mL}\right]^{1/n} \int_\kappa^\xi \left(\frac{\beta^2}{\xi'} - \xi'\right)^{1/n} d\xi' \text{ if } \kappa \leq \xi \leq \beta \\ u_z &= R\left[\frac{\Delta p\, R}{2mL}\right]^{1/n} \int_\xi^1 \left(\xi' - \frac{\beta^2}{\xi'}\right)^{1/n} d\xi' \text{ if } \beta \leq \xi \leq 1\end{aligned} \tag{8.92}$$

where $\xi = r/R$. In order to find the parameter β, the above equations must match at the location of the maximum velocity,

$$\int_\kappa^\xi \left(\frac{\beta^2}{\xi'} - \xi'\right)^{1/n} d\xi' = \int_\xi^1 \left(\xi' - \frac{\beta^2}{\xi'}\right)^{1/n} d\xi' \tag{8.93}$$

This equation is a relation between β, the geometrical parameter κ, and the power-law exponent n. The volumetric flow rate in the annulus becomes

$$\begin{aligned}Q &= 2\pi \int_{\kappa R}^R u_z r dr \\ &= \pi R^3 \left[\frac{\Delta p\, R}{2mL}\right]^{1/n} \int_\kappa^1 \left|\beta^2 - x'^2\right|^{1/n+1} x'^{-1/n} dx' \\ &= \frac{\pi R^{3+1/n}}{1/n+3}\left[\frac{\Delta p\, R}{2mL}\right]^{1/n} \left[\left(1-\beta^2\right)^{1+1/n} - \kappa^{1-1/n}\left(\beta^2 - \kappa^2\right)^{1+1/n}\right]\end{aligned} \tag{8.94}$$

8.4.4 Radial Flow Between two Parallel Discs – Newtonian Model

Radial flow between parallel discs is a very common flow type encountered in polymer processing, particularly during injection mold filling. In this section, we seek the velocity profile, flow rates, and pressure for this type of flow using the notation presented in Fig. 8.21.

Consider a Newtonian fluid that is flowing due to a pressure gradient between two parallel disks that are separated by a distance $2\,h$. The velocity and pressure fields that

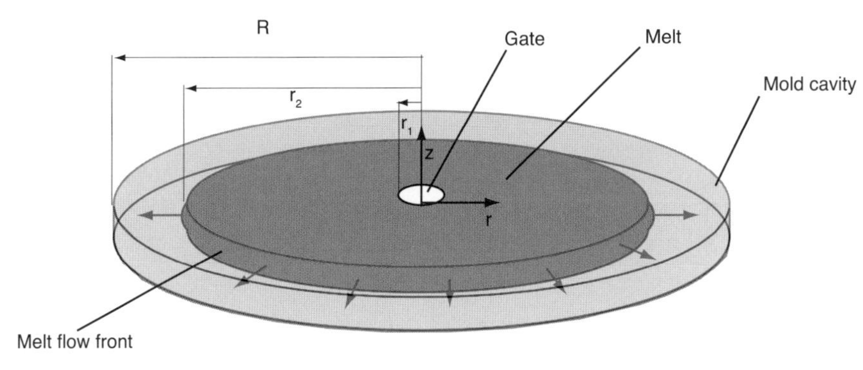

Figure 8.21 Schematic diagram of a center-gated disc-shaped mold during filling

Radial pressure flow:

- Injection molding cavity flow

we will solve for are $u_r = u_r(z, r)$ and $p = p(r)$. According to the Newtonian fluid model, the stress components are

$$\begin{aligned} \tau_{rr} &= -2\mu \frac{\partial u_r}{\partial r} \\ \tau_{\theta\theta} &= -2\mu \frac{u_r}{r} \\ \tau_{rz} = \tau_{zr} &= -\mu \frac{\partial u_r}{\partial z} \end{aligned} \tag{8.95}$$

The continuity equation is reduced to

$$\frac{1}{r}\frac{\partial}{\partial r}(r\, u_r) = 0 \tag{8.96}$$

which indicates that $r\, u_r$ must be a function of z only, $f(z)$. Therefore, from the continuity equation

$$u_r = \frac{f(z)}{r} \tag{8.97}$$

the stresses are reduced to

Order of magnitude analysis is used to reduce the equations. Only shear stresses are significant

$$\begin{aligned} \tau_{rr} &= -2\mu \frac{f(z)}{r^2} \\ \tau_{\theta\theta} &= +2\mu \frac{f(z)}{r^2} \\ \tau_{rz} = \tau_{zr} &= -\frac{\mu}{r}\frac{\partial f(z)}{\partial z} \end{aligned} \tag{8.98}$$

Neglecting the inertia effects, the momentum equation becomes

$$-\frac{1}{r}\frac{\partial}{\partial r}(r\, \tau_{rr}) - \frac{\partial \tau_{zr}}{\partial z} + \frac{\tau_{\theta\theta}}{r} - \frac{\partial p}{\partial r} = 0 \tag{8.99}$$

which is reduced to

$$-\frac{\partial p}{\partial r} + \frac{\mu}{r}\frac{d^2 f(z)}{dz^2} = 0 \tag{8.100}$$

This equation can be integrated, because the pressure is only a function of r. The constants of integration can be solved for by using the following boundary conditions

$$f(\pm h) = 0 \tag{8.101}$$

For the specific case where the gate is at r_1 and the front at r_2, the velocity field is given by

$$u_z(r, z) = \frac{h^2 \Delta p}{2\mu\, r \ln(r_2/r_1)}\left[1 - \left(\frac{z}{h}\right)^2\right] \tag{8.102}$$

The volumetric flow rate is found by integrating this equation over the cross sectional area

$$Q = \int_0^{2\pi}\int_{-h}^{+h} u_z(r, z)\, r\, d\theta\, dz = \frac{4\pi h^3 \Delta p}{3\mu \ln(r_2/r_1)} \tag{8.103}$$

The above equation can also be used to solve for pressure drop from the gate to the flow front

$$\Delta p = \frac{3Q\mu \ln(r_2/r_1)}{4\pi h^3} \tag{8.104}$$

Δp is the boundary condition when solving for the pressure distribution within the disc, by integrating Eq. 8.100

$$p = \frac{\Delta p \ln(r/r_1)}{\ln(r_2/r_1)} \tag{8.105}$$

In order to predict the position of the flow front, r_2, as a function of time, we first perform a simple mass (or volume) balance

$$2h\pi\left(r_2^2 - r_1^2\right) = Q\,t \tag{8.106}$$

which can be solved for r_2 as

$$r_2 = \sqrt{\frac{Q\,t}{2h\pi} + r_1^2} \tag{8.107}$$

The above equations can now be used to plot the pressure requirement, or pressure at the gate, for a given flow rate as a function of time. They can also be used to plot for the pressure distribution within the disc at various points in time or flow front positions. In addition, the same equations can be used to solve for flow rates for given injection pressures.

Example 8.7 Predicting pressure profiles in a disc-shaped mold using a Newtonian model

To show how the above equations are used, let us consider a disc-shaped cavity of $R = 150$ mm, a gate radius, r_1, of 5 mm, and a cavity thickness of 2 mm, i.e., $h = 1$ mm. Assuming a Newtonian viscosity $\mu = 6\,400$ Pa·s and constant volumetric flow rate $Q = 50$ cm^3/s, predict the position of the flow front, r_2, as a function of time, as well as the pressure distribution inside the disc mold.

Equations 8.105 and 8.107 can easily be solved using the given data. Figure 8.22 presents the computed flow front positions with the corresponding pressure profiles.

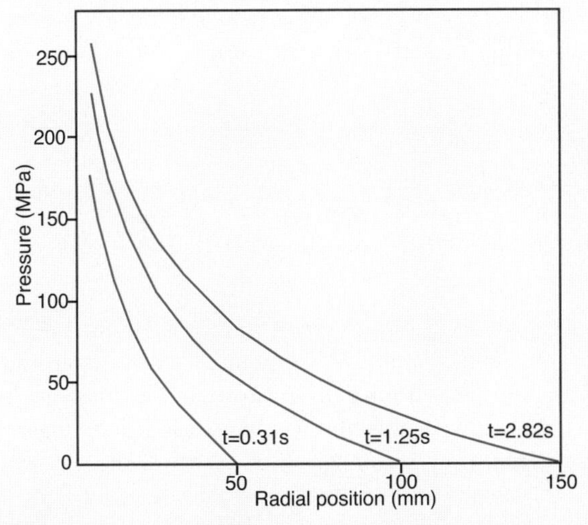

Figure 8.22 Radial pressure profile as a function of time in a disc-shaped mold computed using a Newtonian viscosity model

■

8.4.5 Cooling or Heating in Polymer Processing

Cooling or heating is of great concern in polymer processing. Due to the low thermal conductivity of polymers, the cooling and heating steps of a process control the cycle time. Cooling or heating processes take place inside molds during injection and compression molding, respectively. This heating or cooling process is a contact process during which the polymer melt is pressed against a mold surface, allowing for an effective heat transfer between mold and melt. During some extrusion processes, such as extrusion of fibers or films (during film blowing), cooling takes place in water or air, where we must rely on the heat transfer coefficient between the flowing media and the polymer surface. In other extrusion processes, the polymer soon comes into contact with a metal surface, such as a sizing sleeve during extrusion of plastic pipes.

Cooling time is proportional to the thickness of the part squared

Table 8.6 Penetration thickness and characteristic times in heating and cooling of polymers

L	$T = L^2/\alpha$
100 μm	0.025 s
1 mm	2.5 s
2 mm	10 s
10 mm	250 s

Heating and cooling often take place while the polymer melt flows, making viscous dissipation an influencing factor during the process. However, since most plastic parts are thin, the conduction often occurs only across the thickness and the viscous heating is a result of shear within the narrow gap of a die or mold cavity. For such cases, the equations reduce to

$$\varrho C_{\mathrm{P}} \frac{\partial T}{\partial t} = k \frac{\partial^2 T}{\partial x^2} + \eta \left(\frac{\partial u_z}{\partial x} \right)^2 \tag{8.108}$$

There are various special cases for the above equation, some of which are described in the following.

Cooling or heating of a semi-infinite slab. Although very thick parts are not an issue in polymer processing, we can still solve this problem to illustrate time scales associated with various thicknesses. In a semi-infinite slab, we have a cooling or heating process that takes place in a space from $x = 0$ to $x = \infty$. At $t = 0$, the temperature throughout the slab is T_0 and the surface temperature is suddenly lowered or raised to T_{S}.

For this problem, the above equation reduces to

$$\frac{\partial T}{\partial t} = \alpha \frac{\partial^2 T}{\partial x^2} \tag{8.109}$$

Bird, Stewart, and Lighfoot [10] present a solution to this problem, given by

$$\frac{T - T_{\mathrm{S}}}{T_0 - T_{\mathrm{S}}} = \operatorname{erf}\left(\frac{x}{2\sqrt{\alpha t}} \right) \tag{8.110}$$

where $\operatorname{erf}\left(\frac{x}{2\sqrt{\alpha t}} \right)$ is given by

$$\operatorname{erf}\left(\frac{x}{2\sqrt{\alpha t}} \right) = \frac{2}{\sqrt{\pi}} \int_0^{\frac{x}{2\sqrt{\alpha t}}} e^{-u^2} du \tag{8.111}$$

Figure 8.23 shows the dimensionless temperature as a function of dimensionless time $\left(\frac{x}{2\sqrt{\alpha t}} \right)$.

From the above equations and Fig. 8.24, we can define a heat penetration thickness.

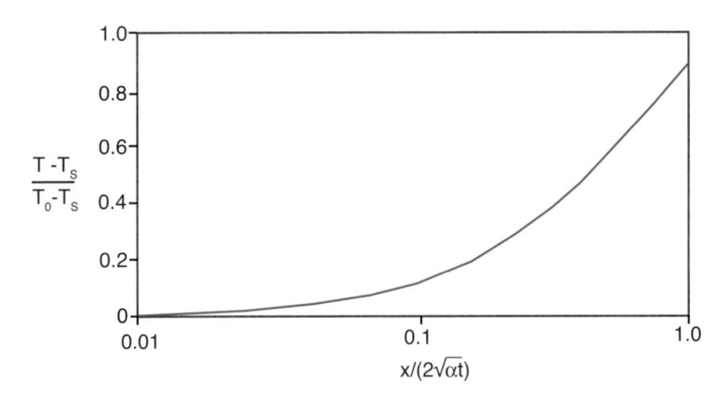

Figure 8.23 Dimensionless temperature as a function of dimensionless time and thickness

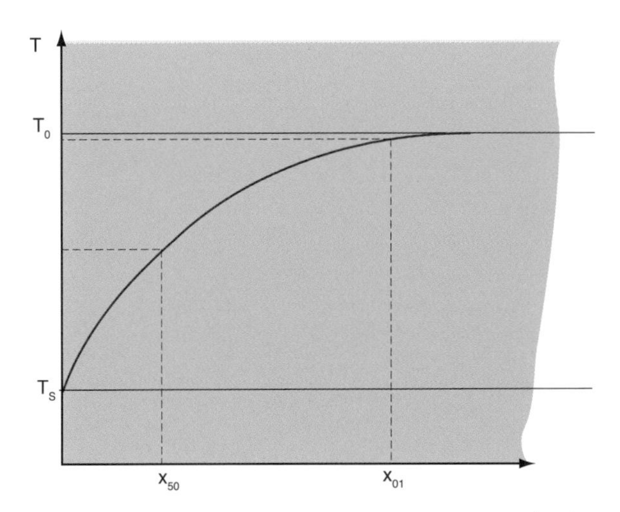

Figure 8.24 Schematic of a semi-infinite cooling body. Denoted are depths at which 50 % and 1 % of the temperature difference is felt

For example, Fig. 8.24 presents two depths, one where 1 % and another where 50 % of the temperature differential is felt. The 1 % temperature differential is defined by

$$T = T_0 + 0.01\,(T_S - T_0) \tag{8.112}$$

or

$$0.99 = \text{erf}\left(\frac{x_{01}}{2\sqrt{\alpha t}}\right) \tag{8.113}$$

which can be used to solve for the given time that leads to a 1 % temperature change for a given depth x_{01}

$$t_{01} = \frac{x_{01}^2}{13.25\alpha} \tag{8.114}$$

The same analysis can be carried out for a 50 % thermal penetration time

$$t_{50} = \frac{x_{50}^2}{0.92\alpha} \tag{8.115}$$

Hence, the time when most of the temperature difference is felt by a part of a given thickness is of the order

$$t = \frac{L^2}{\alpha} \tag{8.116}$$

which can be used as a characteristic time for a thermal event that takes place through diffusion. Because polymers have a thermal diffusivity of about 10^{-7} m^2/s, we can easily compute the characteristic times for heating or cooling as a function of part thickness, $2L$. Some characteristic times are presented as a function of thickness in Table 8.6.

With a characteristic time for heat conduction we can now define a dimensionless time using

The Fourier number is the ratio of the time it takes for a part to reach thermal equilibrium by diffusion to a characteristic process time

$$F_0 = \frac{L^2}{\alpha t} \tag{8.117}$$

which is the well known Fourier number.

Cooling and heating of a finite thickness plate. A more accurate solution of the above problem is to determine the cooling process of the actual part, hence, one of finite thickness $2L$. For the heating process of a finite thickness plate we can solve Eq. 8.109 to give

$$\frac{T - T_S}{T_S - T_0} = 1 - \frac{4}{\pi}\sum_{n=1}^{\infty}\frac{(-1)^{n-1}}{2n-1}\cos\left[\frac{(2n-1)\,\pi x}{L}\right]\exp\left(-\left[\frac{(2n-1)\,\pi^2}{2}F_0\right]\right) \tag{8.118}$$

Figure 8.25 presents the temperature history at the center of the plate and Fig. 8.26 shows a comparison between the prediction and a measured temperature development in an 8 mm thick PMMA plate. As can be seen, the model does a good job of approximating reality.

Cooling and heating of a finite thickness plate using convection. As mentioned earlier, cooling with air or water is very common in polymer processing. For example, the cooling of a film during film blowing is controlled by air blown from a ring located near the die exit. In addition, many extrusion operations extrude into a bath of running chilled water. Here, the controlling parameter is the heat transfer coefficient h, or in dimensionless form the Biot number, Bi, given by

$$Bi = \frac{hL}{k} \tag{8.119}$$

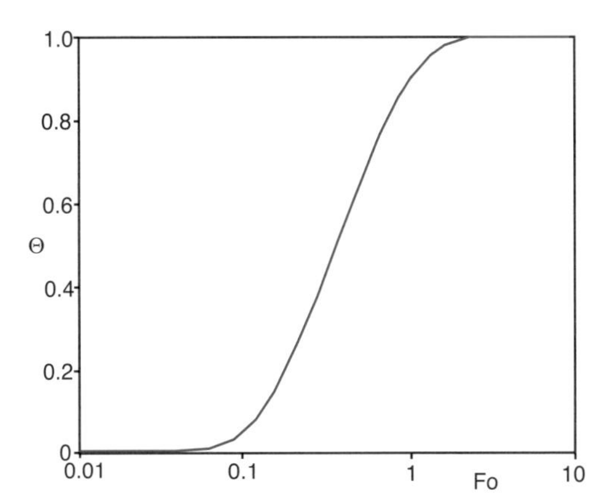

Figure 8.25 Center-line temperature history during heating of a finite thickness plate. Note that cooling is represented by the same curve using $1 - \Theta$ as the dimensionless temperature

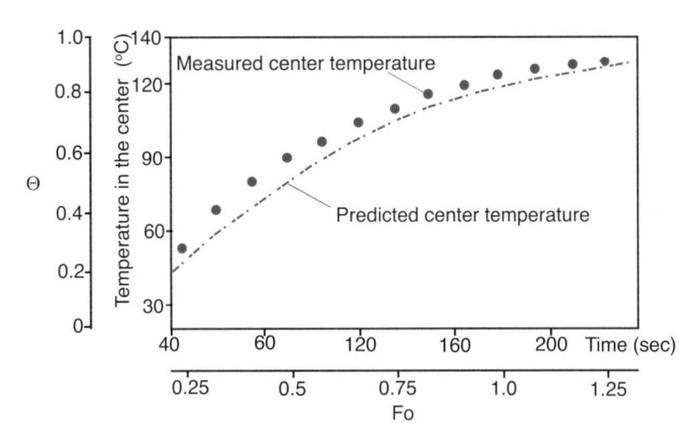

Figure 8.26 Experimental and computed center-line temperature history during heating of an 8 mm thick PMMA plate. The initial temperature $T_0 = 20$ °C and the heater temperature $T_S = 140$ °C [7]

An approximate solution for the convective cooling of a plate of finite thickness is given by Agassant *et al.* [15]

$$\frac{T - T_f}{T_0 - T_f} \approx e^{(-Bi\,Fo)} \cos\left(\sqrt{Bi}\,\frac{x}{L}\right) \tag{8.120}$$

The center-line temperature for plates of finite thickness is given in Fig. 8.26 and a comparison between the prediction and experiments for an 8 mm thick PMMA plate cooled with a heat transfer coefficient, h, of 33 W/m²/K is given in Fig. 8.27. As can be seen, theory and experiment are in relatively good agreement.

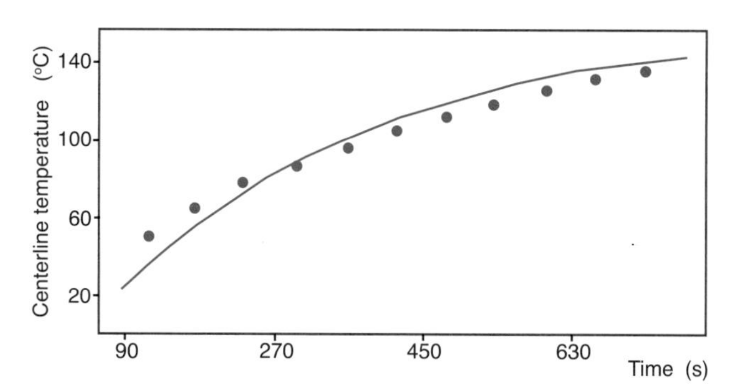

Figure 8.27 Center-line temperature history of an 8 mm thick PMMA plate during convective heating inside an oven set at 155 °C. The initial temperature was 20 °C. The predictions correspond to a Biot number, Bi =1.3 or a corresponding heat transfer coefficient, $h = 33$ W/m²/K [7]

References

1. J. Pawlowski, *Die Ähnlichkeitstheorie in der Physikalisch-Technischen Forschung — Grundlagen und Anwendung*. Springer-Verlag, Berlin, 1971.
2. T. E. Stanton and J. R. Pannell, Similarity of motion in relation to the surface friction of fluids. *Phil. Trans. Roy. Soc. London*, 214:199, 1914.
3. M. Zlokarnik, *Scale-up in Chemical Engineering*, Wiley-VCH,Weinhein, 2002.
4. M. Zlokarnik, Eignung von Rührern zum Homogenisieren von Flüssigkeitsgemischen. *Chem. Ing. Tech.*, 39(9/10):539, 1967.
5. J. Pawlowski, *Chemie-Ing. Techn.*, 38:1229, 1966.
6. J. Pawlowski, *Chemie-Ing. Techn.*, 39:1180, 1967.
7. Z. Tadmor and I. Klein, *Engineering Principles of Plasticating Extrusion*, Van Nostrand Reinhold Company, New York, 1970.
8. C. Rauwendaal, *Polymer Extrusion*, 4th Edition, Hanser Publishers, Munich, 2001.
9. T. Osswald and J. P. Hernández, *Polymer Processing — Modeling and Simulation*, Hanser Publishers, Munich, 2006.
10. R. B. Bird, W. E. Steward, and E. N. Lightfoot, *Transport Phenomena*, John Wiley & Sons, New York, 2nd edition, 2002.
11. W. M. Deen, *Analysis of Transport Phenomena*, Oxford University Press, Oxford, 1998.
12. H. S. Hele-Shaw. *Proc. Roy. Inst.*, 16:49, 1899.
13. F. P. Incropera and D. P. DeWitt, *Fundamentals of Heat and Mass Transfer*, John Wiley & Sons, New York, 1996.
14. H. Potente. Kunststofftechnologie 2. Fachbereich 10 Maschinentechnik I — Technologie der Kunststoffe, Universität-Gesamthochschule Paderborn, 2000.
15. J. F. Agassant, P. Avenas, J.-Ph. Sergent, and P. J. Carreau, *Polymer Processing — Principles and Modeling*, Hanser Publishers, Munich, 1991.

9 Modeling Polymer Processes

Although all polymer processes involve complex phenomena that are non-isothermal, non-Newtonian, and often viscoelastic, most of them can be simplified sufficiently to allow the construction of analytical models. These analytical models involve one or more of the simple flows derived in the previous chapter. These *back of the envelope* models allow us to predict pressures, velocity fields, temperature fields, melting and solidification times, cycle times, etc. The models that are derived will aid the student or engineer to better understand the process under consideration, allowing for optimization of processing conditions, and even geometries and part performance.

This chapter attempts to cover the most important polymer processes. First we derive solutions to non-isothermal approximations of various polymer processes. We begin with a Newtonian analysis of the metering or pumping section of the single screw extruder, followed by the flow in several common extrusion dies, including the analysis using non-Newtonian shear thinning polymer melts. Within this section, we also solve for flow and deformation in a fiber spinning operation using a viscoelastic flow model. The next sections presents a detailed analysis of isothermal, Newtonian, and non-Newtonian flow in two-roll calendering systems. This is followed by the analysis of various injection molding problems. From this point on, non-isothermal problems are introduced, which are exemplified using melting and solidification problems, ending with melting in a plasticating single screw extruder and the curing kinetics of elastomers and thermosetting polymers.

9.1 Single Screw Extrusion – Isothermal Flow Problems

Most flows that take place during polymer processing can be simplified and modeled isothermally. When a system reaches steady state, a polymer melt can be considered isothermal, unless viscous dissipation plays a significant role. As was discussed in Chapter 8, the significance of viscous dissipation during processing is assessed using the Brinkman number given by

$$Br = \frac{\eta\, u_0^2}{k \Delta T} \tag{9.1}$$

where η is a characteristic viscosity, u_0 a characteristic screw speed, k the thermal conductivity of the melt, and ΔT a characteristic temperature variation within the process. In the example below, we attempt to determine if a process can be considered isothermal or if viscous dissipation is significant.

Example 9.1 Determining the effect of viscous dissipation in the metering section of a single screw extruder

Consider a 60-mm diameter extruder with a 4-mm channel depth and a screw speed of 60 rpm. The melt used in this extrusion system is a polycarbonate with a viscosity of 100 Pa·s, a thermal conductivity of 0.2 W/m/K and a heater temperature of 300 °C. To assess the effect of viscous heating, we can choose a temperature difference, ΔT, of 30 K. This simply means that the heater temperature is 30 K above the melting temperature of the polymer. For this system, the Brinkman number becomes

Even a process with Brinkman number that is less than 1 experiences moderate temperature rises due to viscous heating

$$Br = \frac{\eta(\pi Dn)^2}{k\Delta T} = \frac{100\ \text{Pa·s}(0.188\ \text{m/s})^2}{0.2\ \text{W/m/K}\ (30\ \text{K})} = 0.59 \tag{9.2}$$

which means that the heat is conducted out much faster than it is generated by viscous heating, making the isothermal assumption plausible.

This can be easily checked by assuming that the flow inside this section of the screw can be modeled using a simple shear flow, and that most of the conduction occurs through the channel thickness direction. For such a case, the energy equation in that direction, say the y-direction, reduces to

$$0 = k\frac{\partial^2 T}{\partial y^2} + \dot{Q} \tag{9.3}$$

where $\dot{Q}$ is the heat generation by viscous heating, which for simple shear flow reduces to

$$\dot{Q} = \mu\left(\frac{\partial u}{\partial y}\right)^2 = \left(\frac{\pi Dn}{h}\right)^2 \tag{9.4}$$

If we integrate Eq. 9.3 two times and use a boundary condition of $T_0 = 300$ °C on the screw and barrel surfaces, we get

$$T = T_0 + \frac{\mu}{2k}\left(\frac{\pi Dn}{h}\right)^2 (2hy - y^2) \tag{9.5}$$

Setting $y = 0.002$ m, which is located in the middle of the channel and using the above data, we get a temperature of 306.6 °C, a 6.6 K temperature rise. Although there is a measurable temperature rise in this system, it is not significant enough to warrant a non-isothermal analysis.

■

9.1.1 Newtonian Flow in the Metering Section of a Single Screw Extruder

Analyzing the flow in a single screw extruder using analytical solutions can only be done if we assume a Newtonian polymer melt. As can be seen in these sections, the flow inside the screw channel is three-dimensional, made up of a combination of pressure and drag flows.

The geometry of a single screw extruder can be simplified by unwinding or unwrapping the material from the channel, as schematically depicted in Fig. 9.1. By unwinding the channel contents we are assuming that the effects caused by the curvature of the screw are negligible. This is true for most screw geometries where the channel is shallow. Furthermore, if we assume that the barrel rotates instead of the screw, we can model the flow inside the channel using a combination of shear and pressure flow between parallel plates.

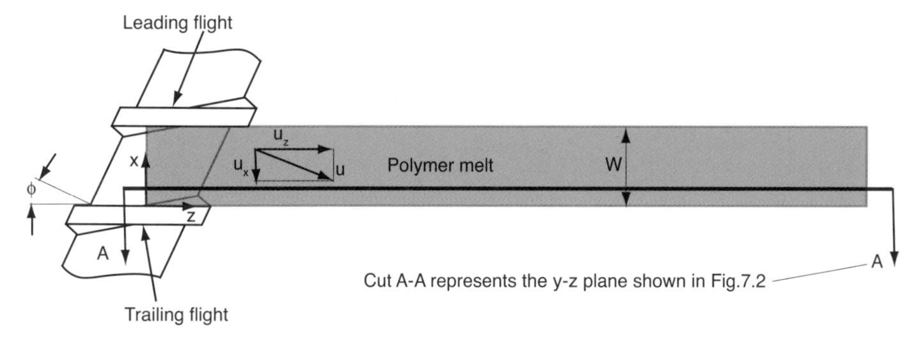

Figure 9.1 Unwrapped screw channel

Figure 9.2 shows a cross-section of the unwrapped channel in the yz-plane. The length L is the helical length of the channel, which for a square pitch screw can be computed using $L = (\text{number of turns})D/\sin(\varphi)$. The lower surface of the channel is the screw root and is assigned a zero velocity and the upper surface is in contact with the barrel, which is given a velocity $u = \pi Dn$. Because of the helical geometry of the screw, this velocity is broken down further into x and z components. The x-component is given by $u_x = u\sin(\varphi)$ and is referred to as the cross-flow component. The z-component, which is given by $u_z = u\cos(\varphi)$, is the down-channel component and is the one that leads to pumping by dragging the polymer down the channel of the screw. Since the polymer is dragged against a die, the pressure will build-up moving in the down-channel direction. This results in a pressure flow that moves in the opposite direction of the drag flow component.

The volumetric flow is given by

$$Q_T = Q_D + Q_P \tag{9.6}$$

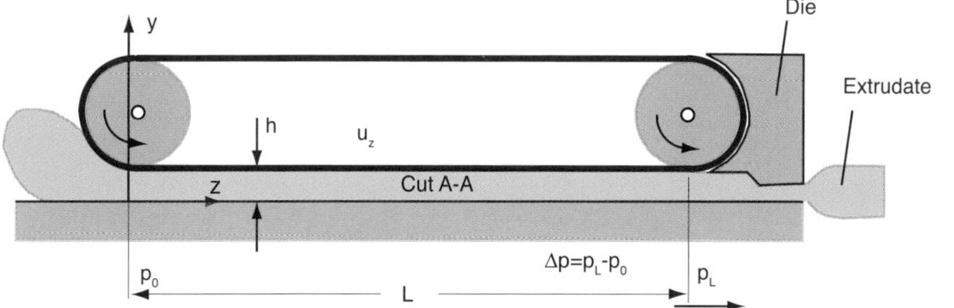

Unwrapped screw channel in the down channel direction

Figure 9.2 Model of a hypothetical viscosity pump

where Q_T is the drag flow due to simple shear, and Q_P is the pressure flow, given by the following equations,

Drag flow

Pressure flow

$$\begin{aligned} Q_D &= \frac{1}{2} u_z h W \\ Q_P &= -\frac{W h^3 \Delta p}{12 \mu L} \end{aligned} \tag{9.7}$$

respectively. The total volumetric flow as a function of Δp is the given by

$$Q_T = \frac{1}{2} u_z h W - \frac{W h^3 \Delta p}{12 \mu L} \tag{9.8}$$

and finally

$$Q_T = \frac{\pi D n h W}{2} \cos \varphi - \frac{W h^3 \Delta p}{12 \mu L} \tag{9.9}$$

As can be seen in the above equations, the resulting total flow is a combination of drag and pressure flows. Depending on the restriction of the die, various types of flows can develop inside the screw channel. Figure 9.3 schematically depicts the different situations that may arise. At closed discharge, which occurs when the die is plugged, the net flow in the down-channel direction is zero, at which point the maximum pressure build-up is achieved. At open discharge, when the die is absent and the extruder is pumping into the atmosphere, the flow in the down-channel direction reduces to a simple shear flow. It should be noted that the above equations neglect the effect of leakage flow over the flight; hence they over-predict the net material throughput as well as the maximum pressure build-up. Furthermore, since there is a no-slip condition

The maximum flow through an extruder occurs at open discharge and equals the drag flow rate

The smallest flow rate through an extruder is zero and it occurs at closed discharge

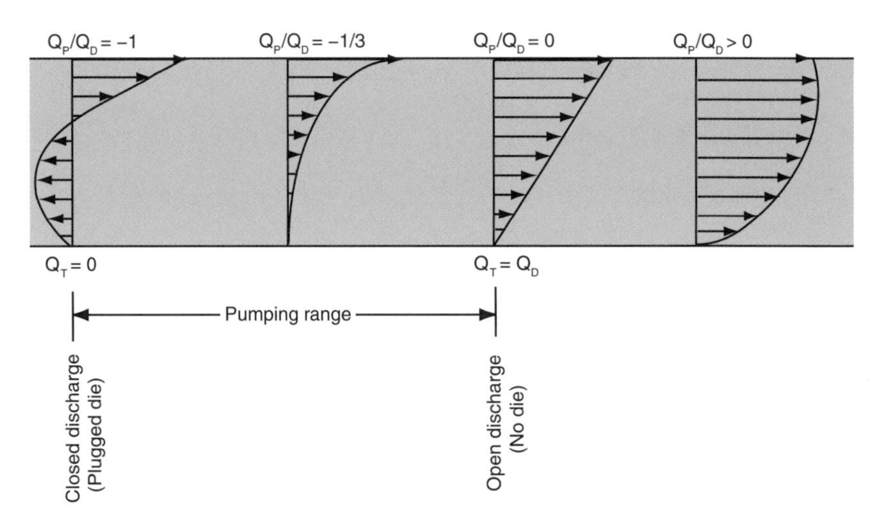

Figure 9.3 Down-channel velocity profiles for different pumping situations with a single screw extruder

between the polymer and the flight walls, the velocity profiles depicted in Fig. 9.3 are only valid away from the flights. This effect further contributes to over-prediction of the volumetric throughput of a single screw extruder.

To correct this effect, which significantly affects extruders with a deep channel screw, Tadmor and Gogos [1] present the following modification

$$Q_T = \frac{\pi D n h W}{2} \cos\varphi F_D - \frac{W h^3 \Delta p}{12\mu L} F_P \tag{9.10}$$

where F_D and F_P are correction factors that account for the flow reduction down the channel of the screw and can be computed using

$$F_D = \frac{16W}{\pi^3 h} \sum_{i=1,3,5}^{\infty} \frac{1}{i^3} \tanh\left(\frac{i\pi W}{2h}\right) \tag{9.11}$$

$$F_P = 1 - \frac{192h}{\pi^3 W} \sum_{i=1,3,5}^{\infty} \frac{1}{i^5} \tanh\left(\frac{i\pi W}{2h}\right) \tag{9.12}$$

It should be noted that the correction is less than 5 % for channels that have an aspect ratio, W/h, larger than 10.

9.1.2 Cross Channel Flow in a Single Screw Extruder

The cross channel flow is derived in a similar fashion as the down channel flow. This flow is driven by the x-component of the velocity, which creates a shear flow in that direction. However, since the shear flow pumps the material against the trailing flight of the screw channel, it results in a pressure increase that creates a counteracting pressure flow, which leads to a net flow of zero.[1] The flow rate per unit depth at any arbitrary position along the z-axis can be defined by

Cross-channel flow rate is zero

$$q_x = -\frac{u_x h}{2} - \frac{h^3}{12\mu} \frac{\partial p}{\partial x} = 0 \tag{9.13}$$

Here, we can solve for the pressure gradient $\frac{\partial p}{\partial x}$ to be

$$\frac{\partial p}{\partial x} = -\frac{6\mu u_x}{h^2} \tag{9.14}$$

Once the pressure is known, we can compute the velocity profile across the thickness of the channel using

$$u_x(y) = -\frac{u_x y}{h} - \frac{1}{2\mu}\left(-\frac{6\mu u_x}{h^2}\right)(hy - y^2) \tag{9.15}$$

1) This assumption is not completely true, because some of the material flows over the screw flight into the regions of lower pressure in the up-channel direction.

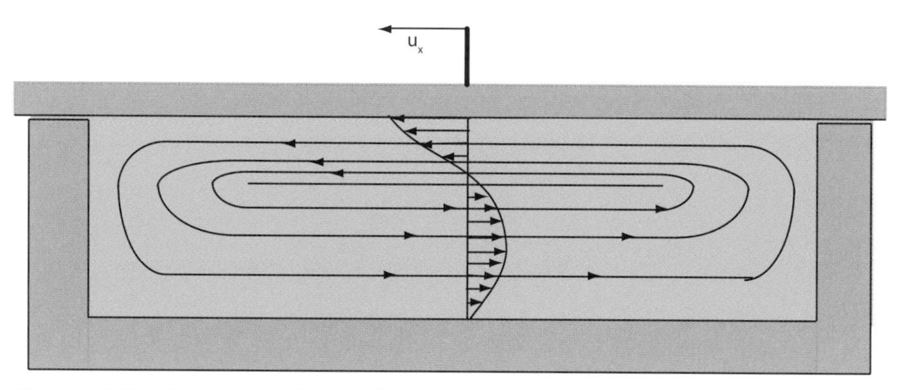

Figure 9.4 View in the down channel direction depicting the resulting cross flow

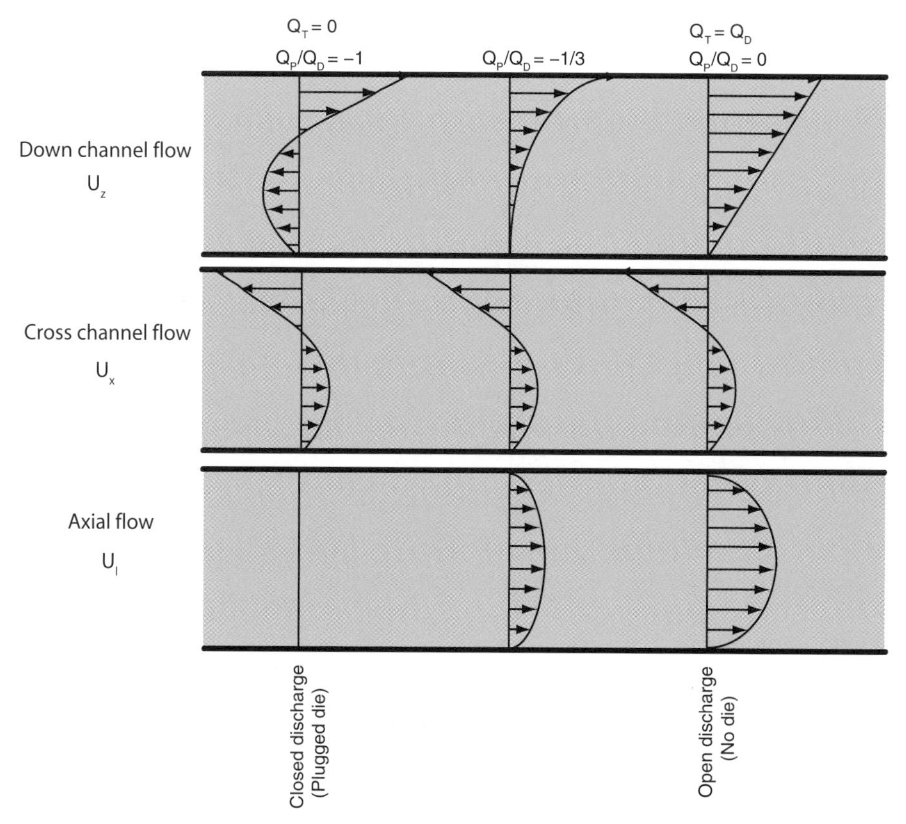

Figure 9.5 Down-channel, cross-channel, and axial velocity profiles for various situations that arise in a single screw extruder

This velocity profile is schematically depicted in Fig. 9.4. As shown in the figure, the cross flow generates a recirculating flow, which performs a stirring and mixing action important in extruders for blending as well as melting.

If we combine the flow generated by the down channel and cross channel flows, a net flow is generated in axial or machine direction (u_1) of the extruder, schematically depicted in Fig. 9.5. As can be seen, the maximum axial flow is generated at open discharge; whereas at closed discharge, the axial flow is zero. From the velocity profiles presented in Fig. 9.5 we can easily deduce which path a particle flowing with the polymer melt will take.

Due to the combination of cross-channel and down-channel flows, peculiar particle paths develop for the various die restrictions. The paths that form for various situations are presented in Fig. 9.6. When the particle flows near the barrel surface of the channel, it moves at its fastest speed and in a direction nearly perpendicular to the axial direction of the screw. As the particle approaches the screw flight, it submerges and approaches the screw root, at which point it travels back at a slower speed, until it reaches the leading flight of the screw, which causes the particle to rise once more and travel in the down-channel direction. Depending on the die restriction, the path may change. For example, for the closed discharge situation, the particle simply travels on a path perpendicular to the axial direction of the screw, recirculating between the barrel surface and screw root.

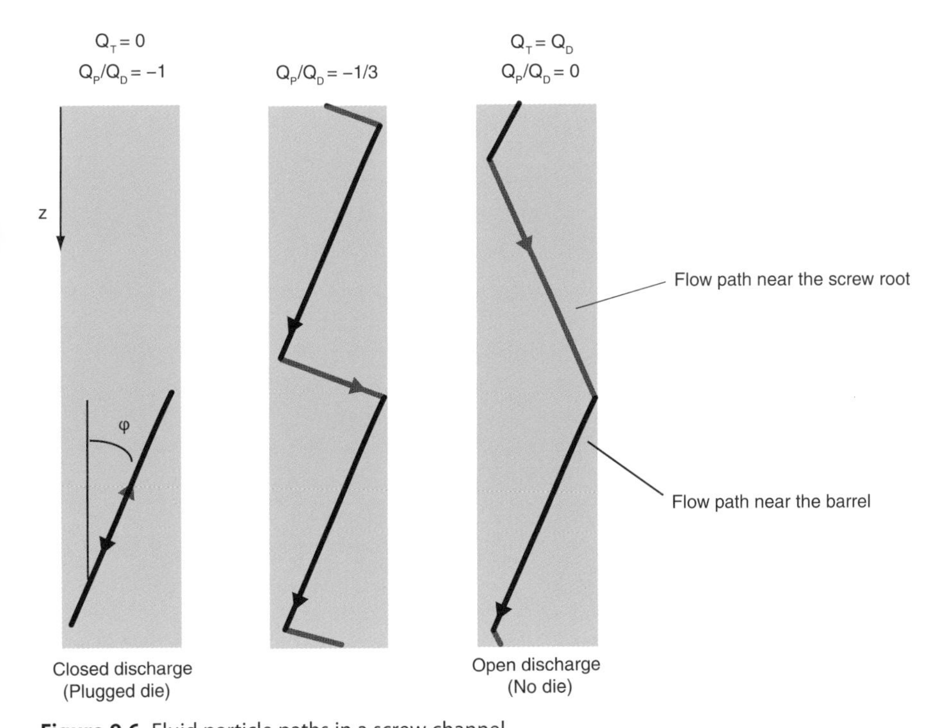

Figure 9.6 Fluid particle paths in a screw channel

9.1.3 Newtonian Isothermal Screw and Die Characteristic Curves

When extruding a Newtonian fluid through a die, the throughput is directly proportional to the pressure build-up in the extruder and inversely proportional to the viscosity as stated in

$$Q_{\text{die}} = \frac{\aleph}{\mu}\Delta p \tag{9.16}$$

where $\aleph$ is a proportionality constant related to the geometry of the die, i. e., for a capillary die of length l and radius R, $\aleph = \pi R^4/8/l$. Equation 9.16 is commonly referred to as the *die characteristic curve.* If we equate the volumetric throughput of the extruder and die, using Eqs. 9.10 and 9.16, we get

The operating point is where the screw characteristic curve and the die characteristic curve cross each other

$$\frac{\pi DnhW}{2}\cos\varphi F_{\text{D}} - \frac{Wh^3\Delta p}{12\mu L}F_{\text{P}} = \frac{\aleph}{\mu}\Delta p \tag{9.17}$$

which can be solved for the pressure build-up, $\Delta p = \Delta p_{\text{D}}$, corresponding to a specific die as

$$\Delta p_{\text{D}} = \frac{6\mu\pi Dn\cos\varphi hWLF_{\text{D}}}{12\aleph L + Wh^3F_{\text{P}}} \tag{9.18}$$

Substituting Eq. 9.18 into Eq. 9.16, we arrive at the volumetric throughput for a single screw pump with a particular die, described by

$$Q = \left[\frac{\pi Dn\cos\varphi hWF_{\text{D}}}{2 + \frac{Wh^3F_{\text{P}}}{6\aleph L}}\right] \tag{9.19}$$

which is commonly referred to as the *operating point.* This concept is more clearly depicted in Fig. 9.7.

As one can imagine, there are numerous types of die restrictions. A die that is used to manufacture a thick sheet of polystyrene is significantly less restrictive than a die that is used to manufacture a thin polyethylene teraphthalate film. To account for the variation of die restrictions, the appropriate screw design for a specific application must be chosen. Figure 9.8 presents two types of dies, a restricted and a less restricted die, along with two screw characteristic curves for a deep channel screw and a shallow channel screw. As can be seen, the deep channel screw has a higher productivity when used with a less restricted die, and the shallow screw works best with a high restriction die. It is obvious that a deep screw carries more material and therefore has a higher productivity at open discharge, whereas a shallow screw carries a smaller overall amount of melt, resulting in lower productivity at open discharge. On the other hand, a shallow screw has higher rates of deformation at the same screw speed, which leads to higher shear stresses. This results in larger pressure build-up, which is needed for the high restriction dies. It is therefore necessary to assess each case on an individual basis and design the screw appropriately.

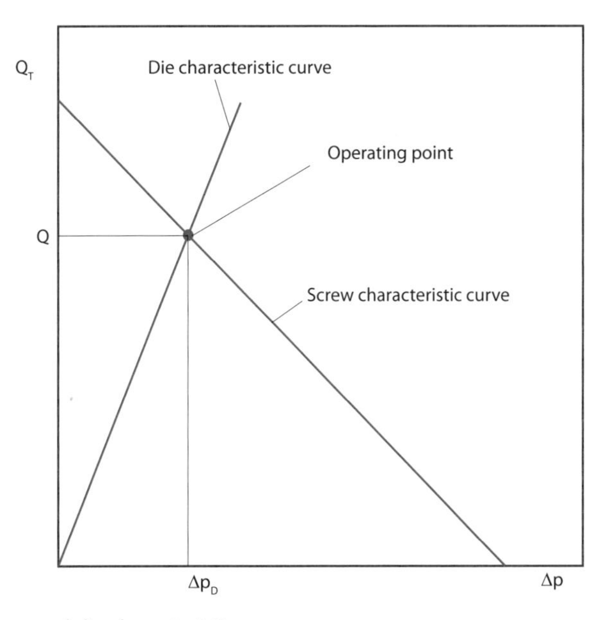

Figure 9.7 Screw and die characteristic curves

The operating point is where the screw characteristic curve and the die characteristic curve cross each other

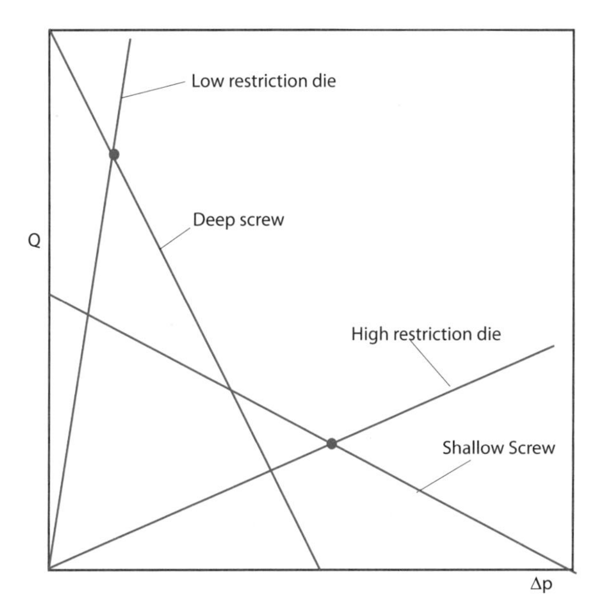

Figure 9.8 Screw and die characteristic curves for various screws and dies

The ideal operating point for a low restriction die lies on the deep channel screw characteristic curve

The ideal operating point for a high restriction die lies on the shallow channel screw characteristic curve

In order to maximize the throughput for a particular screw-die combination, we set the variation of Eq. 9.1 with respect to channel depth, h, to zero as

$$\frac{\partial Q}{\partial h} = 0 \tag{9.20}$$

which results in an optimal channel depth, h^{optimum} for a specific die restriction $\aleph$ of

The optimum channel depth of a screw depends on the flow resistance within the die

$$h^{\text{optimum}} = \left[\frac{6L\aleph}{W}\right]^{1/3} \tag{9.21}$$

Similarly, we can solve for the optimum helix angle, φ, by setting the variation of Eq. 9.19 with respect to the helix angle to zero, $\partial Q/\partial\varphi = 0$. The helix angle is embedded in the channel length, L, term

$$L = \frac{Z}{\sin\varphi} \tag{9.22}$$

where Z is the axial length of the extruder's metering section. After differentiation we get

$$\sin^2\varphi^{\text{optimum}} = \frac{1}{2 + \dfrac{\pi D h^3}{12Z\aleph}} \tag{9.23}$$

Example 9.2 Optimum extruder geometry

You are given the task to find the optimum screw geometry of a 45-mm diameter extruder used for a 3-cm diameter pipe extrusion operation. The pipe's die land length is 100 mm and die opening gap is 2 mm. Determine the optimum channel depth in the metering section and the optimum screw helix angle. Assume a Newtonian isothermal flow and an extrusion metering section that is 5 turns long. Since the die gap is much smaller than the pipe diameter and die length, for the solution of this problem we can assume pressure driven slit flow, which for a Newtonian fluid is governed by

$$Q_{\text{d}} = \frac{W_{\text{d}} h_{\text{d}}^3 \Delta p}{12\mu L_{\text{d}}} \tag{9.24}$$

Where we deduce that the die restriction constant is $\aleph = \frac{W_{\text{d}} h_{\text{d}}^3}{12 L_{\text{d}}}$, which is substituted into Eq. 9.21 assuming a square pitch ($\varphi = 17.65°$) and a channel width of 40 mm (for a 5 mm flight width) to give

The optimum channel depth of a screw is directly proportional to the opening or slit thickness within the die

$$\begin{aligned} h^{\text{optimum}} &= h_{\text{d}}\left[\frac{6(5D/\sin 17.65°)W_{\text{d}}}{12L_{\text{d}}W}\right]^{1/3} \\ &= 2\text{ mm}\left[\frac{6(5\times 45\text{ mm}/\sin 17.65°)\pi(30\text{ mm})}{12(100\text{ mm})(40\text{ mm})}\right]^{1/3} \\ h^{\text{optimum}} &= 4.1\text{ mm} \end{aligned} \tag{9.25}$$

It is interesting to point out that for a die with a pressure flow through a slit, or sets of slits, the optimum channel depth is directly proportional to the die gap. Decreasing the die gap by a certain percentage will result in an optimum channel depth that is reduced by the same percentage. To determine the optimum helix angle, we can re-write Eq. 9.23 for this specific application,

$$\sin^2 \varphi^{\text{optimum}} = \frac{1}{2 + \dfrac{\pi D h^3}{12 Z \aleph}} = \frac{1}{2 + \dfrac{\pi(45\text{ mm})(4.1\text{ mm})^3(100\text{ mm})}{(5 \times 45\text{ mm})(\pi 30\text{ mm})(2\text{ mm})^3}} \qquad (9.26)$$

$$\sin^2 \varphi^{\text{optimum}} = 0.129$$

which results in $\varphi^{\text{optimum}} = 21°$, compared to 17.65° for a square pitch screw. We note that here we used the optimum channel depth.

■

9.2 Extrusion Dies — Isothermal Flow Problems

The flow in many extrusion dies can be approximated with one, or a combination of, simplified models such as slit flow, Hagen Poiseuille flow, annular flow, simple shear flow, etc. A few of these are presented in the following sections using non-Newtonian as well as Newtonian flow models.

9.2.1 End-Fed Sheeting Die

The end-fed-sheeting die, as presented in Fig. 9.9, is a simple geometry that can be used to extrude films and sheets. To illustrate the complexities of die design, we will modify the die, as shown in the figure, in order to extrude a sheet or film with a uniform thickness. In order to achieve this, we must determine the length of the approach zone or die land as a function of the manifold direction, as depicted in the model shown in Fig. 9.10.

For this specific example, the manifold diameter will be kept constant and we will assume a Newtonian isothermal flow, with a constant viscosity μ. The flow of the manifold can be represented using the Hagen-Poiseuille equation as,

The end-fed sheeting die can be simplified to a combination of tube pressure flow and slit pressure flow

$$Q = \frac{\pi R^4}{8\mu}\left(-\frac{dp}{dz}\right) \qquad (9.27)$$

and the flow in the die land (per unit width) can be modeled using the slit flow equation

$$q = \frac{h^3}{12\mu}\left(-\frac{dp}{dl}\right) = \frac{h^3}{12\mu}\left(\frac{p(z)}{L_L(z)}\right) \qquad (9.28)$$

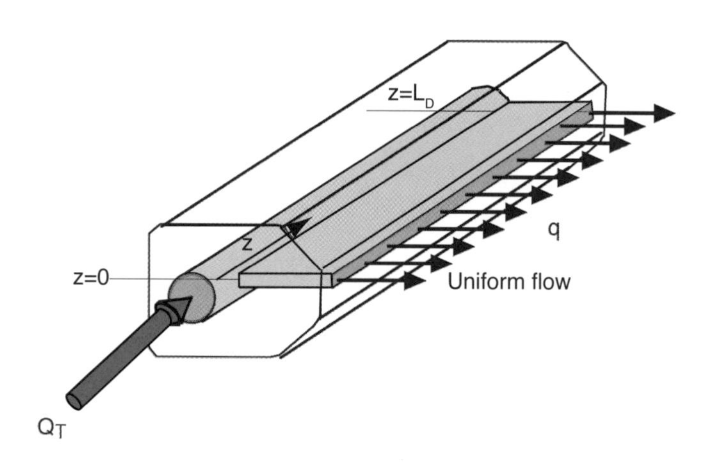

Figure 9.9 Schematic diagram of an end-fed sheeting die

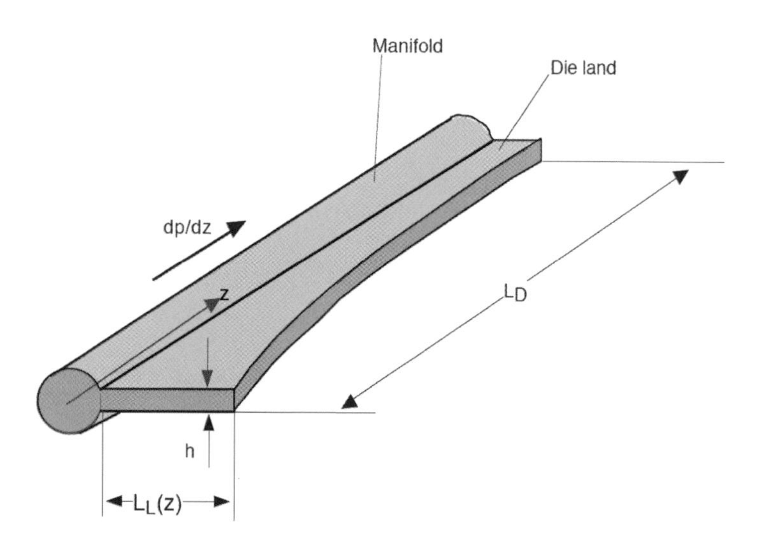

Figure 9.10 Schematic of the manifold and die land in an end-fed sheeting die

The main requirement in a sheeting die is that the flow rate along the slit length is constant (uniform flow), resulting in a uniform film

A manifold that generates a uniform sheet must deliver a constant throughput along the die land. Performing a flow balance within the differential element, presented in Fig. 9.11, results in

$$\frac{dQ}{dz} = -q = \text{constant} \tag{9.29}$$

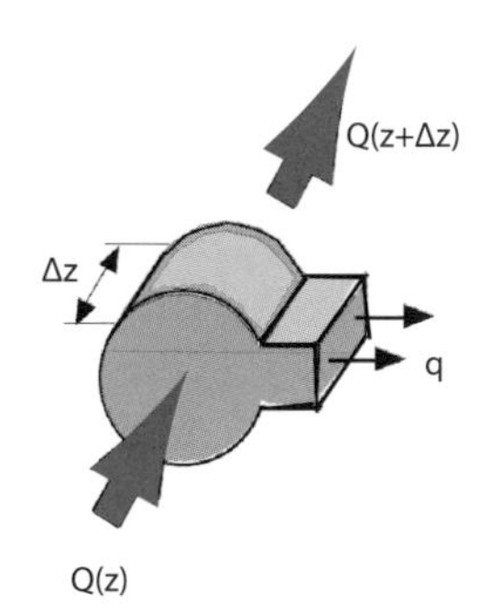

Figure 9.11 Differential element of the manifold in an end-fed sheeting die

Integrating this equation and letting $Q = Q_T$ at $z = 0$ and $Q = 0$ at $z = L_D$, we get,

$$Q(z) = Q_T\left(1 - \frac{z}{L_D}\right) \tag{9.30}$$

A constant flow rate along the slit results in a linearly decreasing flow rate along the manifold length

Therefore,

$$\frac{dQ}{dz} = -\frac{Q_T}{L_D} = -\frac{h^3}{12\mu}\frac{p(z)}{L_L(z)} \tag{9.31}$$

which results in

$$L_L(z) = -\frac{h^3}{12\mu}\frac{L_D}{Q_T}p(z) \tag{9.32}$$

where the pressure as a function of z must be solved for. The manifold equation can be written as

$$\frac{dp}{dz} = -\frac{8\mu}{\pi R^4}Q_T\left(1 - \frac{z}{L_D}\right) \tag{9.33}$$

and integrated from $p = p_0$ at $z = 0$,

$$p(z) = p_0 - \frac{8\mu Q_T L_D}{\pi R^4}\left[\left(\frac{z}{L_D}\right) - \frac{1}{2}\left(\frac{z}{L_D}\right)^2\right] \tag{9.34}$$

which can be substituted into Eq. 9.32 to give

$$L_L(z) = \frac{h^3}{12\mu}\frac{L_D p_0}{Q_T} - \frac{2h^3 L_D^2}{3\pi R^4}\left[\left(\frac{z}{L_D}\right) - \frac{1}{2}\left(\frac{z}{L_D}\right)^2\right] \tag{9.35}$$

Note that the die design equation has pressure, volumetric flow rate, and viscosity embedded inside and can therefore lead to unrealistic results. This is due to the fact that the flow, Q_T, was specified when formulating the equations. However, the die design will be balanced for any volumetric throughput. Hence, during die design it is appropriate to specify the land length at the beginning of the manifold, $L_L(0)$, and pick appropriate combinations of viscosity, flow rate, and pressure,

$$L_L(0) = h^3 L_D \frac{p_0}{12\mu Q_T} \tag{9.36}$$

Example 9.3 End-fed sheeting die

Design a 1000 mm wide end-fed sheeting die with a 1-mm die land gap for a polycarbonate film. For the solution of the problem assume a manifold diameter of 15 mm and the longest portion of the length should be 50 mm. Using the above information, we can write

$$L_L(z) = 50\text{ mm} - \frac{2\,(1\text{ mm})^3\,(1000)^2}{3\pi\,(10\text{ mm})^4}\left[\left(\frac{z}{1000\text{ mm}}\right) - \frac{1}{2}\left(\frac{z}{1000\text{ mm}}\right)^2\right] \tag{9.37}$$

or

$$L_L(z) = 50\text{ mm} - 10.6\left[\left(\frac{z}{1000\text{ mm}}\right) - \frac{1}{2}\left(\frac{z}{1000\text{ mm}}\right)^2\right] \tag{9.38}$$

■

9.2.2 Coat Hanger Die

Perhaps a more common sheeting die is the so-called coat hanger die, presented in detail in Chapter 4 of this book. For a given manifold angle α we must determine the manifold radius profile, $R(s)$, such that a uniform sheet or film is extruded through the die lips.

Using the nomenclature presented in Fig. 9.12 and assuming a land thickness of h we can assume the land length to be described by slit flow and the manifold by the Hagen-Poiseuille flow with a variable radius as

$$q = -\frac{h^3}{12\mu}\left(\frac{dp}{dz}\right) \tag{9.39}$$

and

$$Q(s) = \frac{\pi R(s)^4}{8\mu}\left(-\frac{dp}{ds}\right) \tag{9.40}$$

Equation 9.39 can be rewritten as

$$\frac{Q_T}{2W} = \frac{-h^3}{12\mu}\left(-\frac{dp}{ds}\right) \tag{9.41}$$

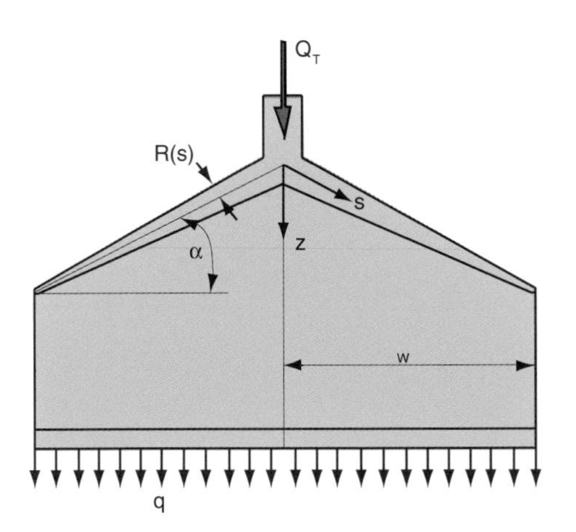

Figure 9.12 Schematic diagram of a coat hanger sheeting die

The coat-hanger sheeting die can be simplified to a combination of tube pressure flow and slit pressure flow

The angle in a coat-hanger sheeting die is fixed and the radius of the manifold is determined

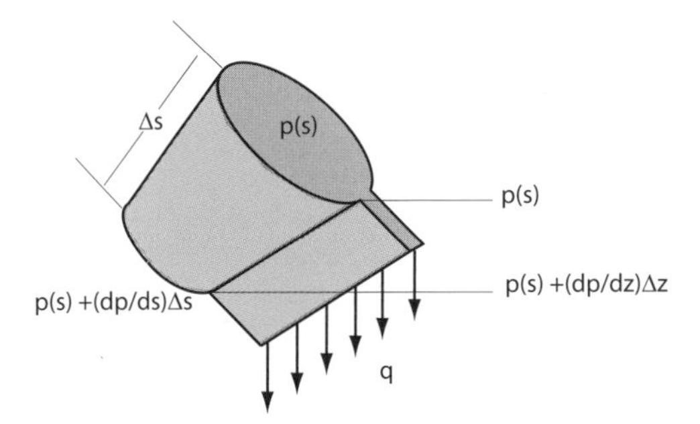

Figure 9.13 Differential element of the manifold in the coat hanger sheeting die

which can be solved for the pressure gradient in the die land

$$\frac{dp}{dz} = -\frac{6\mu Q_{\mathrm{T}}}{Wh^3} \tag{9.42}$$

Here too, we can cut a small element out of the manifold area (see Fig. 9.13) and relate the pressure drop in the s-direction to the drop in the z-direction using

$$p(s) + \frac{dp}{ds}\Delta s = p(s) + \frac{dp}{dz}\Delta z \tag{9.43}$$

Combining the definition of pressure gradient in the die land, Eq. 9.42, with Eq. 9.43 and using geometry we get

$$\frac{dp}{ds} = -\frac{6\mu Q_{\mathrm{T}}}{Wh^3}\sin\alpha \tag{9.44}$$

which can be integrated to become

$$p(s) = p_0 - \frac{6\mu Q_{\mathrm{T}}}{Wh^3}\sin\alpha s \tag{9.45}$$

and a mass balance results in

$$\frac{dQ}{ds} = \frac{Q_\mathrm{T}}{2W}\cos\alpha \tag{9.46}$$

Using the boundary condition that $Q = \frac{Q_\mathrm{T}}{2}$ at $s = 0$ and $Q = 0$ at $s = W/\cos\alpha$, we can integrate Eq. 9.46 to be

$$Q(s) = Q_\mathrm{T}\left(\frac{1}{2} - \frac{\cos\alpha}{2W}s\right) \tag{9.47}$$

We can now set the manifold equation, Eq. 9.40, with the pressure gradient defined in Eq. 9.46, equal to Eq. 9.47

A constant flow rate within the land results in a linearly decreasing flow rate along the manifold length

$$Q_\mathrm{T}\left(\frac{1}{2} - \frac{\cos\alpha}{2W}s\right) = \frac{\pi R(s)^4}{8\mu}\left(\frac{6\mu Q_\mathrm{T}}{Wh^3}\sin\alpha\right) \tag{9.48}$$

which, can be used to solve for the manifold radius profile

$$R(s) = \left(\frac{2(1 - s\cos\alpha/W)}{3\pi\sin\alpha}\left(Wh^3\right)\right)^{1/4} \tag{9.49}$$

A cross-head tubing die is equivalent to the coat hanger die by wrapping it around a cylinder. If we follow the same derivation, but consider a shear thinning power law melt, we get

$$R(s) = \left(\frac{\left[(3 + 1/n)/\pi\right]^n h^{2n+1}\left(W - s\cos\alpha\right)^n}{2^n\left(2 + 1/n\right)^n\left(\sin\alpha\right)}\right)^{1/(3n+1)} \tag{9.50}$$

which for a Newtonian fluid with $n = 1$ reduces to Eq. 9.49.

9.2.3 Extrusion Die with Variable Die Land Thicknesses

When designing plastic parts it is often recommended that the part have uniform thickness. This is especially true for semi-crystalline polymers, where thickness variations lead to variable cooling times, and those in turn lead to variations in the degree of crystallinity in the final part. Variations in crystallinity result in shrinkage variations, which lead to warpage. However, it is often necessary to design parts in which a thickness variation is inevitable, i. e., extrusion profiles with thickness variations, as shown in Fig. 9.14.

The die land thickness differences can be compensated by using different land lengths such that the speed of the emerging melt is constant, resulting in a uniform product. If we assume a power-law viscosity model, a uniform pressure in the manifold and an isothermal die and melt, the average speed of the melt emerging from the die is

$$\overline{u_i} = \frac{h_i}{2\left(s + 2\right)}\left(\frac{h_i \Delta p}{2mL_i}\right)^s \tag{9.51}$$

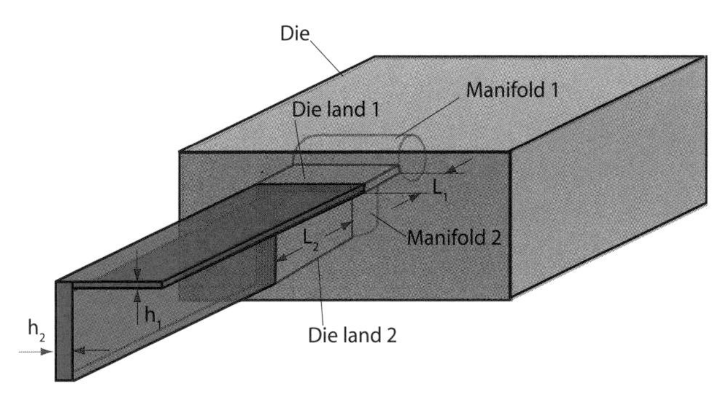

Figure 9.14 Schematic diagram of die with two different die land lengths and thicknesses

where $s = 1/n$. In order to achieve a uniform, product we must satisfy

$$\overline{u_1} = \overline{u_2} \tag{9.52}$$

or

$$\frac{h_1}{2(s+2)}\left(\frac{h_1 \Delta p}{2mL_1}\right)^s = \frac{h_2}{2(s+2)}\left(\frac{h_2 \Delta p}{2mL_2}\right)^s \tag{9.53}$$

which can be rearranged to become

$$\left(\frac{L_1}{L_2}\right) = \left(\frac{h_1}{h_2}\right)^{1+n} \tag{9.54}$$

Example 9.4 Die design with two die land thicknesses

Determine the die land length ratios, L_1/L_2 for a die land thickness ratio, h_1/h_2 of ⅓, for various power-law indices. Using Eq. 9.54, we can easily solve for the land length ratios for several power-law indices. This is presented graphically in Fig. 9.15. Note that while a Newtonian fluid requires a land length ratio of 9, a Bingham fluid, with a power law index of zero, requires a land length ratio of only 3. Hence, die design is very sensitive to the shear thinning behavior of the polymer melt that must always be accounted for.

9.2.4 Fiber Spinning

The process of fiber spinning, described in Chapter 7 and schematically represented in Fig. 9.16, will be modeled in this section using first a Newtonian model, followed by a shear thinning model. To simplify the analysis, it is customary to set the origin of the coordinate system at the location of largest diameter of the extrudate. Because the

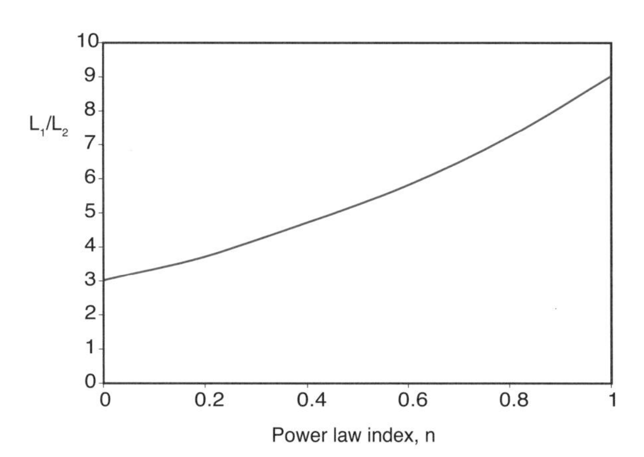

Figure 9.15 Land length ratios as a function of power law index for a die with a land height ratio of 3

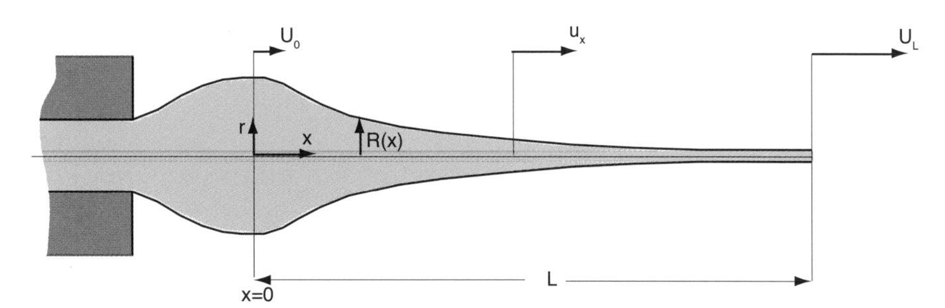

Figure 9.16 Schematic diagram of the fiber spinning process in the post-extrusion die region

distance from the spinnerette to the point of largest swell is very small (only a few die diameters), this simplification will not introduce large problems in the solution.

If we take the schematic of a differential fiber element presented in Fig. 9.17, we can define the fiber geometry by the function $R(x)$ and the unit normal vector $\boldsymbol{n}$. The continuity equation tells us that the volumetric flow rate through any cross-section along the x-direction must be Q

$$Q = \pi R\,(x)^2\, u_x \tag{9.55}$$

We further assume that surface tension is negligible and that in steady state the surface will only move tangentially, which means that

$$\boldsymbol{u} \cdot \boldsymbol{n} = 0 \tag{9.56}$$

The components of the normal vector are described by the geometry of the fiber as

$$n_x = -\frac{dR}{dx}\left[1 + \left(\frac{dR}{dr}\right)^2\right]^{-1/2} \tag{9.57}$$

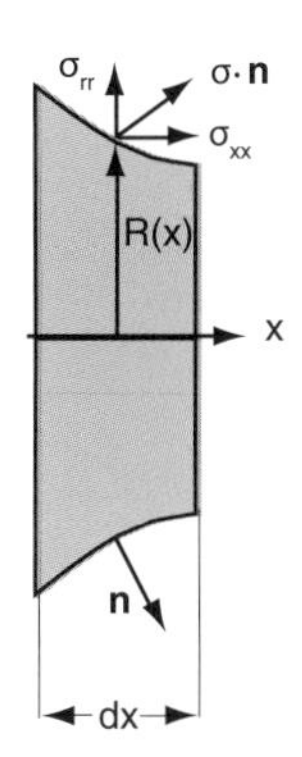

Figure 9.17 Differential element of a fiber during spinning

and

$$n_r = \left[1 + \left(\frac{dR}{dr}\right)^2\right]^{-1/2} \tag{9.58}$$

Due to the negligible effects of surface tension, we can assume that the stress boundary condition is

$$\boldsymbol{\sigma} \cdot \boldsymbol{n} = 0 \tag{9.59}$$

which for each direction can be written as

$$\sigma_{rx} n_r + \sigma_{xx} n_x = 0 \tag{9.60}$$

and

$$\sigma_{rr} n_r + \sigma_{rx} n_x = 0 \tag{9.61}$$

Due to the fact that the fiber is being pulled in the x-direction, we should expect a non-zero σ_{xx} at the free surface. Hence, we can write

$$\sigma_{rx} = -\sigma_{xx} \frac{n_x}{n_r} = \frac{dR}{dr} \sigma_{xx} \tag{9.62}$$

It is clear that only the x-component of the equation of motion plays a significant role in a fiber spinning problem

$$\varrho \left(u_r \frac{\partial u_x}{\partial r} + u_x \frac{\partial u_x}{\partial x} \right) = \frac{1}{r} \frac{\partial}{\partial r} (r \sigma_{rx}) + \frac{\partial \sigma_{xx}}{\partial x} \tag{9.63}$$

The first term in the above equation drops out because u_x is not a function of r. Using Eq. 9.63 and rearranging it somewhat, the equation of motion becomes

$$\varrho u_x \frac{\partial u_x}{\partial x} = \frac{2}{R} \frac{dR}{dr} \sigma_{xx} + \frac{d\sigma_{xx}}{dx} \tag{9.64}$$

For a total stress, σ_{xx}, in a Newtonian approximation we write the constitutive relation

$$\sigma_{xx} = -p + 2\mu \frac{du_x}{dx} \tag{9.65}$$

We can show that the isotropic pressure p is given by

$$p = -(\sigma_{xx} + \sigma_{rr} + \sigma_{\theta\theta})/3 \tag{9.66}$$

However, we can assume that $\sigma_{rr} = 0$ and $\sigma_{\theta\theta} = 0$. Hence, we can write

$$\sigma_{xx} = 3\mu \frac{du_x}{dx} \tag{9.67}$$

Combining the above equation with the continuity equation, Eq. 9.55, the momentum balance presented in Eq. 9.64 becomes

$$\frac{d}{dx}(u_x)^2 = 12 \frac{\mu}{\varrho R} \frac{dR}{dr} \frac{du_x}{dx} + 6 \frac{\mu}{\varrho} \frac{d^2 u_x}{dx^2} \tag{9.68}$$

If we neglect the effect inertia has on the stretching fiber and drop the inertial term in the above equation, it can be solved as

$$u_x = c_1 e^{c_2 x} \tag{9.69}$$

With boundary conditions $u_x = U_0$ at $x = 0$, and $u_x = U_L$ at $x = L$ we get

$$u_x = U_0 e^{\frac{x}{L} \ln(D_\mathrm{R})} = U_0 D_\mathrm{R}^{\frac{x}{L}} \tag{9.70}$$

where D_R is the *draw down ratio* defined by

$$D_\mathrm{R} = \frac{U_L}{U_0} \tag{9.71}$$

Using the continuity equation, we can now write

$$R(x) = R_0 D_\mathrm{R}^{-\frac{x}{2L}} \tag{9.72}$$

The derivation of the fiber spinning equations for a non-Newtonian shear thinning viscosity using a power law model is provided next. For a total stress, σ_{xx}, in a power law fluid, we write the constitutive relation

$$\sigma_{xx} = -p + 2m(3)^{(n-1)/2} \left(\frac{du_x}{dx}\right)^n \tag{9.73}$$

This leads to the velocity distribution

$$u_x = U_0 \left[1 + \left(D_\mathrm{R}^{(n-1)/n} - 1\right) \frac{x}{L}\right]^{n/(n-1)} \tag{9.74}$$

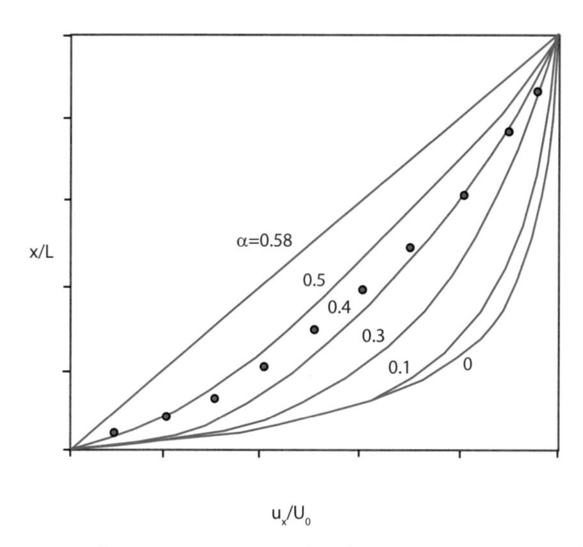

Figure 9.18 Comparison between experimental and computed velocity profiles during fiber spinning using Denn and Fisher's viscoelastic model [2], with $\alpha = \frac{m(3)^{(n-1)/2}}{G}\left(\frac{U_0}{L}\right)^n$ and G a shear storage modulus

Denn and Fisher [2] solved this problem using a viscoelastic analysis. Figure 9.18 presents a plot of their results with various values of their dimensionless viscoelastic parameter α. The graph also presents experimental results for the fiber spinning of polystyrene at 170 °C. The fiber had a value of α between 0.2 and 0.3, but the theoretical prediction compares with the experiments for a value of α between 0.4 and 0.5. Phan-Thien approach showed better agreement between the experiments done with polystyrene and low density polyethylene using the Phan-Thien-Tanner model [3].

9.2.5 Wire Coating Die

One important aspect of wire-coating is the thickness distribution of the polymer on the surface of the wire as well as the velocity distribution within the die. A simplified wire coating process is presented in the Fig. 9.19, where the wire radius is defined by R and the annulus radius by κR. This type of flow is often referred to as an *axial annular Couette flow*.

The most important assumptions when solving this problem are a steady, incompressible, and isothermal flow. Let us now consider a power-law fluid, but neglect the elastic effects. Furthermore, for the solution of this specific problem, let us assume that the flow is primarily driven by drag and that there are no significant pressure drops across the die.

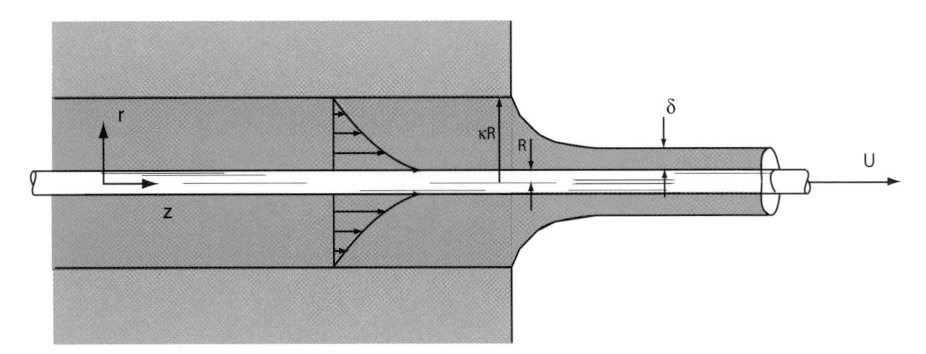

Figure 9.19 Schematic diagram of a wire coating die

The unidirectional flow condition will simplify the momentum equation to

$$\frac{d}{dr}\left(r\tau_{rz}\right)=0 \tag{9.75}$$

using the power-law model and two integrations, we obtain the velocity as

$$u_z=\left(\frac{c_1}{m}\right)^s\frac{r^{1-s}}{s-1}-c_2 \tag{9.76}$$

If we impose the boundary conditions

$$\begin{aligned} u_z\left(R\right)&=U\\ u_z\left(\kappa R\right)&=0 \end{aligned} \tag{9.77}$$

the velocity profile will become

$$\frac{u_z\left(r\right)}{U}=\frac{\xi^{1-s}-1}{\kappa^{1-s}-1} \tag{9.78}$$

where $\xi=r/R$ and $s=1/n$.

In addition, from the definition of the stress tensor, using the power law model for the viscosity, τ_{rz} becomes

$$\tau_{rz}=\frac{mU^n\left(s-1\right)^n}{R^n\left(\kappa^{1-s}-1\right)^n}\left(\frac{R}{r}\right) \tag{9.79}$$

In order to relate the thickness of the coating to R and U, we must perform a mass balance on the region starting at the exit of the die and ending where the fluid has reached the same velocity as the wire. Here, we must assume that the polymer density is constant, although the melt undergoes density changes as it solidifies. The mass balance is written as

$$2\pi\int_{\kappa R}^{R}u_z\left(r\right)rdr=U\pi\left[\left(\kappa R+\delta\right)^2-\left(\kappa R\right)^2\right] \tag{9.80}$$

where δ is the coating thickness. By substituting the velocity profile into the integral in the above equation, an equation for the coating thickness is obtained

$$\frac{2R^2}{\kappa^{1-s}-1}\left[\frac{1-\kappa^{3-s}}{3-s}-\frac{1-\kappa^2}{2}\right]=\left[(\kappa R+\delta)^2-(\kappa R)^2\right] \tag{9.81}$$

which can be written as follows

$$\frac{2R^2\left(1-\kappa^{3-s}\right)}{\left(\kappa^{1-s}-1\right)(3-s)}-\frac{R^2\left(1-\kappa^2\right)}{\left(\kappa^{1-s}-1\right)}-2\kappa R\delta-\delta^2=0 \tag{9.82}$$

This non-linear equation relates R and δ, which can be used to obtain the flow rate and the coating thickness. Finally, the force needed to pull the wire through the die obtained from integrating the stress at the polymer-die surface interface over the area of the die is defined as

$$F=(2\pi\kappa RL)\left[\tau_{rz}\right]_{r=\kappa R}=(2\pi\kappa RL)\left[\frac{mU^n(s-1)^n}{R^n\left(\kappa^{1-s}-1\right)}\frac{1}{\kappa}\right] \tag{9.83}$$

9.3 Processes That Involve Membrane Stretching

There are numerous processes that involve the stretching of a membrane such as film blowing, film casting, extrusion blow molding, injection blow molding, thermoforming, etc. In this section we will address two very important processes: the film blowing process and the thermoforming process.

9.3.1 Film Blowing

Despite the non-isothermal nature of the film blowing process we will develop an isothermal model to show general effects and interactions during the process. In the derivation we follow Pearson and Petrie's approach [4–6]. Even this Newtonian isothermal model requires an iterative solution and numerical integration. Figure 9.20 presents the notation used when deriving the model.

A common form of analyzing film blowing is by setting-up a coordinate system, ξ, that moves with the moving melt on the inner surface of the bubble and that is oriented with the film, as shown in Fig. 9.20. Using the moving coordinates, we can define the three non-zero terms of the local rate of deformation tensor as

$$\begin{aligned}\dot{\gamma}_{11}&=2\frac{\partial u_1}{\partial \xi_1}\\ \dot{\gamma}_{22}&=2\frac{\partial u_2}{\partial \xi_2}\\ \dot{\gamma}_{33}&=2\frac{\partial u_3}{\partial \xi_3}\end{aligned} \tag{9.84}$$

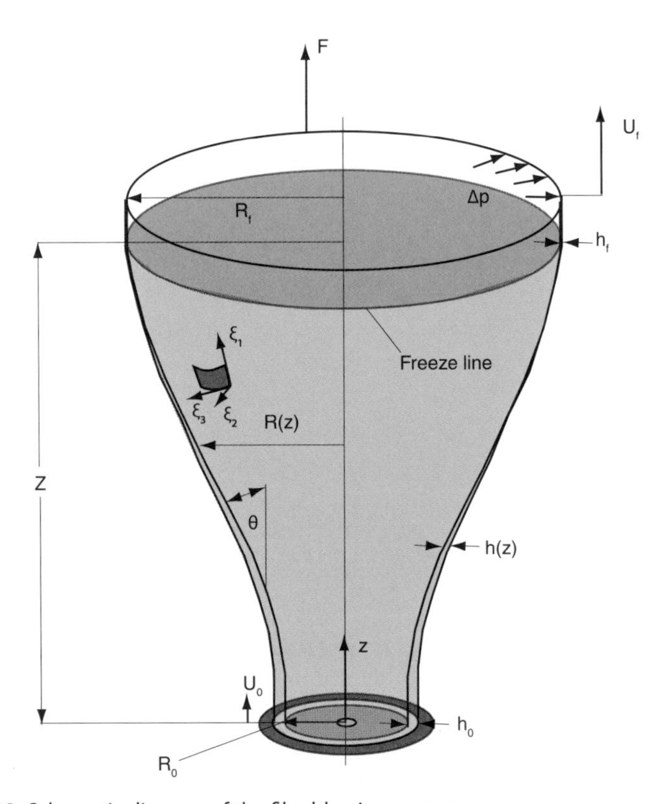

Figure 9.20 Schematic diagram of the film blowing process

For an incompressible fluid, these three components must add up to zero

$$\dot{\gamma}_{11} + \dot{\gamma}_{22} + \dot{\gamma}_{33} = 0 \tag{9.85}$$

For a thin film, where $d\xi_1 = dz/\cos\theta$, the $\dot{\gamma}_{22}$ term can be defined by

$$\dot{\gamma}_{22} = 2\frac{\partial u_2}{h} = \frac{2}{h}\frac{dh}{dt} = \frac{2}{h}\frac{dh}{d\xi_1}\frac{d\xi_1}{dt} = \frac{2u_1}{h}\frac{dh}{d\xi_1} = \frac{2u_1\cos\theta}{h}\frac{dh}{dz} \tag{9.86}$$

The rate of expansion in the circumferential direction is defined by the rate of growth of the circumference

$$u_3 = 2\pi\frac{dR}{dt} \tag{9.87}$$

divided by the local circumference, $2\pi R$, to become

$$\dot{\gamma}_{33} = \frac{1}{R}\frac{dR}{dt} = \frac{2u_1}{R}\frac{dR}{d\xi_1} = \frac{2u_1\cos\theta}{R}\frac{dR}{dz} \tag{9.88}$$

and finally

$$\dot{\gamma}_{11} = -\dot{\gamma}_{22} - \dot{\gamma}_{33} = -\frac{2u_1\cos\theta}{h}\frac{dh}{dz} - \frac{2\pi u_1\cos\theta}{R}\frac{dR}{dz} \tag{9.89}$$

It is possible to relate the total volumetric throughput, Q, to u_1 using

$$Q = 2\pi Rhu_1 \tag{9.90}$$

We can now write

$$\begin{aligned}\dot{\gamma}_{11} &= -\frac{Q\cos\theta}{\pi Rh}\frac{1}{h}\frac{dh}{dz} - \frac{1}{R}\frac{dR}{dz}\\ \dot{\gamma}_{22} &= \frac{Q\cos\theta}{\pi Rh}\frac{1}{h}\frac{dh}{dz}\\ \dot{\gamma}_{33} &= \frac{Q\cos\theta}{\pi Rh}\frac{1}{R}\frac{dR}{dz}\end{aligned} \tag{9.91}$$

The total stress in the ξ-coordinate system is written as

$$\sigma_{ii} = p - \mu\dot{\gamma}_{ii} \tag{9.92}$$

Because surface tension is neglected and no external forces act on the bubble

$$\sigma_{22} = 0 \tag{9.93}$$

Hence

$$p = \mu\dot{\gamma}_{22} = \frac{Q\mu\cos\theta}{\pi Rh^2}\frac{dh}{dz} \tag{9.94}$$

The two stresses become

$$\sigma_{11} = -\frac{\mu Q\cos\theta}{\pi Rh}\left(\frac{2}{h}\frac{dh}{dz} + \frac{1}{R}\frac{dR}{dz}\right) \tag{9.95}$$

and

$$\sigma_{33} = \frac{\mu Q\cos\theta}{\pi Rh}\left(\frac{1}{R}\frac{dR}{dz} - \frac{1}{h}\frac{dh}{dz}\right) \tag{9.96}$$

It is necessary to perform a force balance for the bubble in order to determine the radius, $R(z)$, and the thickness, $h(z)$, of the bubble. The longitudinal force is computed using

$$F_L = 2\pi Rh\sigma_{11} \tag{9.97}$$

and for the small fluid element defined in Fig. 9.21, the transverse force is defined by

$$dF_T = h\xi_1\sigma_{33} \tag{9.98}$$

A force balance about the differential element results in

$$\Delta p = h\left(\frac{\sigma_{11}}{R_L} + \frac{\sigma_{33}}{R_T}\right) \tag{9.99}$$

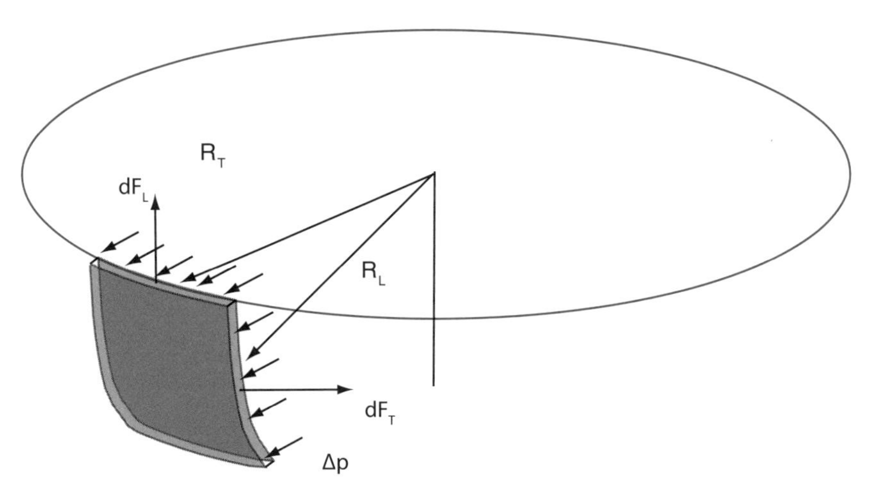

Figure 9.21 Forces acting on a film element

where R_L and R_T are defined in the ξ-coordinate system. In a cylindrical-polar coordinate system we can write

$$R_L = -\frac{\sec^3\theta}{d^2R/dz^2} \tag{9.100}$$

and

$$R_T = R\sec\theta \tag{9.101}$$

The bubble will grow to a maximum (or final)radius, R_f, when it freezes at a position $z = Z$, which is called the *freeze-line*. Because the bubble is pulled by a force F_Z, which is usually referred to as the *draw force*, we can perform a force balance between a position z and Z to give

$$F_Z = 2\pi R\cos\theta h\sigma_{11} + \pi\left(R_f^2 - R^2\right)\Delta p \tag{9.102}$$

For convenience we define the following dimensionless parameters:

- Draw down ratio: $D_R = \frac{U_f}{U_0}$
- Dimensionless pressure: $B = \frac{\pi R_0^3 \Delta p}{\mu Q}$
- Blow-up ratio: $BUR = \frac{R_f}{R_0}$
- Dimensionless stress: $T = \frac{R_0 Fz}{\mu Q} - B\,(BUR)^2$
- Dimensionless take-up force: $\hat{F} = \frac{R_0 Fz}{\mu Q}$

- Dimensionless radius: $\hat{R} = \dfrac{R}{R_0}$
- Dimensionless axial direction: $\hat{z} = \dfrac{z}{R_0}$
- Dimensionless thickness: $\hat{h} = \dfrac{h}{R_0}$
- Thickness ratio: $\dfrac{h_0}{h_f} = D_r BUR$

Using these dimensionless parameters, defining $d\hat{R}/d\hat{z} = \tan\theta$, and combining the above equations yields two dimensionless differential equations

$$2\hat{R}^2\left(T + \hat{R}^2 B\right)\frac{d^2\hat{R}}{d\hat{z}^2} = 6\frac{d\hat{R}}{d\hat{z}} + \hat{R}\left(1 + \left[\frac{d\hat{R}}{d\hat{z}}\right]^2\right)\left(T - 3\hat{R}^2 B\right) \tag{9.103}$$

and

$$\frac{1}{\hat{h}}\frac{d\hat{h}}{d\hat{z}} = -\frac{1}{2\hat{R}}\frac{d\hat{R}}{d\hat{z}} - \frac{\left(1 + \left[\frac{d\hat{R}}{d\hat{z}}\right]^2\right)\left(T + \hat{R}^2 B\right)}{4} \tag{9.104}$$

As boundary conditions we specify that $\hat{R} = 1$ at $\hat{z} = 0$, $d\hat{R}/d\hat{z} = 0$ at $\hat{z} = Z/R_0$ and $\hat{h} = h_0/R_0$ at $\hat{z} = 0$.

Since T depends on *BUR*, we must first specify *BUR* and iterate until a solution of $\hat{R}(\hat{z})$ is found that agrees with the choice of *BUR*. Hence, we must integrate Eq. 9.104 numerically with each choice of *BUR*. After the correct value of *BUR* has been found, we numerically integrate Eq. 9.104. Figures 9.22 and 9.23 present solutions for a fixed value of $\hat{z} = 20$ and a fixed value of $B = 0.1$, respectively.

Example 9.5 Film blowing

A tubular 50 μm thick low density polyethylene film is blown with a draw ratio of 5 at a flow rate of 50 g/s. The annular die has a diameter of 15 mm and a die gap of 1 mm. Calculate the required pressure inside the bubble and draw force to pull the bubble. Assume a Newtonian viscosity of 800 Pa·s, a density of 920 kg/m^3 and a freeze line at 300 mm. Since we know the thickness reduction and the draw ratio of the film, we can compute the blow-up ratio,

$$BUR = \left(\frac{h_0}{h_f}\right)/D_R = \left(\frac{1000\ \mu\text{m}}{50\ \mu\text{m}}\right)/5 = 4 \tag{9.105}$$

Next, we can compute the dimensionless freeze line using,

$$\hat{z} = z/R_0 = 300\ \text{mm}/15\ \text{mm} = 20 \tag{9.106}$$

which allows us to use Fig. 9.22 to get $B = 0.075$ and $\hat{F} = 1.3$, which results in $\Delta p = 307$ Pa and $F_z = 3.77$ N.

■

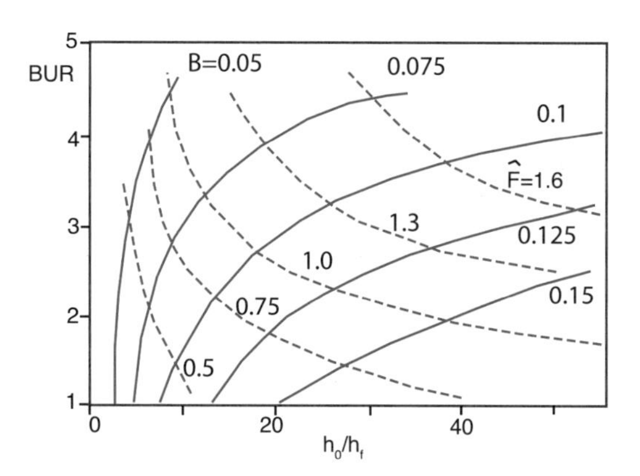

Figure 9.22 Predicted film blowing process using an isothermal Newtonian model for a dimensionless freezing line at $\hat{z}$ =20 [6]

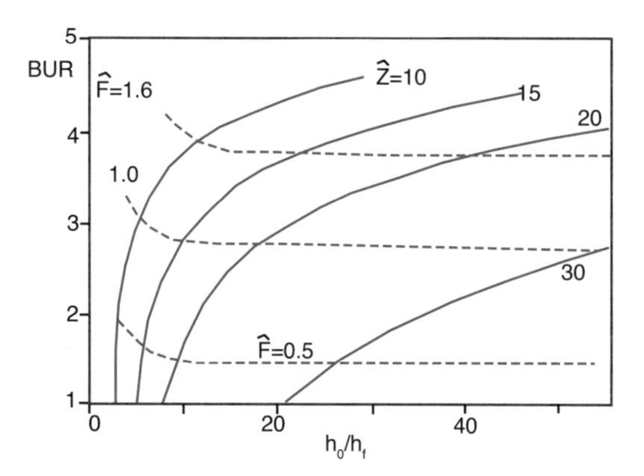

Figure 9.23 Predicted film blowing process using an isothermal Newtonian model for a dimensionless pressure $B = 0.1$ [6]

9.3.2 Thermoforming

A simple approximation of the thermoforming process is based on a mass balance principle. To illustrate this concept, let us consider the thermoforming process of a conical object, as schematically depicted in Fig. 9.24.

For the solution, we assume the notation presented in Fig. 9.24. As shown in the figure, at an arbitrary point in time the bubble will contact the mold at a height z and will have a radius R, which is determined by the mold geometry

$$R = \frac{H - \sin\beta}{\sin\beta \tan\beta} \tag{9.107}$$

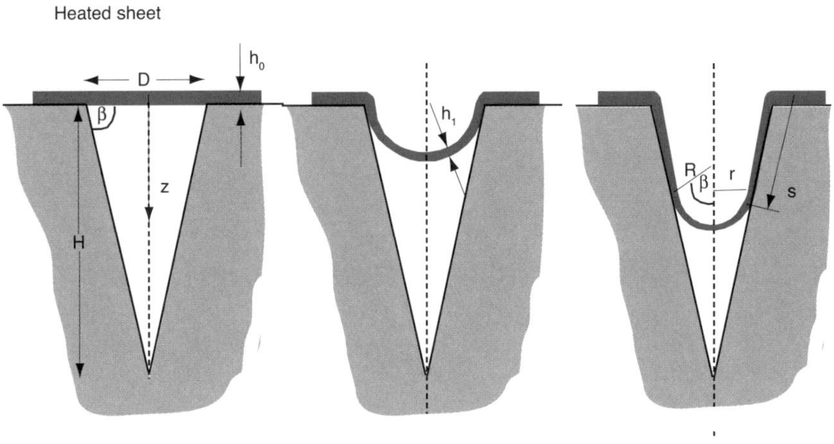

Figure 9.24 Schematic diagram of the thermoforming process of a conical geometry

where H is the depth of the cone, s the contact point along the cone's wall and β the angle described in Fig. 9.24. The surface area of the cone at that point in time is given by

Predicting the thickness distribution using a simple mass balance

$$A = 2\pi R^2 (1 + \cos\beta) \tag{9.108}$$

If we perform a mass balance as the bubble advances a distance Δs, we get

$$2\pi\ R^2 (1 + \cos\beta)\, h\big|_s - 2\pi\ R^2 (1 + \cos\beta)\, h\big|_{s+\Delta s} = 2\pi r h(s) \Delta s \tag{9.109}$$

with $r = R \sin\beta$, the above equation results in

$$-\frac{d}{ds}\left(R^2 h(s)\right) = \frac{R h(s) \sin\beta}{1 + \cos\beta} \tag{9.110}$$

differentiating Eq. 9.110, we get

$$\frac{dR}{ds} = \frac{1}{\tan\beta} \tag{9.111}$$

We can combine the above equations to get

$$\frac{dh}{h(s)} = \left(2 - \frac{\tan\beta \sin\beta}{1 + \cos\beta}\right) \sin\beta \frac{ds}{H - s\sin\beta} \tag{9.112}$$

which can be integrated using $h(0) = h_1$ as a boundary condition

$$\frac{h}{h_1} = \left(1 - \frac{s}{H}\sin\beta\right)^{\sec\beta - 1} \tag{9.113}$$

where h_1 is the thickness of the bubble when it first makes contact with the cone wall. The initial thickness of the sheet, h_0, can be related to h_1 using

$$\frac{\pi D^2 h_0}{4} = \frac{\pi D^2 (1 + \cos\beta)}{2\sin^2\beta} h_1 \tag{9.114}$$

Finally, we can write the thickness distribution using

$$\frac{h}{h_0} = \frac{1+\cos\beta}{2}\left(1 - \frac{s}{H}\sin\beta\right)^{\sec\beta - 1} \tag{9.115}$$

This equation can be extended to simulate the thermoforming process of a truncated cone, which is a more realistic geometry encountered in the thermoforming industry.

9.4 Calendering – Isothermal Flow Problems

As discussed in Chapter 7, the calendering process is used to squeeze a mass of polymeric material through a set of high-precision rolls to form a sheet or film. In this section, we will derive the well known model developed by Gaskell [7] and by McKelvey [8]. For the derivation, let us consider the notation and set-up presented in Fig. 9.25.

The final thickness is found through iteration by finding the point when pressure goes to zero

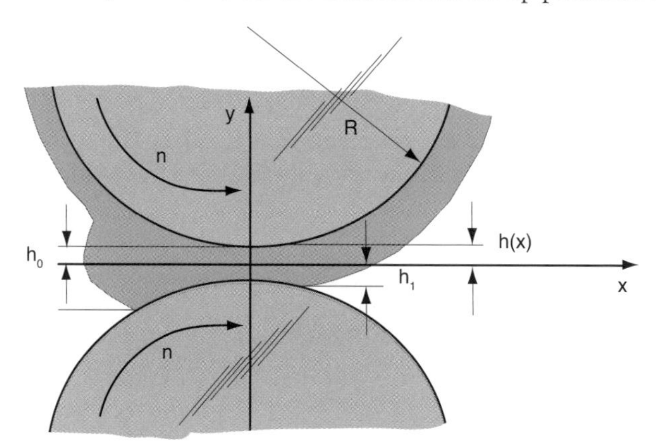

Figure 9.25 Schematic diagram of a two roll calendering system in the nip region

9.4.1 Newtonian Model of Calendering

In Gaskell's treatment, a Newtonian flow was assumed with a very small gap-to-radius ratio, $h \ll R$. This assumption allows us to assume the well known lubrication approximation with only velocity components $u_x(y)$. In addition, Gaskell's model assumes that a very large bank of melt exists in the feed side of the calender. The continuity equation and momentum balance reduce to

$$\frac{du_x}{dx} = 0 \tag{9.116}$$

and

$$\frac{dp}{dx} = \mu\frac{\partial^2 u_x}{\partial y^2} \tag{9.117}$$

respectively. Integrating Eq. 9.117 twice with the boundary conditions $u_x = U$ at $y = h(x)$ and $\partial u_x/\partial y = 0$ at $y = 0$, results in

$$u_x = U + \left[\frac{y^2 - h^2(x)}{2\mu}\right]\frac{dp}{dx} \tag{9.118}$$

where $U = 2\pi nR$ is the speed on the roll surface. Using the velocity profile, we can compute the flow rate per unit width as

$$q = 2\int_0^h u_x dy = 2h\left[U - \frac{h^2}{3\mu}\frac{dp}{dx}\right] \tag{9.119}$$

which will not vary with x. The pressure distribution is unknown and will be solved for next. In order to do this, we require that the velocity at the outlet be uniform and equal to the roll surface speed, $u_x(y) = U$. A uniform velocity implies no shear stress, $\tau_{yx} = 0$, which means that the pressure gradient should also be zero at that point. Hence, at that position the flow rate can be expressed as

$$q = 2h_1 U \tag{9.120}$$

We can combine Eqs. 9.119 and 9.120 to give

$$\frac{dp}{dx} = \frac{3\mu U}{h_1^2}\left(1 - \frac{h_1}{h}\right)\left(\frac{h_1}{h}\right)^2 \tag{9.121}$$

This equation implies that the pressure gradient vanishes at $x = x_1$ as well as at $x = -x_1$, at which point the pressure is at a maximum, as will be shown later.

The half-gap between the rolls is defined by

$$h = h_0 + R - \sqrt{R^2 - x^2} \tag{9.122}$$

but because we can assume that $x \ll R$, the term $\sqrt{R^2 - x^2}$ can be approximated using the first two terms of the binomial series. This results in

$$\frac{h}{h_0} = 1 + \xi^2 \tag{9.123}$$

where $\xi^2 = \frac{x^2}{2Rh_0}$. Now we can integrate Eq. 9.121 to give

$$p = \frac{3\mu U}{4h_0}\sqrt{\frac{R}{2h_0}}\left(\left[\frac{\xi^2 - 1 - 5\lambda^2 - 3\lambda^2\xi^2}{\left(1+\xi^2\right)^2}\right]\xi + \left(1 - 3\lambda^2\right)\tan^{-1}\xi + C(\lambda)\right) \tag{9.124}$$

where $\lambda = \frac{x_1^2}{2Rh_0}$ and $C(\lambda)$ is obtained by letting $p = 0$ at $\xi = \lambda$

$$C(\lambda) = \frac{\left(1 + 3\lambda^2\right)}{\left(1 + \lambda^2\right)}\lambda - \left(1 - 3\lambda^2\right)\tan^{-1}\lambda \tag{9.125}$$

McKelvey [8] approximated $C(\lambda) \approx 5\lambda^3$. The maximum pressure occurs at $x = -x_1(\xi = \lambda)$

$$p_{max} = \frac{3\mu U}{4h_0}\sqrt{\frac{R}{2h_0}}\,(2C(\lambda)) \approx \frac{15\mu U\lambda^3}{2h_0}\sqrt{\frac{R}{2h_0}} \tag{9.126}$$

Figure 9.26 presents a dimensionless pressure, p/p_{max}, as a function of dimensionless x-direction, ξ, for various values of λ, computed using the above equations. Figure 9.27 compares experimental pressure measurements to a curve computed using Gaskell's Newtonian model [7]. The two curves were matched by choosing the best value of λ to match the position of maximum pressure. The predicted pressure is very accurate at values of ξ larger than $-\lambda$ but is not very good before $-\lambda$. Although a shear thinning model would help achieve a better match between experiments and prediction [10], still within that region the accuracy of the models remain poor. Another aspect that should be pointed out at this point is that the maximum pressure, p_{max}, is very sensitive to λ, e.g., doubling λ increases p_{max} 8 times. It is reasonable to assume that $p \rightarrow 0$ when $\xi \rightarrow -\infty$, which means that λ must have a specific value, namely, $\lambda = 0.475$.

It should be noted that the pressure distribution goes to zero in the up-stream position ξ_2, where the material makes contact with both rolls. This position can be determined for any value of λ by letting the pressure in Eq. 9.124 go to zero. Figure 9.28 presents a graph of position of first contact, ξ_2, and the position where the sheet separates from the rolls, λ.

Plug flow exists at the point of highest pressure as well as after the material leaves the rolls

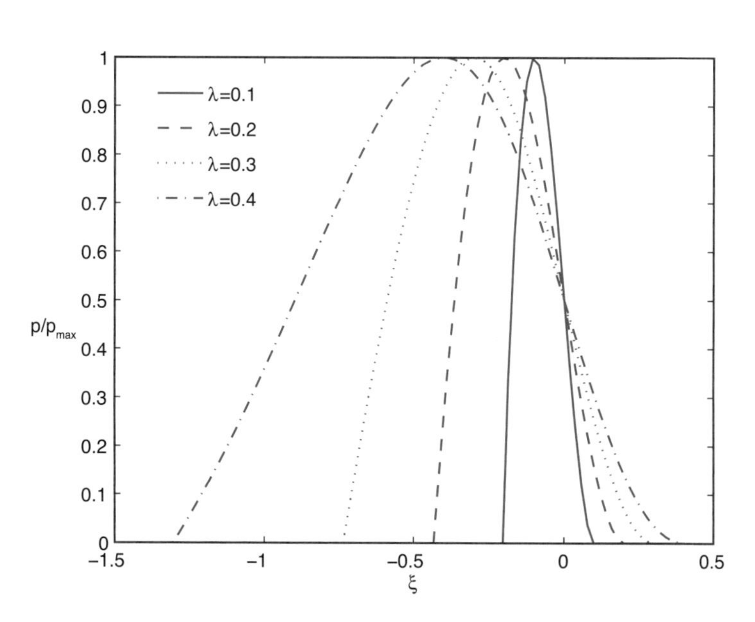

Figure 9.26 Computed pressure distribution between the rolls for various values of λ

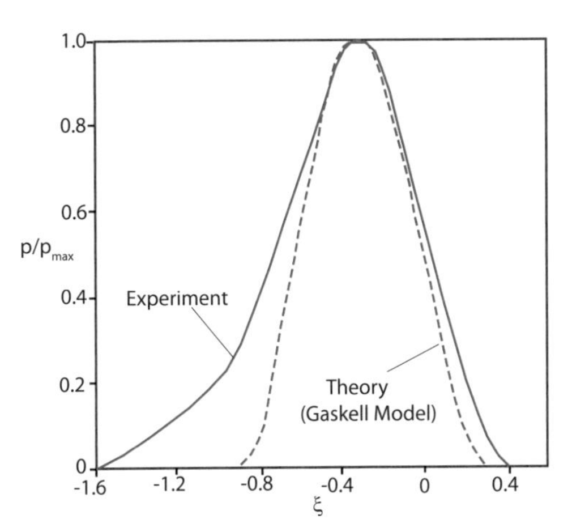

Figure 9.27 Comparison of theoretical and experimental pressure profiles [8]. The experiments were performed by Bergen and Scott [9] with roll diameters of 10 in, a gap in the nip region of 0.025 in and a speed $U = 5$ in/s. The measured viscosity was 3.2×10^9 P (Poise, 1 P = 0.1 Pa·s)

With the above equations, the velocity distribution between the rolls becomes

$$\hat{u}_x = 1 + \frac{3}{2}\frac{\left(1-\eta^2\right)\left(\lambda^2-\xi^2\right)}{\left(1+\xi^2\right)} \tag{9.127}$$

where $\hat{u}_x = u_x/U$ and $\eta = y/h$. Equation 9.127 can be used to determine that a stagnation point, $u_x(0) = 0$, exists at a position $\xi_s = -\sqrt{2+3\lambda^2}$. Figure 9.29 presents the flow pattern that develops in the nip region as predicted by Eq. 9.127. As can be seen, a recirculation pattern develops due to the backflow caused by the pressure build-up as the polymer is forced through the nip region. The calendering process was modeled in 2D using *Radial Function Methods* [11] for a Newtonian as well as power law viscosity models. In comparison to the more complex simulations, the lubrication approximation does an excellent job when modeling the process.

We will calculate the power consumption as well as predict the temperature rise within the material due to viscous heating. In order to compute the power consumption, we need to integrate the product of the shear stress and the roll surface speed over the surface of the roll. The rate of deformation can be computed using Eq. 9.127 as

$$\dot{\gamma}_{yx}(\eta) = \frac{3U\left(\xi^2-\lambda^2\right)}{h_0\left(1-\xi^2\right)^2}\eta \tag{9.128}$$

and the stress

$$\tau_{yx}(\eta) = \mu\frac{3U\left(\xi^2-\lambda^2\right)}{h_0\left(1-\xi^2\right)^2}\eta \tag{9.129}$$

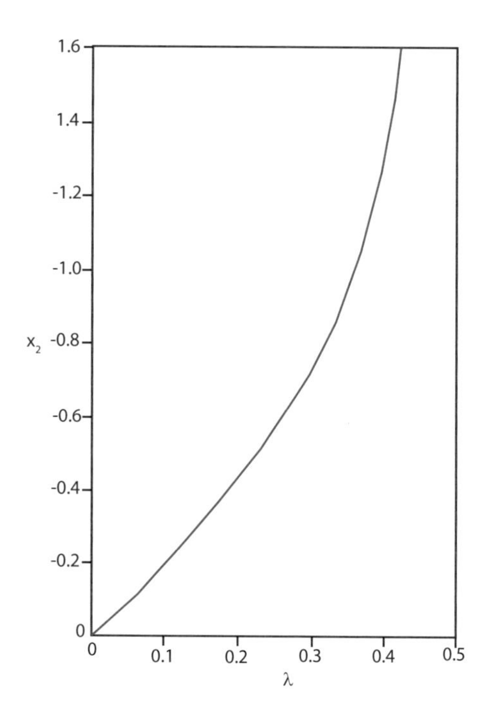

Figure 9.28 Relation between the position of first contact, ξ_2, and the position of sheet separation, λ

Plug flow exists at the point of highest pressure as well as after the material leaves the rolls

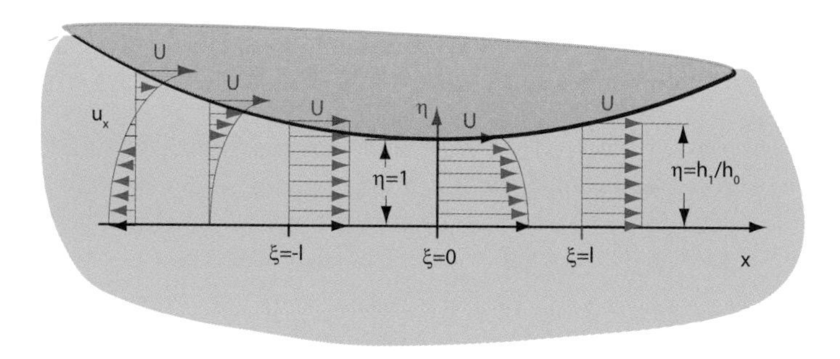

Figure 9.29 Flow pattern that develops in the nip region of a two-roll calendering system

The maximum rate of deformation and shear stress occur at the roll surface, $\eta = 1$, at $\xi = 0$, where the gap is smallest

$$\dot{\gamma}_{\max}(\eta) = \frac{3U\lambda^2}{h_0} \tag{9.130}$$

and

$$\tau_{\max}(\eta) = \mu \frac{3U\lambda^2}{h_0}, \qquad (9.131)$$

respectively. However, the overall maximum of stress and rate of deformation occurs at $\xi = \xi_2$ when $\xi_2 > -\sqrt{1+2\lambda^2}$, and at $\xi = \sqrt{1+2\lambda^2}$, if $\xi_2 < \sqrt{1+2\lambda^2}$. We can compute the overall power requirement by integrating $U\tau_{yx}$ along the surface of the roll, $\eta = 1$

$$P = 3\mu W U^2 \sqrt{\frac{2R}{h_0}} \mathcal{F}(\lambda) \qquad (9.132)$$

where W is the width of the rolls and

$$\mathcal{F}(\lambda) = \left(1-\lambda^2\right)\left[\tan^{-1}\lambda - \tan^{-1}\xi_{\max}\right] - \left[\frac{(\lambda-\xi_{\max})(1-\xi_{\max}\lambda)}{\left(1-\xi_{\max}^2\right)}\right] \qquad (9.133)$$

Of importance to the mechanical design of the calendering system and to the prediction of the film thickness uniformity is the force separating the two rolls, F. This is computed by integrating the pressure over the area of interest on the surface of the roll

$$F = \frac{3\mu URW}{4h_0} \mathcal{G}(\lambda) \qquad (9.134)$$

where $\mathcal{G}$ is given by

$$\mathcal{G}(\lambda) = \left(\frac{\lambda - \xi_2}{1-\xi_2^2}\right)\left[-\xi_2 - \lambda - 5\lambda^3\left(1+\xi_2^2\right)\right] + \left(1-2\lambda^2\right)\left(\lambda\tan^{-1}\lambda - \xi_2\tan^{-1}\xi_2\right) \qquad (9.135)$$

Both functions $\mathcal{F}(\lambda)$ and $\mathcal{G}(\lambda)$ are shown in Fig. 9.30.

Finally, if from an adiabatic energy balance we assume that the power goes into heat generation, we can estimate the temperature rise within the material to be

$$\Delta T = \frac{P}{\varrho Q W C_p} \qquad (9.136)$$

Example 9.6 **Calendering problem with a Newtonian viscosity polymer**

A calender system with $R = 10$ cm, $W = 100$ cm, $h_0 = 0.1$ mm operates at a speed of $U = 40$ cm/s and produces a sheet thickness $h_1 = 0.0218$ cm. The viscosity of the material is given as 1000 Pa·s. Estimate the maximum pressure developed in the material, the power required to operate the system, the roll separating force, and the adiabatic temperature rise within the material.

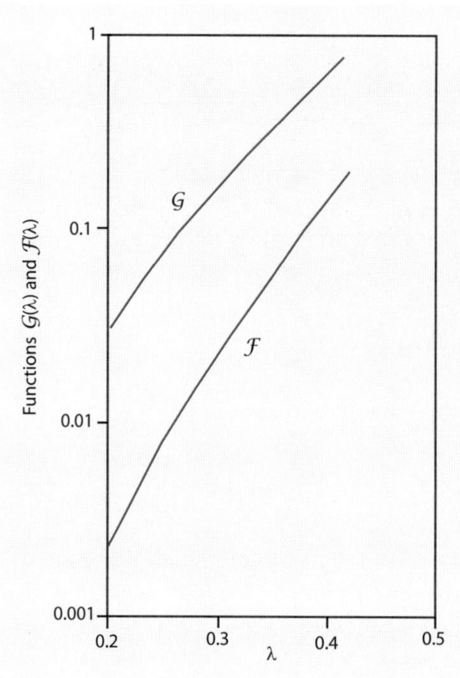

Figure 9.30 Power and force functions $\mathcal{F}(\lambda)$ and $\mathcal{G}(\lambda)$ are used in Eqs. 9.133 and 9.135

Because the final sheet thickness is given we can compute λ using Eq. 9.123 as

$$\frac{h_1}{h_0} = 1 + \lambda^2 \tag{9.137}$$

resulting in $\lambda = 0.3$. Equation 9.126 becomes

$$p_{\max} \approx \frac{15\,(1000\text{ Pa·s})\,(0.40\text{ m/s})\,(0.3)^3}{0.0001\text{ m}} \sqrt{\frac{0.1\text{ m}}{0.0001\text{ m}}} = 18.1\text{ MPa} \tag{9.138}$$

The power is computed using Eq. 9.132 with $\mathcal{F}(0.3) = 0.043$

$$P = 3\,(1000\text{ Pa·s})\,(1\text{ m})\,(0.4\text{ m/s})^2 \sqrt{\frac{2\,(0.1\text{ m})}{(0.0001\text{ m})}}\,\mathcal{F}(0.3) = 923\text{ W} \tag{9.139}$$

The separating force is computed using Eq. 9.134 with $\mathcal{G}(0.3) = 0.16$

$$F = \frac{3\,(1000\text{ Pa})\,(0.4\text{ m/s})\,(0.1\text{ m})\,(1\text{ m})}{4\,(0.0001\text{ m})}\,\mathcal{G}(0.3) = 48\text{ kN} \tag{9.140}$$

Using a volumetric flow rate of $Q = 2Uh_1W = 1.744 \times 10^-4\text{ m}^3/\text{s}$, we can use Eq. 9.136 with a typical specific heat of 1500 J/kg/K and density of 1000 kg/m^3 to compute the adiabatic temperature rise

$$\Delta T = \frac{923\text{ W}}{\left(1000\text{ kg/m}^3\right)(1.744 \times 10^{-4}\text{ m}^3/s)\,(1\text{ m})\left(1500\text{ J/kg/K}\right)} = 3.5\text{ K} \tag{9.141}$$

■

Example 9.7 Calendering problem with floating roll

In a set of calendering rolls, weighing 500 kg each, the upper roll rests on top of the calendered polymer. The calender dimensions are $R = 0.15$ m and $W = 2.0$ m. For a material with a Newtonian viscosity of 1000 Pa·s and a speed of 0.1 m/s, what is the final sheet thickness?

To solve this problem, we begin with Eq. 9.134 and substitute $\mathcal{G}(\lambda)$ with a value of $\lambda = 0.475$

$$F = 1.23\frac{3\mu URW}{4h_0} \tag{9.142}$$

We can solve for h_0 with values of $F = 500 \times 9.81$ N, $U = 0.1$ m/s, $R = 0.15$ m, $W = 2.0$ m and $\mu = 1000$ Pa·s

$$h_0 = 1.23\frac{3\mu URW}{4F} = 0.0022 \text{ m (2.2 mm or 85.8 mils)} \tag{9.143}$$

9.4.2 Shear Thinning Model of Calendering

As with the Newtonian model, we assume a lubrication approximation, where the momentum balance reduces to

$$\frac{dp}{dx} = \frac{\partial \tau_{xy}}{\partial y} \tag{9.144}$$

If we assume a power law model, the shear stress τ_{xy} can be written as

$$\tau_{xy} = m\left|\frac{\partial u_x}{\partial y}\right|^{n-1}\frac{\partial u_x}{\partial y} \tag{9.145}$$

The absolute value in Eq. 9.145 is to avoid taking the root of a negative number. From the Newtonian solution we can see that there are two regions, one where the velocity gradient is positive, $\xi < \lambda$, and one where the velocity gradient is negative, $\xi > -\lambda$. In each region, the above equation must be integrated separately, resulting in two velocity distributions

$$u_x = U + \frac{1}{n/(1+n)}\left(\frac{1}{m}\frac{dp}{dx}\right)^{1/n}\left[y^{n/(1+n)} - h^{n/(1+n)}(x)\right] \tag{9.146}$$

for the region with the negative velocity gradient, and

$$u_x = U - \frac{1}{n/(1+n)}\left(-\frac{1}{m}\frac{dp}{dx}\right)^{1/n}\left[y^{n/(1+n)} - h^{n/(1+n)}(x)\right] \tag{9.147}$$

for the region with the positive velocity gradient. Either equation can be used to solve for the pressure gradient

$$\frac{d\hat{p}}{d\xi} = -\left(\frac{2n+1}{n}\right)^n\sqrt{\frac{2R}{h_0}}\frac{\left(\lambda^2-\xi^2\right)\left|\lambda^2-\xi^2\right|^{n-1}}{\left(1+\xi^2\right)^{2n+1}} \tag{9.148}$$

where $\hat{p}$ is a power law dimensionless pressure defined by

$$\hat{p} = \frac{p}{m}\left(\frac{h_0}{U}\right)^n \tag{9.149}$$

Equation 9.148 can be integrated to become

$$\hat{p}_{\max} = \sqrt{\frac{2R}{h_0}}\left(\frac{2n+1}{n}\right)^n \int_{-\lambda_0}^{\lambda_0} \frac{\left(\lambda_0^2 - \xi^2\right)^n}{\left(1+\xi^2\right)^{1+2n}} d\xi = \sqrt{\frac{2R}{h_0}}\mathcal{P}(n) \tag{9.150}$$

where λ_0 is the position where the integral vanishes

$$0 = \int_{-\infty}^{\lambda_0} \frac{\left(\lambda_0^2 - \xi^2\right)\left|\lambda_0^2 - \xi^2\right|^{n-1}}{\left(1+\xi^2\right)^{1+2n}} d\xi \tag{9.151}$$

Figure 9.31 presents λ_0 as a function of power law index, n. The figure also presents the ratio of final sheet thickness to the nip separation as a function of n.

We can also compute the roll separating force, F, and the power required to drive the system, P,

$$F = RWm\left(\frac{U}{h_0}\right)^n \mathcal{F}(n) \tag{9.152}$$

and

$$P = U^2 Wm\sqrt{\frac{R}{h_0}}\left(\frac{U}{h_0}\right)^{n-1} \mathcal{E}(n) \tag{9.153}$$

Figure 9.32 [11] presents the functions $\mathcal{P}$, $\mathcal{F}$, and $\mathcal{E}$ as a function of the power law index.

Unkrüer [12] performed experimental studies on the flow development during the calendering process of unplasticized polyvinyl chloride to produce thin films. Among

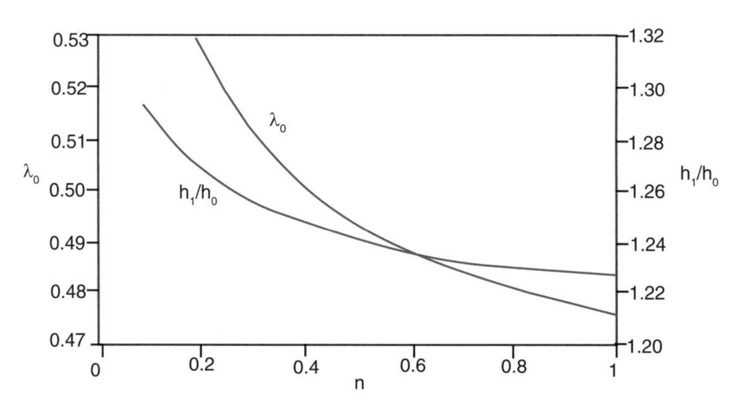

Figure 9.31 Function λ_0 and sheet thickness as a function of power law index, n

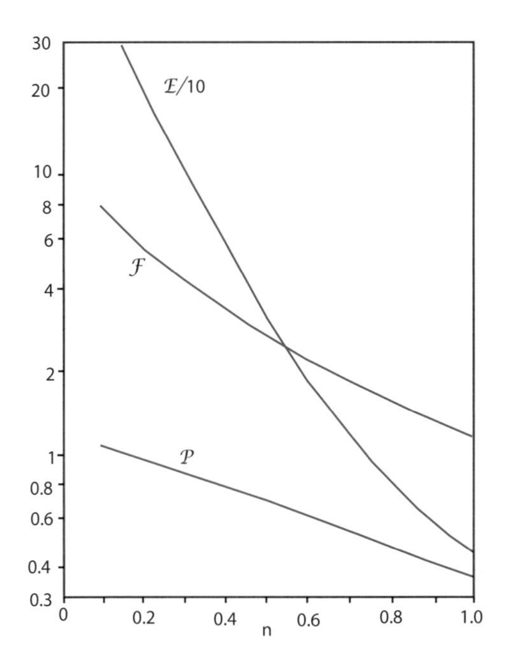

Figure 9.32 Pressure ($\mathcal{P}$ (n)), force ($\mathcal{F}$ (n)), and power ($\mathcal{E}$ (n)) functions as a function of power law index, n

other things he compared the measured maximum pressure that develops between the rolls to the analytical value predicted by the shear thinning model presented above. Figure 9.33 shows this comparison between experiment and theory and their rather good agreement.

9.4.3 Calender Fed with a Finite Sheet Thickness

All the above problems relate to the calendering process where a large mass of polymer melt is fed into the calender. In some industrial applications, a finite polymer sheet of thickness h_f is fed to the calendering rolls, as depicted in Fig. 9.34.

To solve this problem, Eq. 9.150 is replaced by

$$\hat{p}=\sqrt{\frac{2R}{h_0}}\left(\frac{2n+1}{n}\right)^n\int_{-\xi_f}^{\lambda_0}\frac{\left(\lambda_0^2-\xi^2\right)^n}{\left(1+\xi^2\right)^{1+2n}}d\xi=\sqrt{\frac{2R}{h_0}}\mathcal{P}\left(n\right) \tag{9.154}$$

and Eq. 9.150 becomes

$$0=\int_{-\xi_f}^{-\xi}\frac{\left(\lambda_0^2-\xi^2\right)\left|\lambda_0^2-\xi^2\right|^{n-1}}{\left(1+\xi^2\right)^{1+2n}}d\xi \tag{9.155}$$

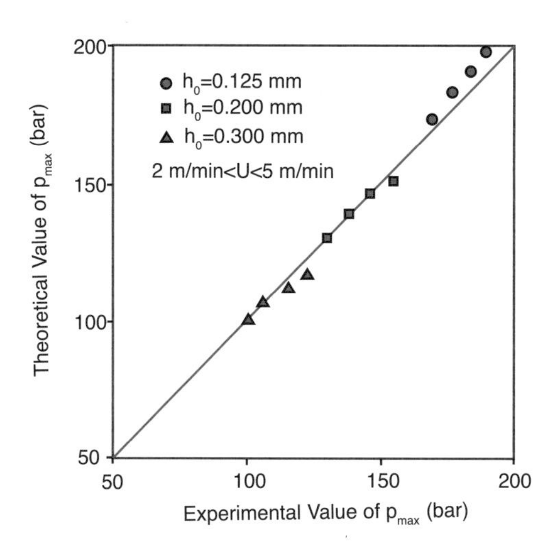

Figure 9.33 Comparison between experiments and theoretical predictions of maximum pressure between the rolls during the calendering process of an unplasticized PVC film. A power law index, $n = 0.1505$ and a consistency index, $m = 155.2$ kPa·s were used in the power law model of the viscosity

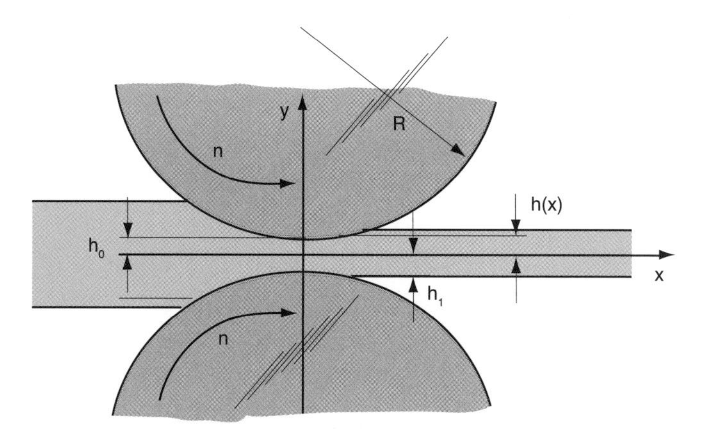

Figure 9.34 Schematic diagram of a two roll calendering system in the nip region fed with a finite sheet

The position where the sheet being fed enters the system can be computed using

$$\xi_f = \sqrt{\frac{h_f}{h_0} - 1} \tag{9.156}$$

Figure 9.35 presents a plot of final sheet thickness as a function of fed sheet thickness for a Newtonian polymer and a shear thinning polymer with a power law index of 0.25.

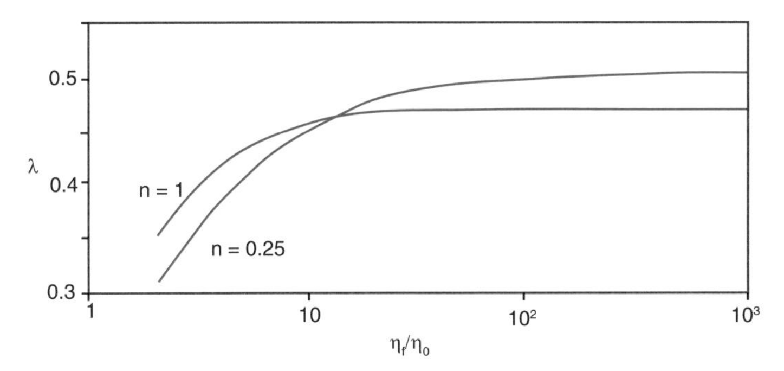

Figure 9.35 Calendered thickness as a function of fed sheet thickness for a Newtonian fluid and a shear thinning fluid with a power law index of 0.25

9.5 Injection Molding — Isothermal Flow Problems

Injection molding is a rather complex process during which non-Newtonian as well as non-isothermal effects play significant roles. Here, we present a couple of problems that are relatively simple to allow an analytical solution. Further information on injection molding models and simulations can be found in the literature [11].

9.5.1 Balancing the Runner System in Multi-Cavity Injection Molds

Inevitably, in multi-cavity injection molds, some of the mold cavity gates are located further than others from the sprue that delivers the melt from the plasticating unit. If the runner system that distributes the melt from the sprue to the individual cavities has a constant diameter, each cavity would receive the melt at a different time and pressure. Differences in pressure will result in variable shrinkage, part weight, and appearance. In order to avoid part inconsistencies, the runner system has to be appropriately designed. For example, let us consider a multi-cavity system, schematically depicted in Fig. 9.36.

In the figure, each portion of the runner is labeled with a number inside a circle and the junctures between the runners are labeled with a number. In order to balance such

A balanced runner system is necessary to have parts of equal weight and properties

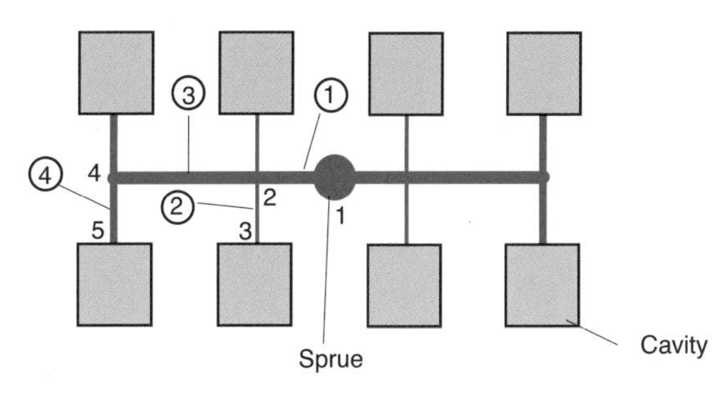

Figure 9.36 Schematic diagram of an eight-cavity mold system with various runner lengths and radii

a runner system, we must assure that the flow-rate into each cavity is the same for the cavities near the gate and the cavities far away. Hence, for the case shown in Fig. 9.36 we must satisfy

$$Q_2 = Q_4 \tag{9.157}$$

The flow through runners 1 and 3 must be

$$Q_3 = 2Q_4 \tag{9.158}$$

and

$$Q_1 = 2Q_2 + 2Q_4 \tag{9.159}$$

respectively. Assuming an isothermal flow inside the runner system,[2)] the flow rate in each runner can be approximated using the Hagen-Poiseuille equation for a power law fluid

$$Q_i = \frac{\pi R_i^3}{s+3}\left(\frac{R_i \Delta p_i}{2mL_i}\right) \tag{9.160}$$

where Δp_i is the pressure drop between the junctures, specific to each runner within the system.

Example 9.8 Sample balancing problem

Let us consider the multi-cavity injection molding process shown in Fig. 9.37. To achieve equal part quality, the filling time for all cavities must be balanced. For the case in question,

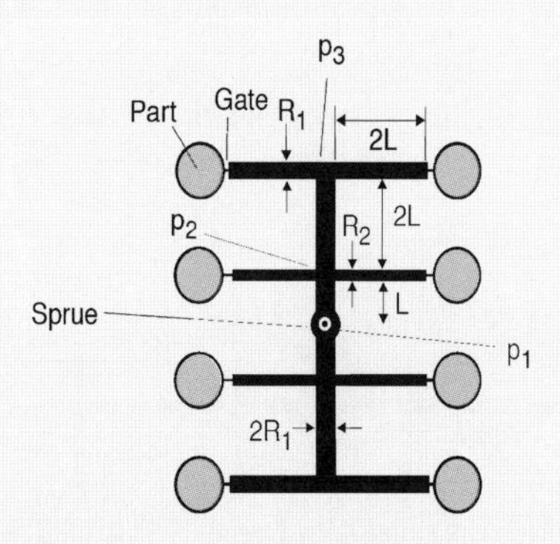

Figure 9.37 Runner system lay out

2) This is only true for a *hot runner system*. However, in many injection molding processes the runner system is directly inside the cooled mold, and the flow is not isothermal.

we need to balance the cavities by solving for the runner radius R_2. For a balanced runner system, the flow rates into all cavities must match. For a given flow rate Q, length L, and radius R_1, solve for the pressures at the runner system junctures. Assume an isothermal flow of a non-Newtonian shear thinning polymer. Compute the radius R_2 for a part molded of polystyrene with a consistency index $m = 2.8 \times 10^4$ Pa·s^n and a power law index $n = 0.28$. Use values of $L = 10$ cm, $R_1 = 3$ mm, and $Q = 20$ cm^3/s.

Assuming a Hagen Poiseuille flow, we can write the following equations for the 4 runner system sections

- Section 1: $4Q = \frac{\pi (1.5R_1)^3}{s+3}\left[\frac{1.5R_1 (p_1 - p_2)}{2mL}\right]^s$
- Section 2: $2Q = \frac{\pi (1.5R_1)^3}{s+3}\left[\frac{1.5R_1 (p_2 - p_3)}{4mL}\right]^s$
- Section 3: $Q = \frac{\pi R_2^3}{s+3}\left[\frac{R_2 (p_2 - 0)}{4mL}\right]^s$
- Section 4: $Q = \frac{\pi R_1^3}{s+3}\left[\frac{R_1 (p_3 - 0)}{4mL}\right]^s$

The unknown parameters, p_1, p_2, p_3, and R_2 can be obtained using the listed equations. For the given values, a radius, $R_2 = 2.34$ mm would result in a balanced runner system, with pressures $p_1 = 562$ bar, $p_2 = 460$ bar, and $p_3 = 292$ bar. For comparison, had we assumed a Newtonian viscosity where $\mu = m$, a radius, $R_2 = 2.76$ mm would have resulted in a balanced system, with much higher pressures of $p_1 = 63\,030$ bar, $p_2 = 49\,120$ bar, and $p_3 = 35\,210$ bar.

■

9.5.2 Radial Flow Between Two Parallel Discs

In Chapter 8 of this book we derived the equations that govern the pressure flow between two parallel discs for a Newtonian fluid. In a similar fashion, we can derive the equations that govern flow rate, gate pressure, and pressure distributions for disc-shaped cavities filling with a shear thinning fluid. For the equations presented in this section, we assumed a power law viscosity. For the velocity distribution we have

$$u_r(r, z) = -\frac{2n+1}{n+1}\frac{Q}{4\pi n h^2}\left(z^{1+1/n} - h^{1+1/n}\right) \tag{9.161}$$

where

$$Q = \frac{4\pi n h^{2+1/n}}{2n+1}\left(-\frac{r^n}{m}\frac{dp}{dr}\right)^{1/n} \tag{9.162}$$

with a pressure distribution of

$$p(r) = \frac{m}{(1-n)h}\left[\frac{(2n+1)Q}{4\pi n h^2}\right]^n \left(R^{1-n} - r^{1-n}\right) \tag{9.163}$$

where r is the radial position in the disc mold. By substituting $r = r_1$, we can compute the pressure at the gate.

Example 9.9 Predicting pressure profiles in a disc-shaped mold using a shear thinning power law model [13]

We can solve the problem presented in Example 8.7 for a shear thinning polymer with power law viscosity model. For this, we will choose the magnitude of the viscosity in Example 8.7 as the power law model consistency index, $m = 6\,400$ Pa·s^n, with a power law index $n = 0.39$. With a constant volumetric flow rate, Q, we get the same flow front location in time as in the previous problem, and we can use Eqs. 9.162 and 9.163 to predict the required gate pressure and pressure profile throughout the disc.

Figure 9.38 presents the pressure profiles within the material for various melt flow front locations. First of all, we can see that the shear thinning behavior of the polymer has caused the pressure requirement to go down significantly (by a factor of 30). The curves presented in Fig. 9.38 also reveal that the shape of the curves was also affected when compared to the Newtonian profiles.

Figure 9.39 presents a comparison of the pressure at the gate for the Newtonian and shear thinning case. The figure also shows the effect of temperature [13]. We can see the effect that cooling has on the pressure requirements. This is caused by a reduction of thickness due to the growth of a solidified layer on the mold surface, as well as an increase in viscosity due to a drop in overall temperature. For a better comparison, Fig. 9.40 presents the pressure requirements for the shear thinning and non-isothermal cases [13].

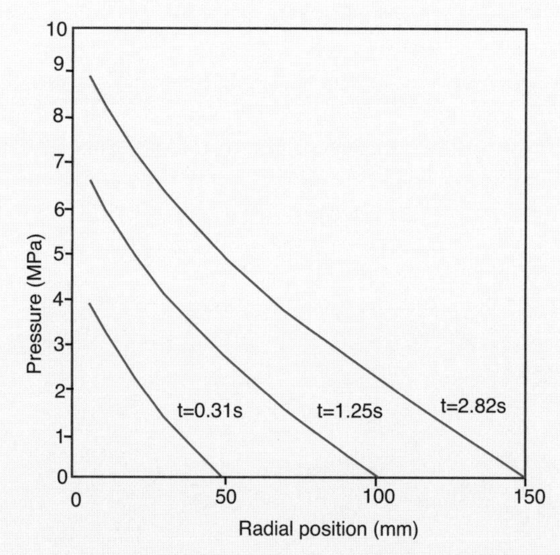

Figure 9.38 Radial pressure profile as a function of time in a disc-shaped mold computed using a shear thinning viscosity model (after Agassant [13])

Shear thinning and non-isothermal effects control flow and pressure during injection molding

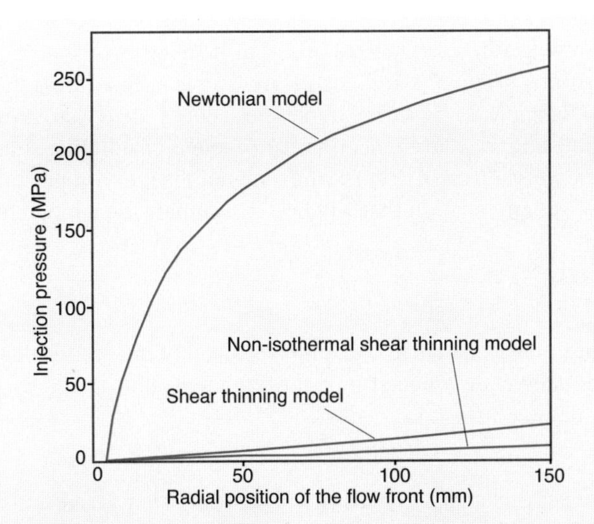

Figure 9.39 Pressure at the gate as a function of melt flow front position in a disc-shaped mold (after Agassant [13])

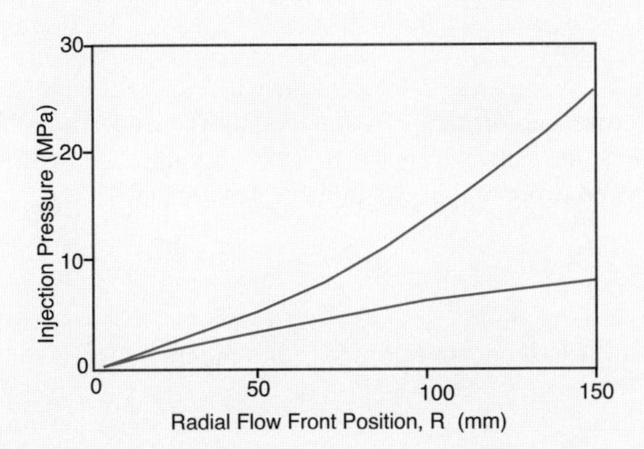

Figure 9.40 Comparison between the analytical shear thinning solution and a non-isothermal shear thinning solution of pressure at the gate as a function of melt flow front position in a disc-shaped mold (after Agassant [13])

9.6 Non-Isothermal Flows

Although we analyze most polymer processes as isothermal problems, many are non-isothermal even at steady state conditions. The non-isothermal effects during flow are often difficult to analyze and make analytical solutions cumbersome or, in many cases impossible. The non-isothermal behavior is complicated further when the energy equation and the momentum balance are fully coupled. This occurs when viscous dissipation is sufficiently high to raise the temperature enough to affect the viscosity of the melt.

Hence, when solving a non-isothermal problem the question arises: is this a problem where the equations of motion and energy are coupled? To address this question we can go back to Example 9.1, where a simple shear flow system was analyzed to decide whether it can be addressed as an isothermal problem or not. In a simple shear flow, the maximum temperature will occur at the center of the melt. By substituting $y = h/2$ into Eq. 9.5, we get an equation that will help us estimate the temperature rise

$$\Delta T_{\text{max}} = \frac{3}{8}\frac{\eta u_0^2}{k} \tag{9.164}$$

When analyzing non-isothermal flow problems, we often assume that the viscosity decays exponentially with temperature following the relation

$$\eta = \eta_0 e^{-a(T-T_0)} \tag{9.165}$$

We can determine the change of viscosity with respect to temperature change by differentiating Eq. 9.165

$$\frac{\partial \eta}{\partial T} = -\eta_0 e^{-a\Delta T} \tag{9.166}$$

Here, we can see that an increase in temperature will reduce the viscosity by an amount controlled by the material constant α (the temperature dependence of the viscosity)and the actual temperature rise, ΔT. Hence, the effect is controlled by the product $a\Delta T$. Taking Eq. 9.164 and dropping the 3/8 term we can say that

$$a\Delta T \propto a\frac{\eta u_0^2}{k} \tag{9.167}$$

Equation 9.167 is the well known Nahme-Griffith number

$$Na = a\frac{\eta u_0^2}{k} \tag{9.168}$$

which is a measure of the degree of coupling between the energy equation and the momentum balance. For the two problems that follow, we will assume that the energy equation and the equation of motion are not coupled, hence $Na \ll 1$.

9.6.1 Non-Isothermal Shear Flow

A common flow problem in polymer processing is a shear flow with a temperature gradient, as depicted in Fig. 9.41. For example, this type of flow occurs within the melt film that develops during melting with drag flow removal, as will be discussed later in this chapter.

In the problem addressed in this section, the two plates are assigned two different temperatures, T_0 on the lower plate and T_1 on the upper moving plate. In addition,

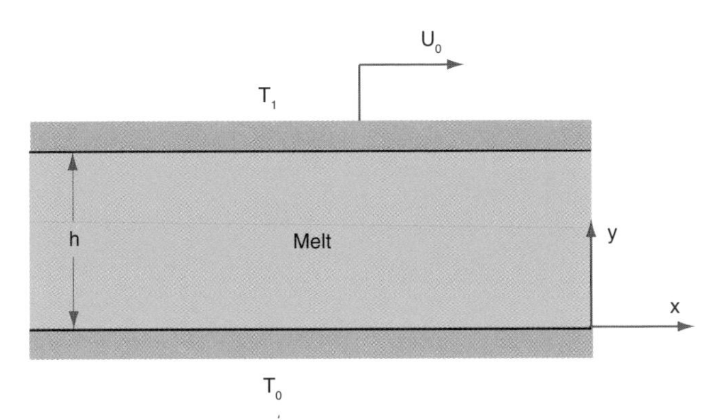

Non-isothermal flows are common within the film during melting, as well as during cavity filling in injection molding

Figure 9.41 Schematic diagram of a shear flow problem with an imposed temperature gradient

we are assuming a Newtonian flow with an exponential temperature dependence, $\mu = \mu_0 e^{-a(T-T_0)}$, constant thermal properties, and negligible viscous heating.

In such a system, the energy balance reduces to

$$0 = k\frac{\partial^2 T}{\partial y^2} \tag{9.169}$$

which can be integrated using the boundary conditions given above to become

$$(T - T_0) = (T_1 - T_0)\left(\frac{y}{h}\right) \tag{9.170}$$

or

$$\Theta = \Theta_1\left(\frac{y}{h}\right) \tag{9.171}$$

The equation of motion reduces to

$$0 = \frac{\partial \tau_{yx}}{\partial y} \tag{9.172}$$

which after integration gives

$$\tau_{yx} = C_1 \tag{9.173}$$

We can now make use of the constitutive equation

$$\tau_{yx} = \mu_0 e^{-a\Theta}\frac{\partial u_x}{\partial y} = C_1 \tag{9.174}$$

which can be rewritten as

$$\frac{\partial u_x}{\partial y} = \frac{C_1}{\mu_0}e^{a\Theta} \tag{9.175}$$

and integrated using the boundary conditions, $u_x = 0$ and $T = T_0$ at $y = 0$, as well as $u_x = U_0$ and $T = T_1$ at $y = h$, to give

$$u_x = U_0 \left(\frac{1 - e^{\Omega (y/h)}}{1 - e^{\Omega}} \right) \tag{9.176}$$

where $\Omega = a\Theta_1$

To illustrate the effect of thermal gradients and temperature dependent viscosity, we can plot a dimensionless velocity, u_x/U_0 as a function of dimensionless position, y/h, for various values of thermal imbalance between the surfaces, Θ_1. Note that Ω, the product between the temperature dependence of the viscosity and the temperature imbalance, is also a dimensionless quantity. This gives a fully dimensionless graph that can be used to assess many case scenarios. Figure 9.42 presents dimensionless velocity distributions across the plates for various dimensionless temperature imbalances, Ω.

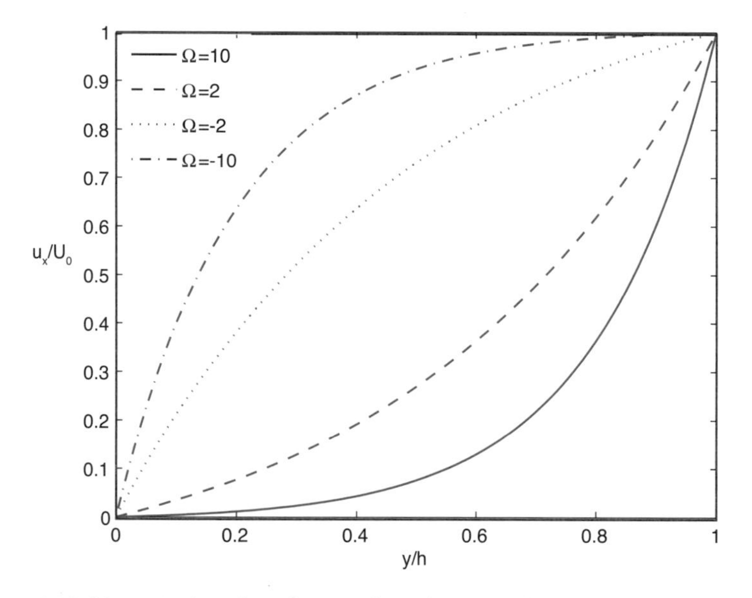

Figure 9.42 Dimensionless shear flow results with various dimensionless temperature imbalances

9.6.2 Non-Isothermal Pressure Flow through a Slit

For this non-isothermal flow, consider a Newtonian fluid between two parallel plates separated by a distance h. Again, we consider the notation presented in Fig. 9.41, however, with both upper and lower plates being fixed. We choose the same exponential viscosity model used in the previous section. We are to solve for the velocity profile between the two plates with an imposed pressure gradient in the x-direction and a temperature gradient in the y-direction.

Again, neglecting viscous heating results in the linear temperature profile, see Eq. 9.171 presented in the previous section. In this case, with the imposed pressure gradient, the equation of motion becomes

$$\frac{\partial p}{\partial x} = \frac{\partial \tau_{yx}}{\partial y} \tag{9.177}$$

which after integration leads to

$$\tau_{yx} = \left(\frac{\partial p}{\partial x}\right) y + C_1 \tag{9.178}$$

The stress can be defined using the constitutive equation $\tau_{yx} = \mu\left(\partial u_x/\partial y\right)$, and letting

$$\frac{\partial p}{\partial x} = \frac{\Delta p}{L}$$

results in

$$\mu_0 e^{(-\Theta_1 a/h)y} \frac{\partial u_x}{\partial y} = \frac{\Delta p}{L} y + C_1 \tag{9.179}$$

Defining $\Omega = \Theta_1 a$ and $\hat{y} = y/h$, we can write

$$\frac{du_x}{dy} = \frac{\Delta p}{\mu_0 L} y e^{\Omega \hat{y}} + \frac{C_1}{\mu_0} e^{\Omega \hat{y}} \tag{9.180}$$

which can be integrated to give

$$u_x = \frac{h^2 \Delta p}{\mu_0 L \Omega^2} e^{\Omega \hat{y}} \left(\Omega \hat{y} - 1\right) + \frac{C_1}{\mu_0 \Omega} e^{\Omega \hat{y}} + C_2 \tag{9.181}$$

Using the boundary conditions $u_x = 0$, at $\hat{y} = 0$ and $\hat{y} = 1$, we get the following velocity distribution

$$u_x = \frac{h^2 \Delta p}{\mu_0 L \Omega} \left[e^{\Omega \hat{y}} \left(\hat{y} + \frac{e^{\Omega}}{1 - e^{\Omega}} \right) - \frac{e^{\Omega}}{1 - e^{\Omega}} \right] \tag{9.182}$$

If we further assume a dimensionless velocity

$$\hat{u}_x = u_x \frac{\mu_0 L \Omega}{h^2 \Delta p} \tag{9.183}$$

Equation 9.182 reduces to

$$\hat{u}_x = \left[e^{\Omega \hat{y}} \left(\hat{y} + \frac{e^{\Omega}}{1 - e^{\Omega}} \right) - \frac{e^{\Omega}}{1 - e^{\Omega}} \right] \tag{9.184}$$

Figure 9.43 presents the dimensionless pressure flow velocity profile for various positive values of Ω. It should be noted that negative values of Ω lead to symmetric velocity profiles as the one generated by the positive values.

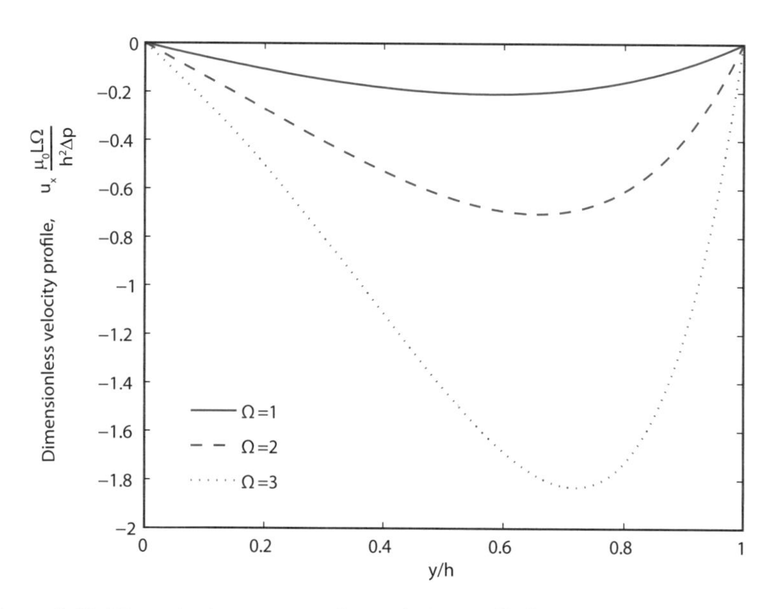

Figure 9.43 Dimensionless pressure flow velocity profile for various dimensionless temperature imbalances

9.7 Melting and Solidification

Melting is an important step in any polymer process. Before the material can be shaped into its final form, it must first be softened or molten. For example, during thermoforming the sheet is heated using radiative heaters to reach a temperature high enough that the sheet can be stretched and formed into, or over, the cavity that will give it its final shape. In extrusion and injection molding the pellets first move from the hopper into the plasticating region of the screw, where they are molten and subsequently pushed out of the die or into the mold cavity.

In the plastication step of the process, melting is critical in controlling cycle time. Also, during thermoforming, the heating of the sheet is the most time consuming step of the whole process. For example, if we consider the melting of an infinite slab, at an initial temperature of T_0, as presented in Fig. 9.44, the heat supplied by the hot wall, set at a heater temperature of T_h, will create a layer of molten polymer of thickness $X(t)$.

For such a case, where the temperature can be computed as a function of time using

$$\frac{T - T_\mathrm{h}}{T_0 - T_\mathrm{h}} = \mathrm{erf}\left(\frac{x}{\sqrt{4\alpha t}}\right) \tag{9.185}$$

where $\alpha = k/\varrho C_\mathrm{p}$ is the thermal diffusivity; the error function, erf, is defined by

$$\mathrm{erf}\left(\frac{x}{\sqrt{4\alpha t}}\right) = \frac{2}{\sqrt{\pi}} \int_0^{\frac{x}{\sqrt{4\alpha t}}} e^{-s^2} ds \tag{9.186}$$

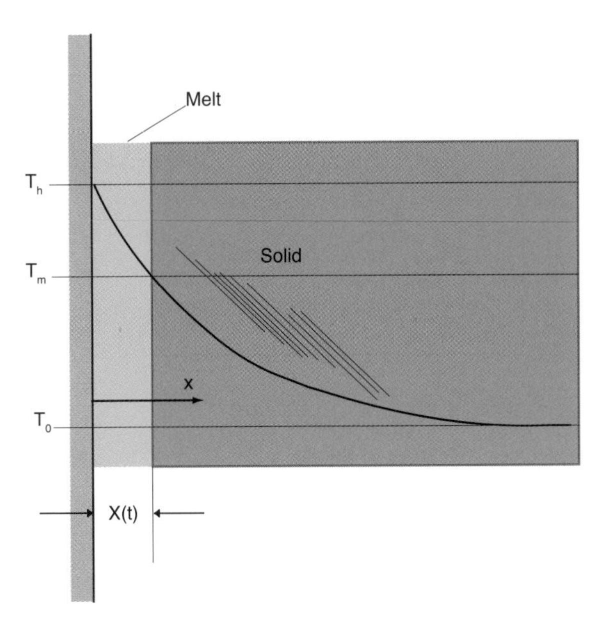

Figure 9.44 Schematic diagram of the melting process of an infinite slab

The equation clearly demonstrates that the molten layer, with temperatures above T_m for semi-crystalline polymers and above T_g for amorphous polymers, will follow the following relation

$$X(t) \propto \sqrt{\alpha t} \tag{9.187}$$

Solidification models predict an infinite solidification growth rate at $t = 0$ – this is not possible for semi-crystalline polymers where the finite growth rate of the solidified front is controlled by crystal nucleation

for amorphous polymers. The melting and solidification process for semi-crystalline materials is more complex because the heat of fusion or heat of crystallization, the nucleation rate, etc. When measuring the specific heat as the material crystallizes, a peak representing the heat of fusion is detected. Although theoretical predictions of melting and solidification in semi-crystalline polymers predict a similar growth rate as with amorphous polymers

$$X(t) \propto f(\lambda, \alpha)\sqrt{t} \tag{9.188}$$

where $f(\lambda, a)$ represents heat of fusion effects, experimental evidence [14] of solidification has demonstrated that the growth rate of the solidified crystallized layer in semi-crystalline polymers is finite at the beginning of the solidification process. This is mainly due to the fact that at the beginning the nucleation can only occur at a finite rate. Hence, the solution presented in Eq. 9.188 as well as the widely used Stefan condition, discussed later, can only be used for melting but do not hold for cooling and solidification of semi-crystalline polymers. This can be seen in Fig. 9.45, which shows the measured thickness of crystallized layers as a function of time for polypropylene plates quenched at three different temperatures. For further reading on this important topic the reader is encouraged to consult the literature [15, 16].

Finite solidification front growth rate for PP plates

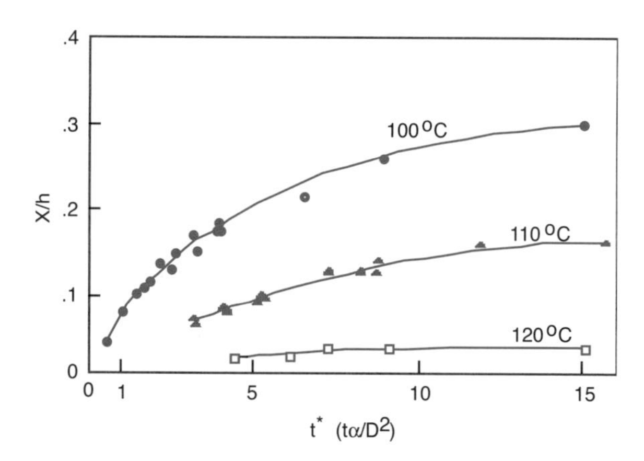

Figure 9.45 Dimensionless thickness of the crystallized layers as a function of dimensionless time for various temperatures of the quenching surface

In either case, the growth rate of the molten layer during conduction melting follows the relation

$$\frac{dX(t)}{dt} \propto \frac{1}{\sqrt{t}} \tag{9.189}$$

which means that the growth rate of the molten layer, $X(t)$ during melting is infinite at $t = 0$, but rapidly decreases as the molten layer increases in thickness. For example, the

Melting with pressure flow melt removal

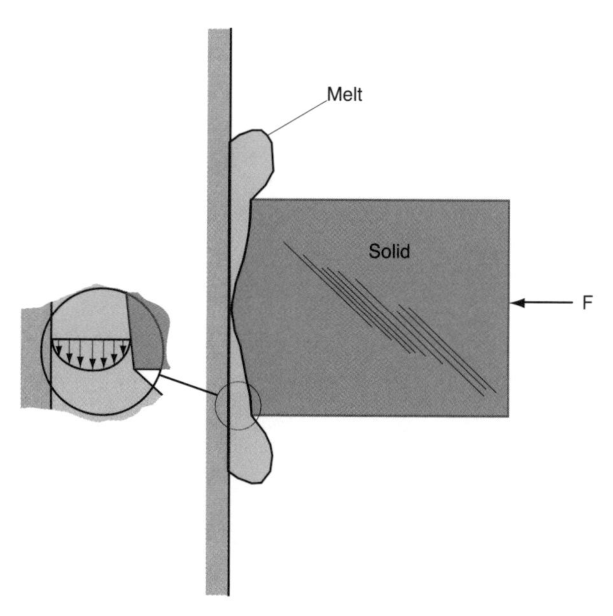

Figure 9.46 Schematic diagram of the melting process with pressure flow melt removal

growth rate at 10 seconds is only 32 % of the growth rate at 1 second, and the growth rate at 1 minute is only 13 % of the growth rate at 1 second. Hence, even after only 10 seconds, the melting rate is quite inefficient, and can only be increased by continuously maintaining a small thickness of the molten layer. This can be achieved by various *melt removal schemes,* namely the *pressure flow melt removal* and the *drag flow melt removal* techniques, presented in Figs. 9.46 and 9.47, respectively.

Melting with drag flow melt removal

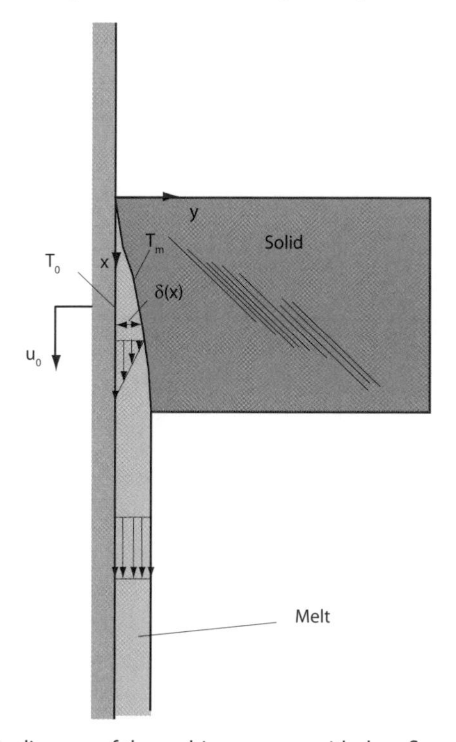

Figure 9.47 Schematic diagram of the melting process with drag flow melt removal

During the pressure flow melt removal scheme a force is applied on the melting solid, resulting in a pressure build-up within the melt film, causing the melt film to be squeezed out from under the solid body. On the other hand, in the drag flow melt removal scheme, the melt is transported out from under the solid by the moving heated surface. This mechanism exists in the melting section of a single screw extruder. Both of these cases are addressed in more detail in the following sections.

9.7.1 Melting with Drag Flow Melt Removal

Let us consider the case described in Fig. 9.47 of an isotropic homogeneous solid infinite slab of width W pushed against a heated moving plate. Here, we want to solve for the temperatures, velocities, and melting rates at steady-state conditions. The problem is essentially two-dimensional, which means that the velocity and temperature fields are functions of x and y. The melt film thickness, $\delta(x)$, is very small at $x = 0$ and it increases in the x-direction; however, its actual shape is unknown. Heat is conducted from the heated plate, T_0, to the solid-melt interface at $T = T_m$. Here, we are considering the more general case of a semi-crystalline polymer and will follow the assumptions and derivation from Tadmor [1, 17]. For simplicity, in his derivation Tadmor assumed constant thermal properties. Furthermore, as a geometric constraint, Tadmor assumed that the film thickness is much smaller than its width, $\delta/W \ll 1$. In addition, all body forces and inertial effects within the film, $Re \ll 1$, are also negligible. Together, these assumptions justify using the lubrication approximation.

The equations of continuity reduce to

$$\begin{aligned} \frac{\partial u_x}{\partial x} + \frac{\partial u_y}{\partial y} &= 0 \\ \frac{\partial \tau_{xy}}{\partial y} &= 0 \end{aligned} \tag{9.190}$$

The energy equation within the melt will include the conductive and the viscous dissipation terms as follows

$$k_m \frac{\partial^2 T}{\partial y^2} - \tau_{xy} \frac{\partial u_x}{\partial y} = 0 \tag{9.191}$$

The boundary conditions for the momentum and energy balance equations are

$$\begin{aligned} &u_x(0) = U_0 \qquad u_x(\delta) = 0 \\ &u_y(0) = 0 \\ &T(0) = T_0 \qquad T(\delta) = T_m \end{aligned} \tag{9.192}$$

The velocity at any point $u_y(\delta)$ is determined by the rate of melting at the interface (see Fig. 9.48), which is obtained from the Stefan condition or heat balance between conduction and the rate of melting at that interface,

$$k_m \left[-\frac{\partial T}{\partial y} \right]_{y=\delta} = \varrho_m \left(-u_y(\delta) \right) \lambda + k_s \left[-\frac{\partial T}{\partial y} \right]_{y=\delta} \tag{9.193}$$

where λ is the heat of fusion and k_s, k_m are the thermal conductivity of the solid and melt, respectively.

In order to reduce Eq. 9.193 we must find the temperature distribution for the solid. The energy balance for the solid is given by

$$\varrho_s C_s u_{sy} \frac{\partial T}{\partial y} = k_s \frac{\partial^2 T}{\partial y^2} \tag{9.194}$$

which can be solved by integration using the boundary conditions $T(\delta) = T_m$ and $T(\infty) = T_{s0}$, which results in the temperature profile

$$T = T_{s0} + (T_m - T_{s0}) e^{\left(\frac{u_{sy}(y\delta)}{\alpha_s} \right)} \tag{9.195}$$

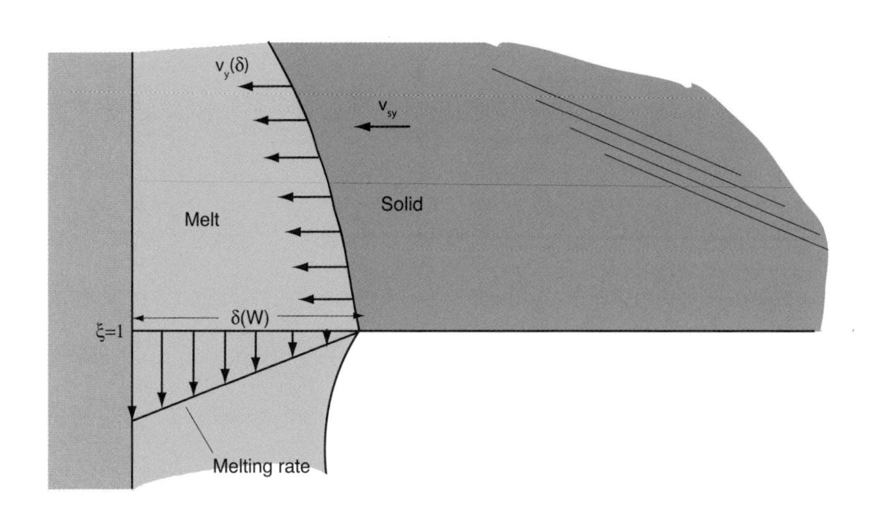

Mass balance – note that the velocity of the solid bed determines the melting rate which is also given by the shear flow profile

Figure 9.48 Schematic diagram of the melting at the interface

The rate of heat conduction out the solid-melt interface will be

$$-k_{\mathrm{s}}\left[\frac{\partial T}{\partial y}\right]_{y=\delta}=-\left(T_{\mathrm{m}}-T_{\mathrm{s}\,0}\right)u_{y}\left(\delta\right)\varrho_{\mathrm{m}}C_{\mathrm{s}} \tag{9.196}$$

where the mass balance in the interface, $u_{\mathrm{sy}}\varrho_{\mathrm{s}}=u_{y}(\delta)\varrho_{\mathrm{m}}$, was used. Equation 9.193 can now be written as

$$k_{\mathrm{m}}\left[\frac{\partial T}{\partial y}\right]_{y=\delta}=\varrho_{\mathrm{m}}u_{y}\left(\delta\right)\lambda' \tag{9.197}$$

where $\lambda'=\lambda+C_{\mathrm{s}}(T_{\mathrm{m}}-T_{\mathrm{s}\,0})$ is the energy required to bring the solid from an initial temperature $T_{\mathrm{s}\,0}$ to T_{m} and to melt it at that temperature.

The variables in the balance Eqs. 9.190 to 9.194 can be scaled into dimensionless variables as

$$\begin{aligned}\Theta&=\frac{T-T_{\mathrm{m}}}{T_{0}-T_{\mathrm{m}}}\\ \xi&=\frac{x}{W} & \zeta&=\frac{y}{\delta}\\ \hat{u}_{x}&=\frac{u_{x}}{U_{0}} & \hat{u}_{y}&=\frac{u_{y}}{U_{0}\left(\delta_{0}/W\right)}\end{aligned} \tag{9.198}$$

where the characteristic film thickness, δ_0, is determined from a scaling analysis of Eq. 9.197, using characteristic values, and reduces to

$$\delta_{0}=\left(\frac{k_{\mathrm{m}}\left(T_{0}-T_{\mathrm{m}}\right)W}{\lambda'\varrho_{\mathrm{m}}U_{0}}\right) \tag{9.199}$$

The dimensionless form of the boundary conditions Eq. 9.192 is

$$\begin{aligned} &\Theta(0) = 1 \qquad \hat{u}_x(0) = 1 \qquad \hat{u}_y(0) = 0 \\ &\Theta(1) = 1 \qquad \hat{u}_x(1) = 0 \end{aligned} \tag{9.200}$$

Melting model using a Newtonian fluid with temperature independent viscosity μ. For a Newtonian fluid the equation of motion reduces to

$$\frac{\partial^2 \hat{u}_x}{\partial \zeta^2} = 0 \tag{9.201}$$

which with the boundary conditions, Eq. 9.200, has the solution

$$\hat{u}_x = 1 - \zeta \tag{9.202}$$

The energy equation, which can be solved independently from the equation of motion, reduces to

$$\frac{\partial^2 \Theta}{\partial \zeta^2} + Br\left(\frac{\partial \hat{u}_x}{\partial \zeta}\right)^2 = 0 \tag{9.203}$$

where the Brinkman number, *Br*, is defined by

$$Br = \frac{\mu U_0^2}{k_m (T_0 - T_m)} \tag{9.204}$$

Integrating Eq. 9.203 with the corresponding boundary conditions yields the temperature profile

$$\Theta = (1 - \zeta) + \frac{Br}{2}\zeta(1 - \zeta) \tag{9.205}$$

This temperature profile can be used to find the y-velocity component at the interface, i. e.,

$$\left(\frac{\partial \Theta}{\partial \zeta}\right)_{\zeta=1} = \frac{\delta}{\delta_0}\hat{u}_y(1) \tag{9.206}$$

to obtain

$$\hat{u}_y(1) = -\frac{\delta}{\delta_0}\left(1 + \frac{Br}{2}\right) \tag{9.207}$$

Integrating the continuity equation and using Eq. 9.202, it is found that

$$\hat{u}_y(1) = -\frac{1}{2}\frac{\delta}{\delta_0} \tag{9.208}$$

Combining these two equations and integrating results in the film thickness profile

$$\delta(\xi) = \delta_0 \sqrt{(4 + 2Br)\,\xi} \tag{9.209}$$

The rate of melting, per unit width, is given by

$$w_L(x) = \varrho_m U_0 \delta \int_0^1 \hat{u}_x d\zeta = \frac{U_0 \delta}{2} \varrho_m \tag{9.210}$$

which results in

$$w_L = \left[\frac{U_0 \varrho_m \left(k_m (T_0 - T_m) + \mu U_0^2/2\right)}{\lambda'} W \right]^{1/2} \tag{9.211}$$

In this analysis, we neglected the convection in the film. Tadmor *et al.* [1, 17] accounted approximately for convection by including in λ' the heat needed to bring the melt from T_m to the mean temperature, i. e.,

$$\lambda'' = \lambda + C_s (T_m - T_{s0}) + C_s (T_0 - T_m)\,\bar{\Theta} \tag{9.212}$$

where the mean temperature is defined by

$$\bar{\Theta} = \frac{\int_0^1 \hat{u}_x \Theta d\zeta}{\int_0^1 \hat{u}_x d\zeta} = \frac{2}{3} + \frac{Br}{12} \tag{9.213}$$

In addition, w_L must be reduced by a factor of 2 because the newly melted material must be removed from the interface, allowing it to flow into the film at that point, keeping the film thickness constant, i. e.,

$$w_L = \left[\frac{U_0 \varrho_m \left(k_m (T_0 - T_m) + \mu U_0^2/2\right)}{2\lambda''} W \right]^{1/2} \tag{9.214}$$

Power law model fluid with temperature dependent viscosity $m_0 = e^{(-a(T-T_m))}$. The rate of melting is strongly dependent on the shear thinning behavior and the temperature dependent viscosity of the polymer melt. However, we can simplify the problem significantly by assuming that the viscous dissipation is low enough that the temperature profile used to compute the viscosity is linear, i. e.,

$$\Theta = (1 - \zeta) \tag{9.215}$$

such that the equation of motion reduces to

$$\frac{\partial}{\partial \zeta} \left(e^{b(1-\zeta)} \left(-\frac{\partial \hat{u}_x}{\partial \zeta} \right)^n \right) = 0 \tag{9.216}$$

Solving this equation, the local velocity profile (derived in the previous sections) becomes

$$\hat{u}_x = \frac{e^{b'\zeta} - e^{b'}}{1 - e^{b'}} \tag{9.217}$$

where $b' = b/n = -a\,(T_0 - T_m)\,/n$. The energy equation without convection, will be

$$\frac{\partial^2 \Theta}{\partial \zeta^2} + Br\left(\frac{\delta_0}{\delta}\right)^{n-1} e^{b(1-\zeta)}\left(-\frac{\partial \hat{u}_x}{\partial \zeta}\right)^{n+1} = 0 \tag{9.218}$$

where the Brinkman number is defined by

$$Br = \frac{m_0 U_0^{(3n+1)/2} \varrho_m^{(n-1)/2} \lambda'^{(n-1)/2}}{(T_0 - T_m)^{(n+1)/2}\, k_m^{(n+1)/2} W^{(n-1)/2}} \tag{9.219}$$

Integrating the equation will give the temperature profile

$$\Theta = (1-\zeta) + Br\left(\frac{\delta_0}{\delta}\right)^{n-1}\left(\frac{b'}{1-e^{-b'}}\right)^{n+1}\frac{e^{-b'}}{b'^2}\left[1 - e^{b'} - \zeta\left(1 - e^{b'}\right)\right] \tag{9.220}$$

with these equations, and similar to the Newtonian case, the film profile can be obtained

$$\delta(\xi) = \delta_0\left[\frac{4\left[1 + Br\left(\frac{\delta_0}{\delta}\right)^{n-1}\left(\frac{b'}{1-e^{-b'}}\right)^{n+1}\left(\frac{b'-1+e^{-b'}}{b'^2}\right)\right]\xi}{U_2}\right]^{1/2} \tag{9.221}$$

where $\bar{\delta}$ is an assumed value of the mean melt thickness and

$$U_2 = 2\frac{1 - b' - e^{-b'}}{b'\left(e^{-b'} - 1\right)} \tag{9.222}$$

By substitution of the parameters we find

$$\delta = \left[\frac{2\,(2k_m\,(T_0 - T_m) + U_1)\,\xi}{U_2 \varrho_m U_0 \lambda'}\right] \tag{9.223}$$

where

$$U_1 = \frac{2m_0 U_0^{n+1}}{\bar{\delta}^{n-1}}\left(\frac{b'}{1-e^{-b'}}\right)^{n+1}\frac{b' - 1 + e^{-b'}}{b'^2} \tag{9.224}$$

And the rate of melting, per unit width, is given by

$$w_L(\xi) = \left[\frac{\varrho_m U_0 U_2\,(k_m\,(T_0 - T_m) + U_1/2)\,\xi}{\lambda'}\right]^{1/2} \tag{9.225}$$

Convection can be accounted for, using the same approximation used for the Newtonian case [1, 17], then λ' is replaced by λ'' (Eq. 9.211), and $w_L(\xi)$ is reduced by a factor of $\sqrt{2}$, i.e.,

$$w_L(\xi) = \left[\frac{\varrho_m U_0 U_2 \left(k_m \left(T_0 - T_m\right) + U_1/2\right)\xi}{2\lambda''}\right]^{1/2} \tag{9.226}$$

where the mean temperature of the film is

$$\bar{\Theta} = \frac{b'/2 + e^{-b'}\left(1 + 1/b'\right) - 1/b'}{b' + e^{-b'} - 1} \tag{9.227}$$

9.7.2 Melting Zone in a Plasticating Single Screw Extruder

One of the most important aspects when designing the screw geometry for a single screw extruder is to control the melting of the pellets within the screw. Here, we will present the model developed by Tadmor [17], which is based on observations of the state of the material along the screw channel. If we unwrap the channel contents, several characteristics can be recognized, as schematically represented in Fig. 9.49. The first section of the screw, the solids conveying zone, compacts and pressurizes the polymer pellets and, as discussed in Chapter 4 of this book, is responsible for starting the motion in the down-channel direction. The first portion of the solids conveying zone is cooled with chilled water in order to avoid melt formation on the barrel surface. However, after the chilled region, due to friction of the pellets against each other and against the barrel surface, as well as due to the effect of the nearby heaters, a melt film forms between the bed of solid pellets and the barrel. One to three turns later, a melt pool forms against the trailing flight of the screw. These few turns, between the melt film formation and the start of the melt pool is often referred to as the *delay zone*. The delay zone has been described by various researchers in the past, and was quantitatively described by Tadmor and Klein [17]. Noriega *et al.* [18] were able to clearly discern its existence and attributed it to a *solid bed saturation process*, where the melt pool does not form until

Solid bed profile

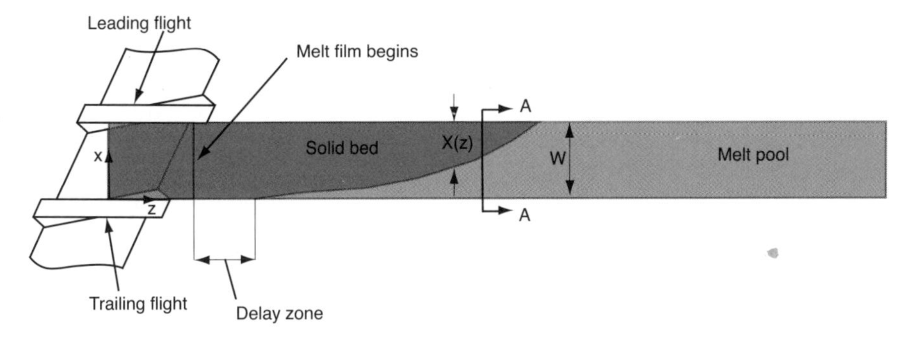

Figure 9.49 Schematic diagram of unwrapped channel contents of a single screw extruder

the melt has completely filled the gaps between the pellets. From that point on, a solid bed profile develops, as schematically depicted in Fig. 9.49.

Cutting a cross-section A—A of the channel contents in Fig. 9.49 leads to a cross-sectional view that is shown Fig. 9.50. Furthermore, to help visualize and to aid in the mass balances, we can break the channel contents in the melting section down to a differential element, shown in Fig. 9.51.

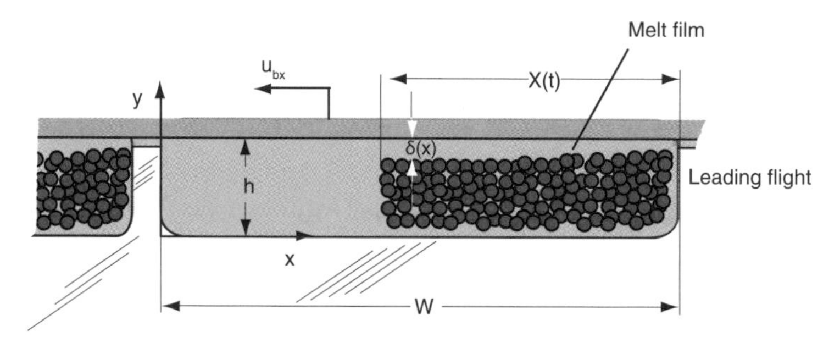

Figure 9.50 Schematic diagram of a channel cross-section in the melting zone of a single screw extruder

Mass and energy balance

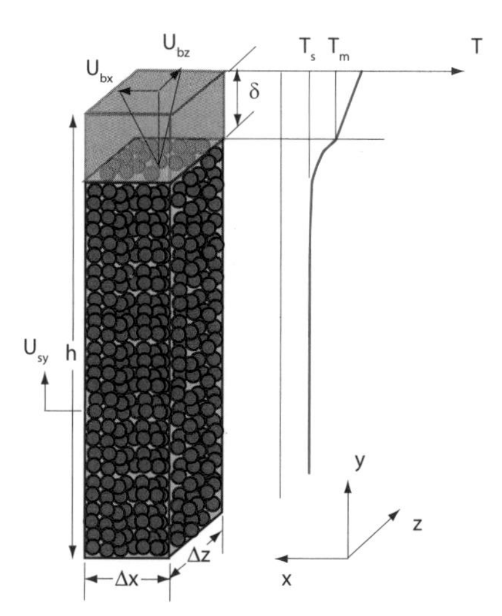

Figure 9.51 Differential element of the channel contents in the melting section of a single screw extruder. The element includes the temperature distribution across the channel as well as isothermal velocity distributions

Once the melt film forms, the conveying mechanism changes at the barrel surface, where viscous drag is now dominant, but frictional drag is still important at the root of the screw and the flights. The thickness of the melt film continues to increase as the plug proceeds down the channel until it attains a value of several times the flight clearance. At this point, the melt film thickness stays nearly constant and the melt is scraped off and accumulated on the side of the trailing or pushing flight. The axial distance from where the melt film appears until melt begins to accumulate at the pushing flight is referred to as the delay zone.

The Tadmor model assumes Newtonian fluids and shallow channels. The channel cross section and that of the solid bed are assumed to be rectangular. The width of the solid bed profile is denoted by $X(z)$, which is the main objective that we are seeking with the model. The solid bed that develops at steady state conditions is the focal interest here. Furthermore, Tadmor assumed that melting only occurs at the barrel surface and that the solid bed is homogeneous, continuous, and deformable.

From observation, it is clear that the solid bed has a local down-channel velocity u_{sz} and a local velocity component into the melt film of u_{sy}. As before, we resolve the barrel surface velocity U_b into a down-channel, u_{bz}, and a cross-channel component, u_{bx}. Using Tadmor's notation, we define the relative velocity between barrel surface and solid bed as

$$|U_j| = \sqrt{u_{bx}^2 + \left(u_{bz} - u_{sz}\right)^2} \tag{9.228}$$

which controls the rate of viscous dissipation. The rate of melting per unit down-channel distance is obtained directly from the Newtonian model of melting with drag removal

$$w_L = \left[\frac{U_0 \varrho_m \left(k_m \left(T_0 - T_m\right) + \mu U_0^2/2\right)}{2\lambda''} W\right]^{1/2} \tag{9.229}$$

with X replacing W, T_b replacing T_0, U_{bx} replacing U_0 in the first term, and U_j replacing U_0 in the second term, we have

$$w_L(z) = \left[\frac{u_{bx}\varrho_m \left(k_m \left(T_b - T_m\right) + \mu U_j^2/2\right)}{2\left[\lambda + C_s \left(T_m - T_{s0}\right) + C_m \bar{\Theta} \left(T_b - T_m\right)\right]} X\right]^{1/2} \tag{9.230}$$

The change in the solid width is obtained by a differential mass balance as follows

$$\varrho_s u_{sz} \left(h - \delta\right) X|_z - \varrho_s u_{sz} \left(h - \delta\right) X|_{z+\Delta z} = w_L \Delta z \tag{9.231}$$

Neglecting the film thickness change in the down-channel direction and taking the limit as $\Delta z \rightarrow 0$, we get

$$-\frac{d\left(HX\right)}{dz} = \frac{w_L(z)}{\varrho_s u_{sz}} \tag{9.232}$$

By substituting the definition of $w_L(z)$, it reduces to

$$-\frac{d\left(HX\right)}{dz} = \frac{\Phi\sqrt{X}}{\varrho_s u_{sz}} \tag{9.233}$$

where

$$\Phi = \left[\frac{u_{bx}\varrho_m \left(k_m \left(T_b - T_m \right) + \mu U_j^2/2 \right)}{2\left[\lambda + C_s \left(T_m - T_{s0} \right) + C_m \bar{\Theta} \left(T_b - T_m \right) \right]} \right]^{1/2} \tag{9.234}$$

For a constant channel depth, Eq. 9.233 can be integrated to give

$$\frac{X_2}{W} = \frac{X_1}{W} \left[1 - \frac{\psi \left(z_2 - z_1 \right)}{2H} \right]^2 \tag{9.235}$$

where X_1 and X_2 are the widths of the solid bed at locations z_1 and z_2, respectively and

$$\psi = \frac{\Phi}{\varrho_s u_{sz} \sqrt{X}} \tag{9.236}$$

Hence, for a constant channel depth we can determine the length of the channel required to melt the solid bed from Eq. 9.235. For a tapered channel of constant taper, Eq. 9.233 can be written as

$$-\frac{d\left(HX \right)}{dz} = \frac{\Phi\sqrt{X}}{A\varrho_s u_{sz}} \tag{9.237}$$

Where

$$A = -\frac{dH}{dz} \tag{9.238}$$

which integrated will give

$$\frac{X_2}{W} = \frac{X_1}{W} \left[\frac{\psi}{A} - \left(\frac{\psi}{A} - 1 \right) \sqrt{\frac{H_1}{H_2}} \right]^2 \tag{9.239}$$

where X_1 and X_2 are the widths of the solid bed at down-channel locations corresponding to heights H_1 and H_2. Equations 9.235 and 9.239 represent the basic equations for the melting model of an extruder. The total length of melting for a channel of constant depth is

$$z_T = \frac{2H}{\psi} \tag{9.240}$$

and for the tapered channel is

$$z_T = \frac{2H}{\psi} \left(2 - \frac{A}{\psi} \right) \tag{9.241}$$

Example 9.10 Solid bed profile prediction for a plasticating single screw extruder

In this example we would like to use Tadmor's model to predict the solid bed profile of a low density polyethylene in a plasticating single screw extruder, based on experiments published by Tadmor and Klein [17]. In their experiments they used the following screw geometry:

- Square pitch screw, $D = 63.5\text{mm}$, $L/D = 26.5$ and $W = 54.16$
- Feed zone -12.5 turns and $h_1 = 9.4$ mm
- Transition zone -9.5 turns and constant taper from $h_1 = 9.4$ mm to $h_2 = 3.23$ mm
- Metering zone -4.5 turns and $h_2 = 3.23$ mm

The processing conditions used in the experiments were:

- Initial temperature of pellets: $T_0 = 24$ °C
- Barrel or heater temperature: $T_b = 149$ °C
- Pressure buildup: $\Delta p = 204$ bar
- Screw speed: $N = 60$ rpm
- Mass throughput: $\dot{m} = 61.8$ kg/hr

The material properties of the PE-LD to be used in the calculations are for a power law viscosity model and thermal properties independent of temperature and pressure, except for the melt density:

- Viscosity: $m_0 = 56\,000$ Pa·s^n, $n = 0.345$, $a = 0.01$/K and $T_m = 110$ °C
- Thermal properties: $k_m = 0.1817$ W/m/K, $C_m = 2,596$ J/kg/K, $C_s = 2,763$ J/kg/K, $\varrho_{bulk} = 595$ kg/m^3, $\varrho_s = 915.1$ kg/m^3, $\varrho_m = 852.7 + 5.018 \times 10^{-7}p - 0.4756T$, and $\lambda = 129.8$ kJ/kg.

From the above data, we can first compute the barrel velocity to be $U_b = \pi DN = 0.1995$ m/s. Note that this speed is on the low end of realistic speeds used in industry for low density polyethylene. PE-LD usually can have screw speeds that lead to velocities up to of 1 m/s. Other polymers can take up to 0.5 m/s, and PVC about 0.2 m/s. The down-channel and cross channel velocities become

$$u_{bz} = U_b \cos\varphi \tag{9.242}$$

and

$$u_{bx} = U_b \sin\varphi \tag{9.243}$$

respectively. The solid bed velocity is computed using

$$u_{bz} = \frac{\dot{m}}{\varrho_{bulk} W h} = 0.0567 \text{ m/s} \tag{9.244}$$

The relative speed between the solids bed and the barrel is calculated using

$$|U_j| = \sqrt{u_{bx}^2 + (u_{bz} - u_{sz})^2} = 0.1465 \text{ m/s} \tag{9.245}$$

We can now compute

$$b' = -a\,(T_0 - T_m)\,/\,n = -1.1304 \tag{9.246}$$

$$U_2 = 2\frac{1 - b' - e^{-b'}}{b'\left(e^{-b'} - 1\right)} = 2\left[\frac{1 + 1.1304 - e^{1.1304}}{-1.1304\,(e^{1.1304} - 1)}\right] = 0.8155 \tag{9.247}$$

and

$$\bar{\Theta} = \frac{b'/2 + e^{-b'}\,(1 + 1/b') - 1/b'}{b' + e^{-b'} - 1} = 0.7002 \tag{9.248}$$

Neglecting the pressure effects, we can compute the melt density using an average temperature of 129.5 °C as

$$\varrho_m = 852.7 + -0.4756T = 852.7 - 0.4756\,(129.5) = 791.1 \text{ kg/m}^3 \tag{9.249}$$

The rest of the equations depend from each other and will reduce to:

$$U_1 = \frac{2m_0 U_0^{n+1}}{\bar{\delta}^{n-1}}\left(\frac{b'}{1 - e^{-b'}}\right)^{n+1}\frac{b' - 1 + e^{-b'}}{b'^2} = 2786.8\,(\delta)^{0.655} \tag{9.250}$$

$$\delta = \left[\frac{2\,(2k_m\,(T_0 - T_m) + U_1)\,\xi}{U_2 \varrho_m U_0 \lambda'}\right] = 2.418 \times 10^{-4}\left[(14.7 + U_1)\,\xi\right]^{1/2} \tag{9.251}$$

$$\Phi = \left[\frac{u_{bx}\varrho_m\left(k_m\,(T_b - T_m) + \mu U_j^2/2\right)}{2\left[\lambda + C_s\,(T_m - T_{s0}) + C_m\bar{\Theta}\,(T_b - T_m)\right]}\right]^{1/2}$$
$$= 4.72 \times 10^{-3}\,[14.17 + U_1]^{1/2} \tag{9.252}$$

$$\psi = \frac{\Phi}{\varrho_s u_{sz}\sqrt{X}} = 6.01 \times 10^{-4}\,[14.17 + U_1]^{1/2} \tag{9.253}$$

For the constant channel depth region in the solids section, we compute with the initial position of the melt film of 1.26 m, we can write

$$\frac{X}{W} = [1 - 53.19\psi\,(z - 1.26)]^2 \tag{9.254}$$

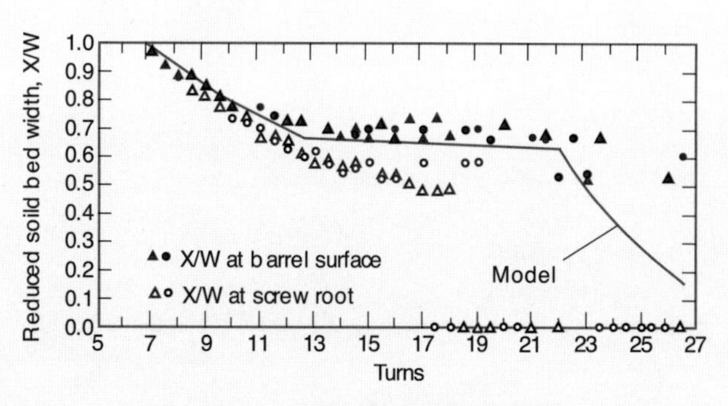

Figure 9.52 Predicted and experimental solids bed profile [17]

Table 9.1 Thermal properties for selected polymeric materials

Turns	z	δ (m)	U_1 (N/s)	ψ	X (m)	X/W
7.0	1.466	0.0002967	13.63	0.003169	0.05416	1.000
7.5	1.571	0.0002905	13.44	0.003158	0.05227	0.965
8.0	1.675	0.0002844	13.26	0.003147	0.05227	0.931
8.5	1.780	0.0002783	13.07	0.003137	0.05043	0.898
9.0	1.885	0.0002723	12.88	0.003126	0.04863	0.866
9.5	1.990	0.0002664	12.70	0.003115	0.04688	0.834
10.0	2.094	0.0002605	12.52	0.003105	0.04517	0.803
10.5	2.199	0.0002548	12.33	0.003094	0.04350	0.773
11.0	2.304	0.0002490	12.15	0.003083	0.04188	0.744
11.5	2.408	0.0002434	11.97	0.003073	0.04030	0.716
12.0	2.513	0.0002378	11.79	0.003062	0.03876	0.688
12.5	2.618	0.0002323	11.61	0.003052	0.03580	0.661

Equations 9.250, 9.251, 9.253, and 9.254 must now be solved simultaneously starting at the end of the 6th turn. The results are presented in Table 9.1.

The remaining two sections are computed in a similar fashion. Figure 9.52 presents a comparison between the measured and the predicted solid bed profiles.

■

9.8 Curing reactions during processing

While thermoplastics solidify and melt by lowering and raising the temperature below and above transition temperatures, thermosets and elastomers undergo an irreversible cross-linking or curing reaction during processing. When processing these materials it is important to predict the cycle time, in order to know when to de-mold a part. In addition, predicting the curing reaction allows us to optimize, control, and understand the process. In this section, we follow the procedure used by Enns and Gillham [19].

A general cure kinetic equation that describes the reaction can have the form

$$\frac{dc}{dt} = ae^{-E/RT} f(c) \tag{9.255}$$

where the function $f(c)$ describes the nature and order of the reaction. For a zero order reaction $f(c) = 1$, a first order reaction $f(c) = 1 - c$, and a second order reaction $f(c) = (1-c)^2$. A first order autocatalytic reaction is described by $f(c) = (B+c)(1-c)$, and a second order autocatalytic reaction by $f(c) = (B+c)(1-c)^2$. For the sake of the analysis here, let us assume a first order reaction such that

First order curing model

$$\frac{dc}{dt} = ae^{-E/RT}(1-c) \tag{9.256}$$

We can rearrange Eq. 9.256 and form the integral

$$t = \int_0^{c(t)} \frac{e^{E/RT}}{a\,(1-c)} dc \tag{9.257}$$

which after integration results in

$$t = -\frac{e^{E/RT}}{a} \ln (1-c) \tag{9.258}$$

This equation can be used to solve for the time to gelation

$$t_{gel} = -\frac{e^{E/RT}}{a} \ln \left(1-c_g\right) \tag{9.259}$$

and using DiBenedetto's equation,[3] which relates the glass transition temperature of a curing resin as a function of the degree of cure [20],

$$\hat{T}_g - 1 = \frac{T_g - T_{g0}}{T_{g0}} = \frac{(E_x/E_m - F_x/F_m)\,c}{1-(1-F_x/F_m)\,c} \tag{9.260}$$

where E_x/E_m is the ratio of lattice for cross-linked and uncross-linked resins and F_x/F_m is the ratio of molecular mobilities. The ratio $\frac{E_x/E_m}{F_x/F_m}$ is determined from measurements of the minimum and maximum glass transition temperature as

$$\frac{T_{g1}}{T_{g0}} = \frac{E_x/E_m}{F_x/F_m} \tag{9.261}$$

The individual ratios, E_x/E_m and F_x/F_m are then determined by fitting Eq. 9.260 to measured data of glass transition temperatures as a function of cure. We can take Eq. 9.260 and solve for the degree of cure

$$c = \frac{\hat{T}_g}{E_x/E_m - 1 + (1-F_x/F_m)\,\hat{T}_g} \tag{9.262}$$

which in turn can be substituted into Eq. 9.257 to give

$$t_{vitr} = -\frac{e^{E/RT}}{a} \ln \left(\frac{(F_x/F_m)\,\hat{T}_g}{E_x/E_m - 1 + (1-F_x/F_m)\,\hat{T}_g} \right) \tag{9.263}$$

We are now in the position to construct a time-temperature-transformation (TTT) diagram [11]. Figure 9.53 presents a TTT diagram generated for an epoxy with $E_x/E_m = 0.34$, $F_x/F_m = 0.19$, $T_{g0} = 254$ K, $T_{g1} = 439$ K, $a = 4.5 \times 10^6$ min^{-1} and $R = 1.987$ cal/mol/K.

3) Another form of this equation is $T_g = T_{g0} + \frac{\left(T_{g1}-T_{g0}\right)\lambda c}{1-(a-\lambda)c}$.

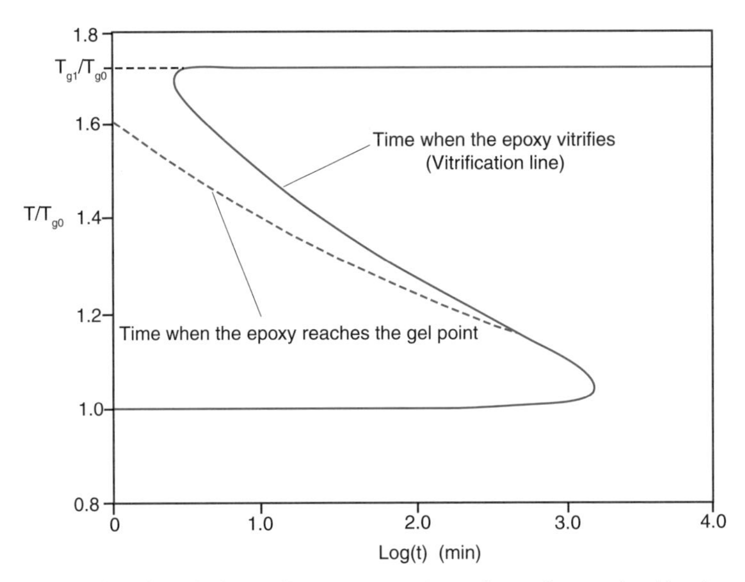

Figure 9.53 Predicted TTT behavior for an epoxy using a first order reaction kinetic model

Processing must occur below the gel point line and to the left of the vitrification line

9.9 Estimating Injection Pressure and Clamping Force

To aid the polymer processing student and engineer in finding required injection pressures and clamping forces, Stevenson [22] derived the non-isothermal non-Newtonian equations for the flow in a disc and solved them using finite difference techniques. The outcome was a set of dimensionless groups and graphs that can be applied to any geometry after a lay-flat approximation. In his analysis, Stevenson represented the flow inside the cavity with radial flow between two parallel plates. In order to use this representation, we must first *lay flat* the part to find the longest flow path, as schematically depicted in Fig. 9.54. Since the longest flow path may exceed the radius of the projected area that causes mold separating pressures, one must also find the radius of an equivalent projected area, R_p, to compute a more accurate mold clamping force. However, to perform the calculations necessary to predict velocities and pressure fields, Stevenson assumed a disc geometry of radius R and thickness h. As a constitutive model for the momentum balance, Stevenson chose a temperature dependent power law model represented by $\eta = m_0 e^{-a(T-T_{\text{ref}})}|\dot{\gamma}|^{n-1}$, and plotted his results in terms of four dimensionless groups defined below.

- The dimensionless temperature β determines the intensity of the coupling between the energy equation and the momentum balance. It is defined by

$$\beta = a\,(T_i - T_m) \tag{9.264}$$

where T_i and T_m are the injection and mold temperatures, respectively.

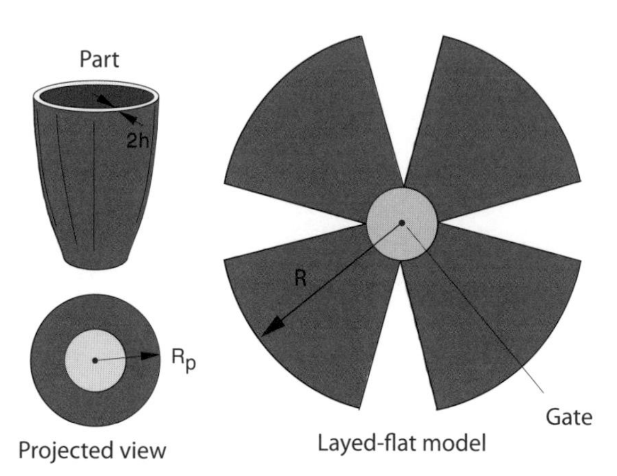

Figure 9.54 Schematic diagram of an injection molding item with its projected area and its lay-flat representation

- The dimensionless time τ is the ratio of the filling time, t_{fill}, and the time for thermal equilibrium via conduction, defined by

$$\tau = \frac{t_{\text{fill}} k}{h^2 \varrho C_{\text{p}}} \tag{9.265}$$

- The Brinkman number Br is the ratio of the energy generated by viscous dissipation and the energy transported by conduction. For a non-isothermal, non-Newtonian model it is

$$Br = \frac{m_0 e^{-aT_{\text{i}}} h^2}{k(T_{\text{i}} - T_{\text{m}})} \left(\frac{R}{t_{\text{fill}} h} \right)^{n+1} \tag{9.266}$$

- The power-law index n of power law model reflects the shear thinning behavior of the polymer melt.

Once the dimensionless parameters are calculated, the dimensionless injection pressures $\left(\frac{\Delta p}{\Delta p_{\text{I}}}\right)$ and dimensionless clamping forces $\left(\frac{F}{F_{\text{I}}}\right)$ are read from Figs. 9.55 to 9.58, where the isothermal pressure and force are computed using

$$\Delta p_{\text{I}} = \frac{m_0 e^{-aT_{\text{i}}}}{1-n} \left[\frac{1+2n}{2n} \frac{R}{t_{\text{fill}} h} \right]^n \left(\frac{R}{h} \right) \tag{9.267}$$

and

$$F_{\text{I}} = \pi R^2 \left(\frac{1-n}{3-n} \right) \Delta p_{\text{I}} \tag{9.268}$$

Since the part area often exceeds the projected area, Fig. 9.59 can be used to correct the computed clamping force.

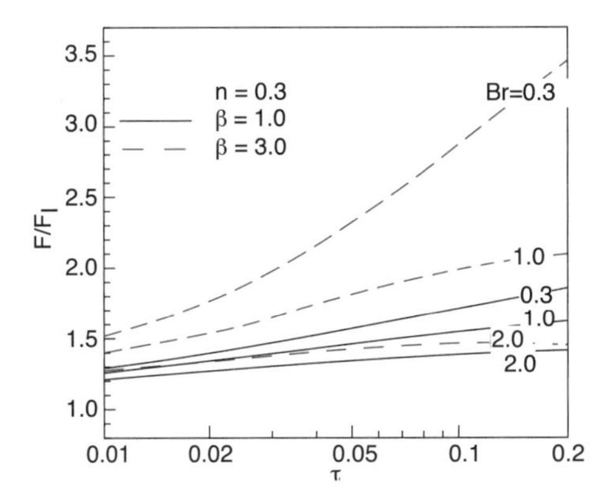

Figure 9.55 Dimensionless injection pressure as a function of β, Br and τ for $n = 0.3$

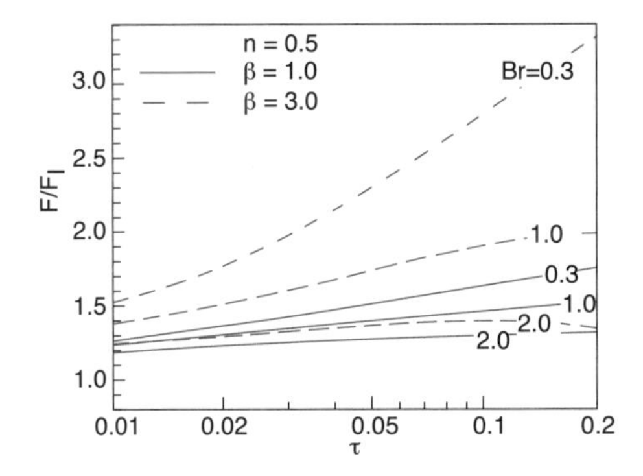

Figure 9.56 Dimensionless injection pressure as a function of β, Br and τ for $n = 0.5$

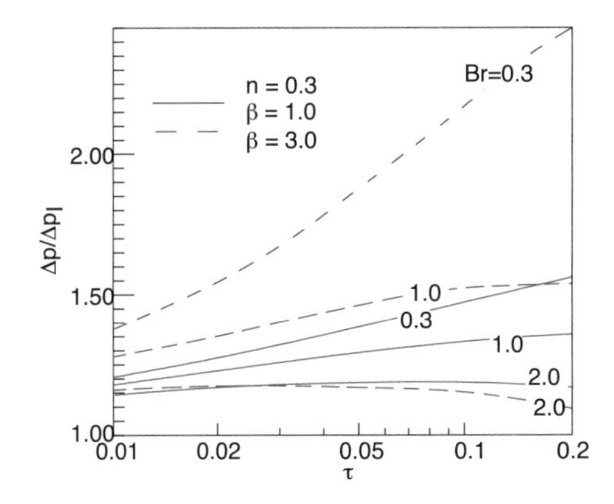

Figure 9.57 Dimensionless clamping force as a function of β, Br and τ for $n = 0.3$

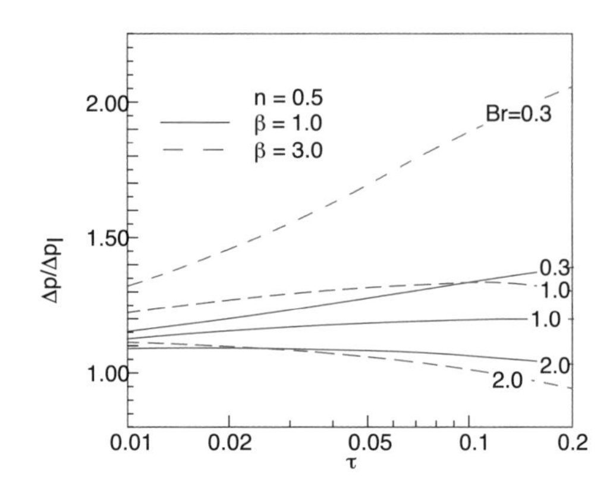

Figure 9.58 Dimensionless clamping force as a function of β, Br and τ for $n = 0.5$

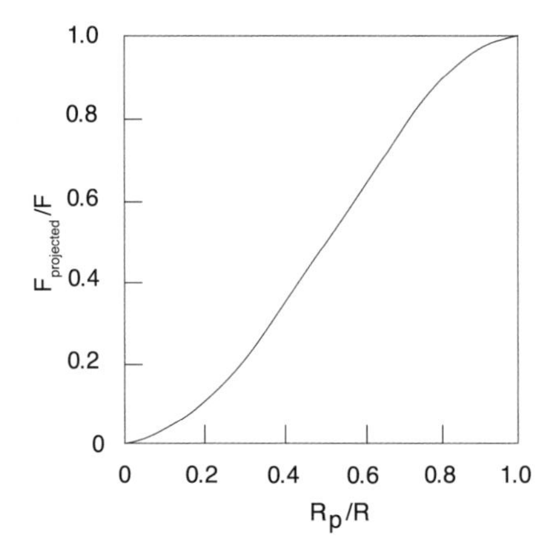

Figure 9.59 Clamping force correction for the projected area

Example 9.11 Injection pressure and clamping force example

You are to determine the maximum clamping force and injection pressure required to mold an ABS suitcase with a filling time, $t_f = 2.5$ seconds. Use the dimensions shown in Fig. 9.60, an injection temperature $T_i = 227$ °C (500 K), and a mold temperature $T_m = 27$ °C (300 K). The properties necessary for the calculations are given below.

Properties for ABS

$n = 0.29$,	$\varrho = 1020\ \text{kg/m}^3$
$m_0 = 29 \times 10^6\ \text{Pa·s}^n$,	$C_p = 2343\ \text{J/kg/K}$
$a = 0.01369/\text{K}$,	$k = 0.184\ \text{W/m/K}$

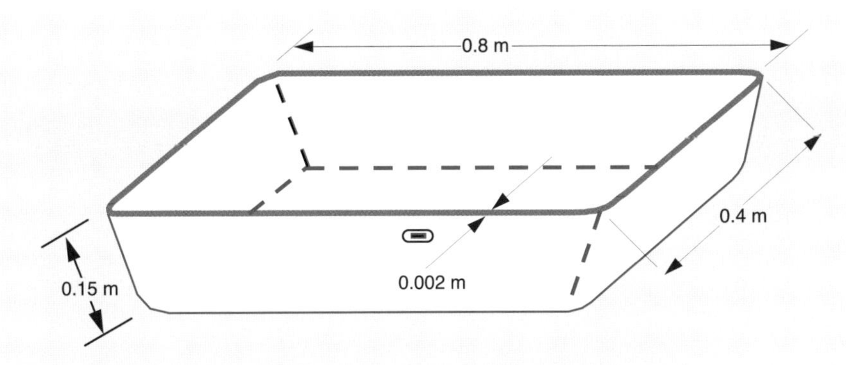

Figure 9.60 Suitcase

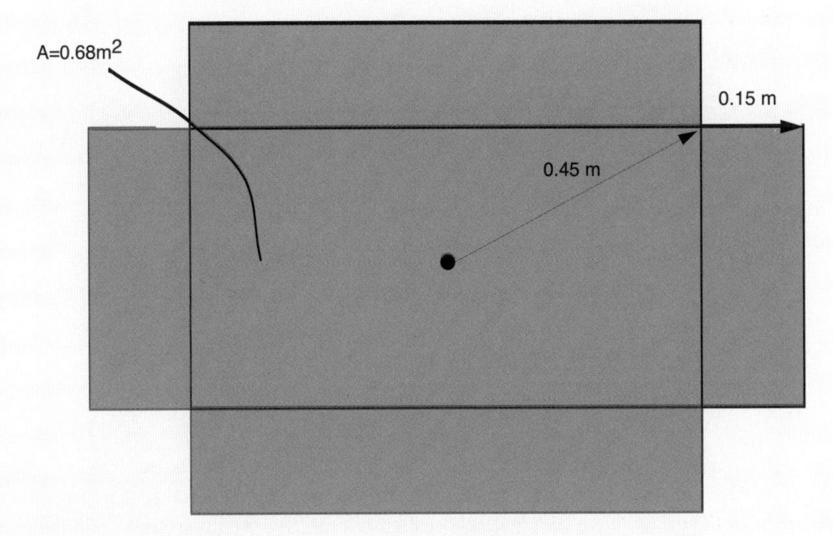

Figure 9.61 Layed-flat suitcase

We start this problem by first laying the suitcase flat and determining the required geometric factors (Fig. 9.61). From the suitcase geometry, the longest flow path, R, is 0.6 m and the radius of the projected area, R_p, is 0.32 m. We can now compute the dimensionless parameters given in Eqs. 9.264 to 9.266

$$\beta = 0.01369(500 - 300) = 2.74$$

$$\tau = \frac{2.5(0.184)}{(0.001)^2(1020)(2343)} = 0.192$$

$$Br = \frac{(29 \times 10^6)e^{-0.01369(500)}(0.001)^2}{0.184(500 - 300)} \left(\frac{0.6}{2.5(0.001)}\right)^{0.29+1} = 0.987$$

The isothermal injection pressure and clamping force are computed using Eqs. 9.267 and 9.268, respectively

$$\Delta p_{\mathrm{I}} = \frac{29 \times 10^6 e^{-0.01369(500)}}{1-0.29}\left[\frac{1+2(0.29)}{2(0.29)}\frac{0.6}{2.5(0.001)}\right]^{0.29}\left(\frac{0.6}{0.001}\right) = 171\ \mathrm{MPa}$$

$$F_{\mathrm{I}} = \pi(0.6)^2\left(\frac{1-0.29}{3-0.29}\right)(17.1 \times 10^7) = 50.7 \times 10^6\ \mathrm{N}$$

We now look up $\frac{\Delta p}{\Delta p_{\mathrm{I}}}$ and $\frac{F}{F_{\mathrm{I}}}$ in Figs. 6.8 to 6.11. Since little change occurs between $n = 0.3$ and $n = 0.5$, we choose $n = 0.3$. However, for other values of n we can interpolate or extrapolate. For $\beta = 2.74$, we interpolate between 1 and 3 as

$$\beta = 1 \rightarrow \frac{\Delta p}{\Delta p_{\mathrm{I}}} = 1.36 \text{ and } \frac{F}{F_{\mathrm{I}}} = 1.65$$

$$\beta = 3 \rightarrow \frac{\Delta p}{\Delta p_{\mathrm{I}}} = 1.55 \text{ and } \frac{F}{F_{\mathrm{I}}} = 2.1$$

$$\beta = 2.74 \rightarrow \frac{\Delta p}{\Delta p_{\mathrm{I}}} = 1.53 \text{ and } \frac{F}{F_{\mathrm{I}}} = 2.04$$

$$\Delta p = \left(\frac{\Delta p}{\Delta p_{\mathrm{I}}}\right)\Delta p_{\mathrm{I}} = 262\ \mathrm{MPa} = 2\,620\ \mathrm{bar}$$

$$F = \left(\frac{F}{F_{\mathrm{I}}}\right)F_{\mathrm{I}} = 10.3 \times 10^7\ \mathrm{N} = 10\,300 \text{ metric tons}$$

The clamping force can be corrected for an $R_{\mathrm{p}} = 0.32$ m using Fig. 6.12 and $\frac{R_{\mathrm{p}}}{R} = 0.53$.

$$F_{\mathrm{projected}} = (0.52)10\,300 = 5\,356 \text{ metric tons}$$

For our suitcase cover,where the total volume is 1 360 cc and total part area is 0.68 m^2, the above numbers are too high. A useful rule-of-thumb is a maximum allowable clamping force of 2 tons/in^2. Here, we have greatly exceeded that number. Normally, around 3 000 metric tons/m^2 are allowed in commercial injection molding machines. For example, a typical injection molding machine with a shot size of 2 000 cc has a maximum clamping force of 630 metric tons, with a maximum injection pressure of 1 400 bar. A machine with much larger clamping forces and injection pressures is suitable for much larger parts. For example, a machine with a shot size of 19 000 cc allows a maximum clamping force of 6 000 metric tons, with a maximum injection pressure of 1 700 bar. For this example we must reduce the pressure and clamping force requirements. This can be accomplished by increasing the injection and mold temperatures or by reducing the filling time. Recommended injection temperatures for ABS range between 210 and 240 °C and recommended mold temperatures range between 40 and 90 °C. As can be seen, there is room for adjustment in the processing conditions, so one must repeat the above procedure using new conditions.

■

References

1. Z. Tadmor, I. J. Duvdevani, and I. Klein. Melting in plasticating extruders — theory and experiments. *Polym. Eng. Sci.*, 7:198–217, 1967.
2. M. M. Denn and R. J. Fisher. The mechanics and stability of isothermal melt spinning. *AIChE J.*, 22:236, 1979.
3. N. Phan-Thien. A nonlinear network viscoelastic model. *J. Rheol.*, 31(8):259–283, 1978.
4. J. R. Pearson and C. J. S. Petrie. A fluid-mechanical analysis of the film-blowing process. *Plast. Polym.*, 38:85, 1970.
5. J. R. Pearson and C. J. S. Petrie. The flow of a tubular film. part i. formal mathematical representation. *J. Fluid Mech.*, 40(1), 1970.
6. J. R. A. Pearson and C. J. S. Petrie. The flow of a tubular film. part ii. interpretation of the model and discussion of solutions. *J. Fluid Mech.*, 42:609, 1970.
7. R. E. Gaskell. The calendering of plastic materials. *J. Appl. Mech.*, 17:334–336, 1950.
8. J. M. McKelvey. *Polymer Processing*. John Wiley and Sons, Inc., New York, 1962.
9. J. T. Bergen and G. W. Scott. Pressure distribution in the calendering of plastic materials. *J. Appl. Mech.*, 18:101, 1951.
10. C. Kiparissides and J. Vlachopoulos. Finite element analysis of calendering. *Polym. Eng. Sci.*, 16:712–719, 1979.
11. T. Osswald and J. P. Hernández. *Polymer Processing — Modeling and Simulation*. Hanser Publishers, Munich, 2006.
12. W. Unkrüer. *Beitrag zur Ermittlung des Druckverlaufes und der Fließvorgänge im Walzenspalt bei der Kalenderverarbeitung von PVC-hart zu Folien*. PhD thesis, RWTH-Aachen, 1970.
13. J. F. Agassant, P. Avenas, J.-Ph. Sergent, and P. J. Carreau. *Polymer Processing — Principles and Modeling*. Hanser Publishers, Munich, 1991.
14. G. Krobath, S. Liedauer, and H. Janeschitz-Kriegl. *Polymer Bulletin*, 14:1, 1985.
15. G. Eder and H. Janeschitz-Kriegl. *Polymer Bulletin*, 11:93, 1984.
16. H. Janeschitz-Kriegl and G. Krobath. *Intern. Polym. Proc.*, 3:175, 1988.
17. Z. Tadmor and I. Klein. *Engineering Principles of Plasticating Extrusion*. Van Nostrand Reinhold Company, New York, 1970.
18. M.d.P. Noriega, PhD thesis, University of Wisconsin-Madison, 2003.
19. J. B. Enns and J. K. Gillman. Time-temperature-transformation (ttt) cure diagram: Modeling the cure behavior of thermosets. *J. Appl. Polym. Sci.*, 28:2567–2591, 1983.
20. A. T. DiBenedetto and L. E. Nielsen. *J. Macromol. Sci., Rev. Macromol. Chem.*, C3:69, 1969.

Subject Index